# Flash Floods: Forecasting, Monitoring and Mitigation Strategies

# Flash Floods: Forecasting, Monitoring and Mitigation Strategies

Editors

**Xiekang Wang**
**Philippe Gourbesville**
**Changjun Liu**

MDPI • Basel • Beijing • Wuhan • Barcelona • Belgrade • Manchester • Tokyo • Cluj • Tianjin

*Editors*
Xiekang Wang
Sichuan University
China

Philippe Gourbesville
Université Côte d'Azur
France

Changjun Liu
China Institute of Water Resources and Hydropower Research
China

*Editorial Office*
MDPI
St. Alban-Anlage 66
4052 Basel, Switzerland

This is a reprint of articles from the Special Issue published online in the open access journal *Water* (ISSN 2073-4441) (available at: https://www.mdpi.com/journal/water/special_issues/Flash_Floods_Mitigation).

For citation purposes, cite each article independently as indicated on the article page online and as indicated below:

| LastName, A.A.; LastName, B.B.; LastName, C.C. Article Title. *Journal Name* **Year**, *Volume Number*, Page Range. |
|---|

**ISBN 978-3-0365-7818-7 (Hbk)**
**ISBN 978-3-0365-7819-4 (PDF)**

# Contents

# About the Editors

**Xiekang Wang**

Xiekang Wang, Professor. He has served as a professor of State Key Laboratory of Hydraulics and Mountain River Engineering, Sichuan University since 2006. His research focuses on basic theoretical issues and engineering techniques of hydraulics and river dynamics, water-sediment disaster, flash flood etc.

**Philippe Gourbesville**

Philippe Gourbesville, Professor. He has served as professor of Hydroinformatics at Université Côte d'Azur, Nice, France since 1997. His research activities are concentrated on deterministic modelling methods for hydrology and hydraulics, including extreme conditions such as flash floods, real-time simulation, water resources management and natural disaster management.

**Changjun Liu**

Changjun Liu, Professor. He has served as professor at the China Institute of Water Resources and Hydropower Research. His research interests include flash flood prevention and control, hydrology and hydraulic modelling, remote science, flood risk management, and water and soil conservation.

water

MDPI

*Editorial*

# Flash Floods: Forecasting, Monitoring and Mitigation Strategies

**Xiekang Wang [1,*], Philippe Gourbesville [2] and Changjun Liu [3]**

1 State Key Laboratory of Hydraulics and Mountain River Engineering, Sichuan University, Chengdu 610065, China
2 Université Côte d'Azur, Polytech Nice Sophia, 06903 Sophia Antipolis, France; gourbesv@unice.fr
3 China Institute of Water Resources and Hydropower Research, Beijing 100038, China; lcj2005@iwhr.com
* Correspondence: wangxiekang@scu.edu.cn

In recent decades, flash floods have become a major natural disaster and show a continuously increasing trend on a worldwide scale. The magnitude of the damages associated with flash floods requires forecasting and monitoring strategies to understand the vulnerability factors, analyze the mechanisms of flash floods, and mitigate disasters.

Research efforts are needed to improve early warning mechanisms, risk control, and hazard prevention that could obviously aim at a reduction in casualties, social impacts, and economic losses. New technical approaches such as surface monitoring and combined hydrologic–hydrodynamic models are in development and are offering useful information for field managers.

The main causes of flash flood disasters can be investigated by analyzing the hydrological and hydrodynamic process of flash flood disaster events. The simultaneous flooding of tributaries and the main river has been identified as one of the main causes to amplify the discharge peak in the main river. Therefore, the monitoring and forecasting of tributaries' discharge may be an effective solution to issuing a flash flood early warning. Meanwhile, subgrade water damming that serves to mitigate flash flood impact, however, may exaggerate the risk of a flash flood disaster if a subgrade water damming structure collapses due to extreme weather conditions. Thus, artificial efforts to a mitigate flash flood disaster should be discussed and assessed.

Recently, the sediment-transport-related geomorphological evolution of mountain rivers drew considerable attention when assessing flash flood propagation. Due to sediment deposition in local mild reaches, the flash flood stage under a significant change in the riverbed morphology essentially differs from that under relatively stable river morphology, which is likely to result in the incorrect warning of flash flood disasters.

Debris flows might be triggered by storms and flash floods, which have not yet been understood. Well-designed flume experiments may help researchers to deepen the understanding of the triggering mechanisms of debris flows, as well as the dynamic characteristics of debris flows under different impact factors.

Data-driven approaches are considered appropriate to predict flood behavior due to the availability of an increasing number of high-quality data. Models with different complexities from simple regression to complex machine learning can be applied for the hydrological prediction and susceptibility map establishment of flash floods. Specific treatments, such as parameter regionalization, have a good potential to improve the modeling performance.

Flash flood disaster prevention and mitigation have not only been studied with the development of different approaches, but they have also been practiced with these appropriate approaches. However, the relationship between different socio-economic effects and flash flood disaster prevention and control should be discussed, which can offer information for good planning and policies for a further support of flash flood disaster prevention and mitigation.

**Citation:** Wang, X.; Gourbesville, P.; Liu, C. Flash Floods: Forecasting, Monitoring and Mitigation Strategies. *Water* **2023**, *15*, 1700. https://doi.org/10.3390/w15091700

Received: 14 April 2023
Accepted: 17 April 2023
Published: 27 April 2023

The Special Issue *Flash Floods: Forecasting, Monitoring and Mitigation Strategies* includes major flash flood disaster event analysis; the key factors for flash floods and monitoring strategies; field observations for flash flood processes; the modeling and forecasting of flash flood events; a risk assessment for flash floods; and the prevention and mitigation measures for flash floods. It includes twelve research papers.

Three papers describe flash-flood-caused hazard events which have recently happened in small mountain catchments in China. Different mechanisms triggering flash flood disasters occurred across different regions. One paper is about discharge peak amplification caused by simultaneous flooding for tributaries and main river. The second paper addresses the impact of the collapse of subgrade water damming on the severity of the disasters. The third paper details the coupling effect of flow and sediment transport to exaggerate the flash flood dynamics, which caused a disaster. With the insights of these three papers, common causes for mountain basin flash flood disasters can be checked.

Post-event reanalysis can be effective and necessary for identifying the causes of flash flood disasters which can be implemented using hydrologic and hydrodynamic approaches.

By developing a new distributed hydrological model named China-FFMS that simulates the evolution of natural disasters and make an assessment, Hao et al. [1] simulated the 8.12 flash flood disaster that occurred in the Liulin county of Hubei province on 12 August ("8.12") and the paper includes the data collected from the national flash flood disaster investigation and assessment. They identified that the main factor leading to the disaster was the overlapped peak flow where the Dunne flood peak of three different tributaries from the upper reach met together at the same time, and the peak flow of the Lianhua river at the downstream of Liulin County also arrived at the same time as the upstream peak, which obstructed the flood's progress and increased the damage of the disaster.

When a hydrodynamic model is coupled with a hydrological model, the simulation can provide not only flood discharge routing but also the evolution of flood stage, which, once overbank, a flooding disaster is caused in theory. Hao et al. [2] numerically investigated the "7.20" flash flood in the Wangzongdian river basin by establishing a hydrological–hydrodynamic model using limited measured data. The extreme rainstorm accounting for flooding in mountainous areas and the collapse of subgrade water damming led to the high-level flood quickly flowing into Wangzongdian Village over a short distance, causing a serious disaster.

In mountain basins, the sediment supply and its delivery have a significant impact on the geomorphological change in steep rivers with sediment deposition and erosion processes, thus determining the geometric boundaries where flash floods evolve. Therefore, Yang et al. [3] used a depth-averaged two-dimensional hydrodynamic model to simulate a flash flood that occurred in Sanjiang Town, Sichuan, China, on 20 August 2019. Inflows and sediment deposition were the main factors that contribute to flash flood enlargement in confluence and bifurcation streams. This study deepened the understanding that flow in the supercritical slope runs at a very fast velocity and seldom deposits sediment in the steep channel, while most sediment is transported to the streams with flat hydraulic slopes.

Currently, an increasing sense of awareness has been realized that the triggering of flash floods may cause other kinds of natural water-sediment-related disasters such as debris flows. Two papers, therefore, addressed the importance of mountain gully debris flows in a strong earthquake area and the periglacial debris flow in Southeast Tibet. Zhang et al. [4] designed a lateral erosion flume model experimental device to explore the erosion characteristics of debris flow. In total, 18 groups of incomplete orthogonal experiments were conducted to investigate the effects of the unit weight of debris flow, the content of fine particles, and the longitudinal slope gradient of the gully. The major finding demonstrates that the erosion width, depth, and volume decrease with the increase in the fluid bulk density and increase with the increase in the gully slope and the unit weight of debris flow has the greatest impact on the erosion degree of the side slope. Du et al. [5] analyzed the importance of potential indicators to the development of periglacial debris flows in the Parlung Zangbo Basin of southeast Tibet and introduced three machine learning approaches combined with the borderline resampling technique for predicting debris flow occurrences. They found that temperature, precipitation, and vegetation coverage are closely related to the development of periglacial debris flow in the study area. By testing and comparing several scenarios, the best model of predicting debris flow events was suggested.

Precipitation data as the input driving the hydrological process, as well as the quality and representation and data source, control the performance of hydrological modeling. Guo et al. [6] Evaluated the performance of three precipitation products (rain gauge observations, CMADS, and TRMM) in the hydrological modeling of the Danjiang River Basin (DRB) with the SWAT model at monthly, daily, and spatial scales. Simulation calibration and validation were performed at three hydrological stations using the SWAT Calibration Uncertainties Program (SWAT-CUP). This study finally provides a reference for choosing precipitation datasets in watersheds like the DRB where ground-based rain gauge data are unavailable.

Apart from precipitation, the surface condition of the catchment may largely influence the runoff and sediment generation. Ding et al. [7] numerically assessed the impact of grass coverage degrees and spatial arrangements on the runoff and sediment yield both on the hillslope and gully slope. The relative contribution of hillslope and gully side account for the total erosion on the entire hillslope gully system was analyzed, which helps to understand the validity of the policy on soil and water conservations in Loess Plateau. This study highlighted the importance of vegetation coverage in reducing soil erosion and the need for further research on the impacts of different vegetation coverage and arrangement patterns on the slope gully system.

Data-driven approaches are becoming increasingly popular to predict flood behavior. As the quantity of historical data is becoming larger and the quality is better due to systematical monitoring techniques, different data-driven approaches, such as logistic regression and machine learning, can be applied. El-Rawy et al. [8] provided a method for flood risk assessment by incorporating principal component analysis and logistic regression in the Sinai Peninsula, Egypt, using hydro-morphometric parameters. Cross-validation of the model was conducted to ensure reliability and robustness. A flash flood susceptibility map with four categories of risks (low, moderate, high, and very high) was established for the Sinai Peninsula that can be useful for authorities and decision makers in impact assessment, flash flood management, and the planning and implementation of mitigation measures.

Wang et al. [9] studied the influence of different methods on the parameter regionalization of distributed hydrological model parameters in hilly areas of Hunan Province, China. Shortest distance, attribute similarity, support vector regression, generative adversarial networks, classification and regression tree, and random forest methods were evaluated to create parameter regionalization schemes, with 426 floods of 25 catchments for calibration and 136 floods of 8 catchments for verification. The study showed that the random forest model is the most stable solution and significantly outperforms other methods and can improve the accuracy of flood simulation in ungauged areas with parameter regionalization, which is of great significance for flash flood forecasting and early warning.

The regionalization method can be also applied in the process-based modeling of flash floods. Williams et al. [10] applied a rainfall regionalization method to construct flash flood hyetographs with several return periods using the flash flood shape of the historical event that occurred in the Tlalnepantla River basin, Mexico. A semi-distributed model in HEC-HMS was used to obtain the outflow hydrograph and hydrodynamic model in Iber, and Hec-Ras 2D to simulate free surface flow was used for a hydrological–hydrodynamic two-dimensional analysis. This study successfully estimated the potential consequences of synthetic design storms on the site and provided insights into the study of flash floods at the global level, highlighting the need for a methodology for threat assessment.

With effective modeling approaches, a flash flood early warning system can be designed and established for disaster mitigation, such as for a reduction in causality and economic loss. Arganis et al. [11] developed a pilot impacts-based flood early warning decision support system for the Vaisigano River in Samoa. The flood early warnings decision support system alerted the hazard monitoring and emergency responders of imminent flooding with up to 24–48 h lead time, with information of a specific context and real-time automated river monitoring and forecast. It includes a web-based information portal which enables interaction with the decision support information tools, which was conducted in a practice of operational testing during the 2020/2021 tropical cyclone season in Samoa.

Finally, to mitigate the hazard impact of flash flood on mountain area residents and the environment, China has implemented a two-decadal flash flood prevention and mitigation project, which is still under operation currently. Therefore, whether these projects with recent advanced technology and methods are useful for the prevention and mitigation of flash flood disasters in a mountain area is important for sustainable investment, policy, and research. Zhang et al. [12] applied the Kaya identity and a Logarithmic Mean Divisia Index (LMDI) approach to quantitatively measure the driving effects of interannual changes in economic loss related to flood disasters in China. Five flood-related driving effects, including demographic effect, economic effect, flash flood disaster control effect, capital efficiency effect, and loss-rainfall effect, were evaluated. This paper shows that the flash flood disaster control effect most obviously reduced flood-related economic losses, and non-engineering measures for flash flood prevention and control have been implemented since 2010, achieving remarkable results. The discussion of the relationship between flood-related economic loss and flash flood disaster prevention and control in China adds value for the adjustment and formulation of future flood disaster prevention policies.

This collection of papers highlights the efforts of researchers in mitigating flash flood disasters, involving deepening the understanding of the causes triggering the disasters with existing cases, finding appropriate modeling approaches, and practicing disaster mitigation. This topic is a good example, which not only inspires the future research direction for flash floods but also supports the current practice of flash flood disaster prevention and mitigation.

**Author Contributions:** Conceptualization, X.W., P.G. and C.L.; writing—original draft preparation, X.W.; writing—review and editing, X.W., P.G. and C.L.; supervision, X.W. All authors have read and agreed to the published version of the manuscript.

**Funding:** This research was supported by the National Natural Science Foundation of China (Nos. 52239006, 51579163).

**Data Availability Statement:** Not Available.

**Acknowledgments:** As Guest Editors, we sincerely appreciate the authors for their contributions to this Special Issue. Our thanks also go to the editors and referees because their professional contributions obviously improved the quality of the twelve published papers.

**Conflicts of Interest:** The authors declare no conflict of interest.

## References

1. Hao, S.; Wang, W.; Ma, Q.; Li, C.; Wen, L.; Tian, J.; Liu, C. Analysis on the Disaster Mechanism of "8.12" Flash Flood in Liulin River Basin. *Water* **2022**, *14*, 2017. [CrossRef]
2. Hao, S.; Wang, W.; Ma, Q.; Li, C.; Wen, L.; Tian, J.; Liu, C. Model-Based Mechanism Analysis of "7.20" Flash Flood Disaster in Wangzongdian River Basin. *Water* **2023**, *15*, 304. [CrossRef]
3. Yang, Q.; Wang, X.; Sun, Y.; Duan, W.; Xie, S. Numerical Investigation on a Flash Flood Disaster in Streams with Confluence and Bifurcation. *Water* **2022**, *14*, 1646. [CrossRef]
4. Zhang, J.; Luo, D.; Li, H.; Pei, L.; Yao, Q. Experimental Study on Gully Erosion Characteristics of Mountain Torrent Debris Flow in a Strong Earthquake Area. *Water* **2023**, *15*, 283. [CrossRef]
5. Du, J.; Zhang, H.-y.; Hu, K.-h.; Wang, L.; Dong, L.-y. Prediction of the Periglacial Debris Flow in Southeast Tibet Based on Imbalanced Small Sample Data. *Water* **2023**, *15*, 310. [CrossRef]
6. Guo, Y.; Ding, W.; Xu, W.; Zhu, X.; Wang, X.; Tang, W. Assessment of an Alternative Climate Product for Hydrological Modeling: A Case Study of the Danjiang River Basin, China. *Water* **2022**, *14*, 1105. [CrossRef]

7. Ding, W.; Wang, X.; Zhang, G.; Meng, X.; Ye, Z. Impacts of Grass Coverage and Arrangement Patterns on Runoff and Sediment Yield in Slope-Gully System of the Loess Plateau, China. *Water* **2023**, *15*, 133. [CrossRef]
8. El-Rawy, M.; Elsadek, W.M.; De Smedt, F. Flash Flood Susceptibility Mapping in Sinai, Egypt Using Hydromorphic Data, Principal Component Analysis and Logistic Regression. *Water* **2022**, *14*, 2434. [CrossRef]
9. Wang, W.; Zhao, Y.; Tu, Y.; Dong, R.; Ma, Q.; Liu, C. Research on Parameter Regionalization of Distributed Hydrological Model Based on Machine Learning. *Water* **2023**, *15*, 518. [CrossRef]
10. Williams, S.; Griffiths, J.; Miville, B.; Romeo, E.; Leiofi, M.; O'Driscoll, M.; Iakopo, M.; Mulitalo, S.; Ting, J.C.; Paulik, R.; et al. An Impacts-Based Flood Decision Support System for a Tropical Pacific Island Catchment with Short Warnings Lead Time. *Water* **2021**, *13*, 3371. [CrossRef]
11. Arganis, M.; Preciado, M.; Luna, F.D.; Cruz, L.; Domínguez, R.; Santana, O. Application of a Regionalization Method for Estimating Flash Floods: Cuautepec Basin, Mexico. *Water* **2023**, *15*, 303. [CrossRef]
12. Zhang, Z.; Li, Q.; Liu, C.; Ding, L.; Ma, Q.; Chen, Y. Driving Effects and Spatial-Temporal Variations in Economic Losses Due to Flood Disasters in China. *Water* **2022**, *14*, 2266. [CrossRef]

water

MDPI

Article

# Research on Parameter Regionalization of Distributed Hydrological Model Based on Machine Learning

**Wenchuan Wang [1], Yanwei Zhao [1,*], Yong Tu [2], Rui Dong [3], Qiang Ma [2] and Changjun Liu [2]**

1 College of Water Resources, North China University of Water Resources and Electric Power, Zhengzhou 450046, China
2 China Institute of Water Resource and Hydropower Research, Beijing 100038, China
3 Beijing Tianzhixiang Information Technology Co., Ltd., Beijing 100191, China
* Correspondence: zyw_1987@sina.com

**Abstract:** In the past decade, more than 300 people have died per year on average due to mountain torrents in China. Mountain torrents mostly occur in ungauged small and medium-sized catchments, so it is difficult to maintain high accuracy of flood prediction. In order to solve the problem of the low accuracy of flood simulation in the ungauged areas, this paper studies the influence of different methods on the parameter regionalization of distributed hydrological model parameters in hilly areas of Hunan Province. According to the terrain, landform, soil and land use characteristics of each catchment, we use Shortest Distance, Attribute Similarity, Support Vector Regression, Generative Adversarial Networks, Classification and Regression Tree and Random Forest methods to create parameter regionalization schemes. In total, 426 floods of 25 catchments are selected to calibrate the model parameters, and 136 floods of 8 catchments are used for verification. The results showed that the average values of the Nash–Sutcliffe coefficients of each scheme were 0.58, 0.64, 0.60, 0.66, 0.61 and 0.68, and the worst values were 0.27, 0.31, 0.25, 0.43, 0.35 and 0.59. The random forest model is the most stable solution and significantly outperforms other methods. Using the random forest model to regionalize parameters can improve the accuracy of flood simulation in ungauged areas, which is of great significance for flash flood forecasting and early warning.

**Keywords:** mountain torrents; distributed hydrological model; parameters regionalization; machine learning

**Citation:** Wang, W.; Zhao, Y.; Tu, Y.; Dong, R.; Ma, Q.; Liu, C. Research on Parameter Regionalization of Distributed Hydrological Model Based on Machine Learning. *Water* **2023**, *15*, 518. https://doi.org/10.3390/w15030518

Academic Editor: Renato Morbidelli

Received: 3 January 2023
Revised: 22 January 2023
Accepted: 25 January 2023
Published: 28 January 2023



## 1. Introduction

Hunan Province is located in the southeast inland of China, with abundant rainfall but extremely uneven temporal and spatial distribution. Due to frequent and high-intensity rainfall and short confluence time in hilly areas, the flood rises and falls steeply, which can very easily cause mountain torrents. The climate, underlying surface and geomorphic types in hilly areas are diverse, and most of them are areas without data. This is an important challenge for flood forecasting and early warning in hilly areas.

The hydrological model is an important tool for understanding the laws of hydrological science, analyzing hydrological processes and studying hydrological cycle mechanisms [1]. How to identify hydrological parameters in ungauged areas accurately is an important area of research for PUB (Prediction in Ungauged Basins). The regionalization method is usually used to determine the parameters of hydrological models for ungauged basins at present, and the commonly used methods include shortest distance, attribute similarity, regression, average, machine learning, etc. The main idea of the regionalization method is to analyze the relationship between model parameters and characteristic attributes of basins, and the parameters of the hydrological model for ungauged basins are deduced from the calibration results of gauged basins [2].

The parameter transplant method includes the shortest distance method and the attribute similarity method. Among them, the distance approach refers to finding one or more basins adjacent to the research object in the geographical location. The attribute similarity method is used to find a basin that is similar to the research basin in attributes. Young achieved the ideal result of parameter transplant by computing the spatial distance between 260 catchments in the UK [3]. Parajka et al. selected indicators such as watershed area, average slope, watershed latitude, river network density, vegetation coverage, drought index, etc., to analyze the similarity of watershed attributes and complete parameter transplantation. The results show that attribute selection plays a decisive role in the performance of transplantation [4]. Li et al. compared the shortest distance method with the attribute similarity method and pointed out that the performance of transplantation results is affected by the density of hydrological stations, and it is easier to achieve better results in areas with dense hydrological stations [5].

The parameter regression method is mainly used to establish the functional relationship between watershed characteristics and model parameters. Yokoo et al. established a multiple linear regression equation between the Tank model parameters and soil, geology and land use data [6]. Cheng et al. established a regression equation between the SCS model parameter CN, concentration time and soil, land use, average slope and river length [7]. Based on the parameter regionalization method combining spatial proximity and stepwise regression analysis, Yao et al. found that stepwise regression analysis can effectively deduce the sensitive parameters [8]. Sun et al. pointed out that the parametric regression method is prone to the phenomenon of "the same effect of different parameters", and the basin properties screening is highly subjective, which is not suitable for small samples [9].

Machine learning research mainly includes SOM classification and the CART decision tree method. Yi et al. used hierarchical clustering analysis HCA and unsupervised neural network SOM methods to divide the sub basins of Dianchi Lake basin into 7 groups based on 16 physical characteristics, and they believed that the basin parameters of the same group can be transplanted to each other [10]. Ragettli et al. took 35 basins in different regions of China as the research object, comprehensively considering the physical properties of watersheds and the spatial distance of watersheds; the CART tree model was used to optimize the parameter transplantation rules, and the results show that the CART tree has better parameter adaptability [11]. Liu et al. conducted a parametric zoning study on 19 small catchments in Henan Province; the success rate of parameter transplantation based on the CART tree is about 20% higher than that of random transplantation [12].

The advantage of the CART tree is that it is easy to interpret and the mapping between basin characteristics and transplantation rules is intuitive. In recent years, with the advent of machine learning algorithms, more and more models have been used to create parameter transplantation schemes. However, many machine learning algorithms usually require a large number of samples, and data showing that hydrological model modeling can be used for parameter calibration is often very limited, so it is necessary to reasonably build a large number of learning samples, or to study intelligent algorithms suitable for small sample research. In this study, 33 small and medium-sized catchments in Hunan Province were taken as examples. We constructed distributed hydrological models of these catchments and selected four machine learning models—Support Vector Regression, Generative Adversarial Networks, Classification and Regression Tree, and Random Forest—to create different parametric regionalization schemes and compared them with two traditional methods—Shortest Distance and Attribute Similarity. By analyzing the transplantation results of different schemes, it can provide a reference for determining the parameters of the distributed hydrological model in ungauged areas, which is very valuable for flash flood forecasting and early warning.

## 2. Materials and Methods

### 2.1. Study Area

Hunan Province is located on the South Bank of the middle reaches of the Yangtze River. The general geomorphological characteristics are that it is surrounded by mountains in the east, south and west, hills in the middle, plains and lakes in the north, and an asymmetric horseshoe basin that was high in the southwest and low in the northeast. XueFeng mountain runs through the central part of the province from southwest to northeast, which divides the whole province into two parts: mountainous area and hilly area. Due to the comprehensive influence of monsoon circulation and the geomorphic conditions, the mid subtropical monsoon humid climate with obvious continental characteristics is formed. Mountain torrents occur frequently because of the complex topography, developed water system and abundant rainfall. The average annual precipitation in Hunan Province is 1450 mm, but the distribution of precipitation is uneven in time and space, and the interannual variation is large, with an average annual variation of 1200–1800 mm. The province's annual average water surface evaporation is 736.5 mm, with a variation range of 600–900 mm.

### 2.2. Data Collection

Taking 33 hydrological stations with observation data from 1979 to 2020 in Hunan Province as examples, we collected the ASTER GDEM V2 dataset, land use layer and soil type layer in Hunan Province. At the same time, a distributed hydrological model of all hydrological stations was established with 30 min as the simulation step. In this study, 426 floods in 25 catchments were selected to calibrate the model parameters, and the regionalization scheme was determined by comparing the simulation results of the other 8 catchments. Figure 1 shows the distribution of hydrological stations.

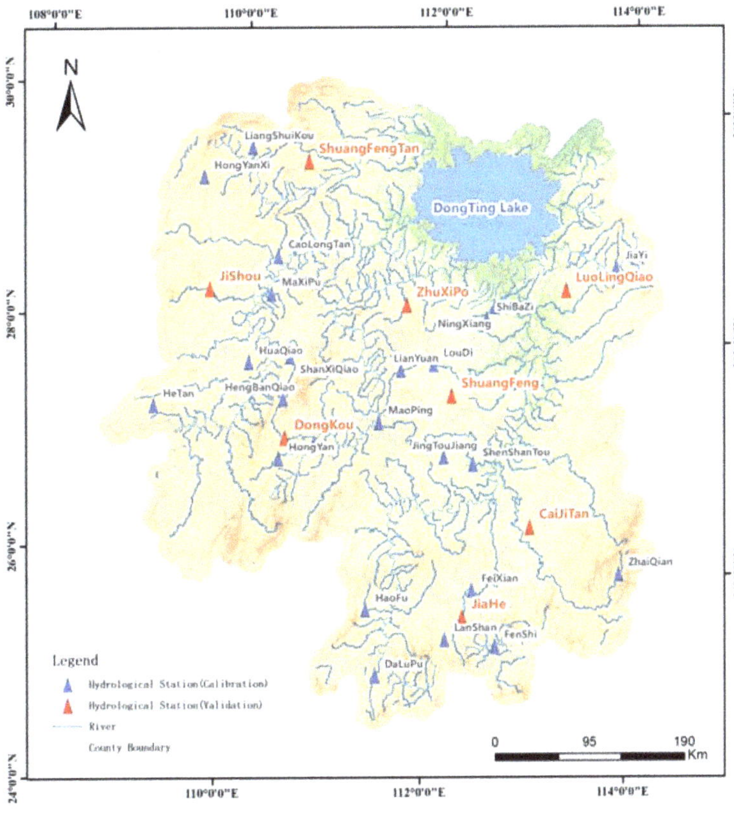


**Figure 1.** Distribution of hydrological stations.

The smallest catchment is HengBanQiao, with a catchment area of 31 km$^2$, and the largest catchment is FeiXian, with a catchment area of 3659 km$^2$. The hydrological data collection is shown in Table 1.

**Table 1.** Information of study stations.

| Station Name | Area ($km^2$) | Data Years | Number of Floods | Number of Rain Stations | Number of Sub Basins | Type |
|---|---|---|---|---|---|---|
| NingXiang | 2250 | 2013–2020 | 18 | 65 | 174 | calibration |
| ShiBaZi | 564 | 2013–2020 | 18 | 17 | 50 | calibration |
| FeiXian | 3659 | 2013–2020 | 28 | 157 | 257 | calibration |
| FenShi | 923 | 2013–2020 | 24 | 27 | 69 | calibration |
| ZhaiQian | 392 | 2015–2020 | 8 | 11 | 31 | calibration |
| JingTouJiang | 173 | 2014–2019 | 4 | 9 | 14 | calibration |
| ShenShanTou | 2930 | 2014–2019 | 5 | 71 | 227 | calibration |
| CaoLongTan | 350 | 2013–2015 | 6 | 7 | 22 | calibration |
| HeTan | 445 | 2014–2020 | 7 | 13 | 34 | calibration |
| HengBanQiao | 40 | 2014–2020 | 13 | 5 | 2 | calibration |
| HuaQiao | 81 | 2013–2020 | 22 | 6 | 5 | calibration |
| MaXiPu | 342 | 2012–2020 | 24 | 4 | 25 | calibration |
| ShanXiQiao | 1211 | 2013–2020 | 12 | 24 | 82 | calibration |
| LianYuan | 154 | 1979–2020 | 39 | 17 | 11 | calibration |
| LouDi | 1556 | 2014–2020 | 17 | 58 | 112 | calibration |
| HongYan | 711 | 2014–2019 | 8 | 17 | 55 | calibration |
| HuangQiao | 2689 | 2012–2019 | 14 | 76 | 211 | calibration |
| SheBu | 1434 | 2013–2020 | 9 | 38 | 109 | calibration |
| MaoPing | 2114 | 2014–2020 | 9 | 54 | 163 | calibration |
| HongYanXi | 190 | 2012–2020 | 20 | 4 | 11 | calibration |
| DaLuPu | 635 | 2013–2020 | 29 | 18 | 47 | calibration |
| HaoFu | 440 | 2013–2020 | 26 | 10 | 35 | calibration |
| LanShan | 305 | 2013–2020 | 32 | 25 | 19 | calibration |
| JiaYi | 1475 | 2013–2020 | 16 | 32 | 96 | calibration |
| LiangShuiKou | 865 | 2012–2020 | 18 | 17 | 65 | calibration |
| LuoLingQiao | 340 | 2012–2020 | 16 | 21 | 30 | verification |
| JiaHe | 1501 | 2012–2020 | 31 | 58 | 103 | verification |
| CaoJiTan | 387 | 2013–2020 | 13 | 9 | 30 | verification |
| ShuangFeng | 1552 | 2014–2020 | 10 | 36 | 115 | verification |
| DongKou | 928 | 2013–2020 | 13 | 18 | 66 | verification |
| JiShou | 788 | 2012–2020 | 26 | 30 | 56 | verification |
| ZhuXiPo | 699 | 2013–2020 | 15 | 18 | 53 | verification |
| ShuangFengTan | 444 | 2013–2020 | 12 | 20 | 35 | verification |

### 2.3. *Modeling Approaches*

#### 2.3.1. Distributed Hydrological Model

Based on the ASTER GDEM V2 dataset, the sub basin and river are extracted by GIS tools. The resolution of the DEM data grid is 30 m, and the area of the sub basin is controlled within 10–30 $km^2$. At the same time, the attributes of sub basins and rivers are extracted, including basin area, slope, longest concentration path, average altitude, average drop (average elevation minus outlet elevation), river length, river section gradient, geomorphic unit hydrograph, etc.

The Xinanjiang model is adopted for runoff generation computation [13–15]. A three-layer evaporation model is used to calculate watershed evaporation. The total runoff produced by rainfall is computed according to the concept of saturated runoff, and the influence of the uneven underlying surface on runoff yield area is considered by the water storage curve of the basin. In the aspect of runoff component division, according to the runoff production theory of "hillside hydrology", the total runoff is divided into saturated surface runoff, soil water runoff and groundwater runoff by a reservoir with limited volume, a side hole and a bottom hole. The unit hydrograph is used to convert the surface runoff into the overland flow, and the linear reservoir model is used to calculate the interflow and groundwater flow, and is finally incorporated into the river network. Figure 2 shows the structure of the Xinanjiang model.

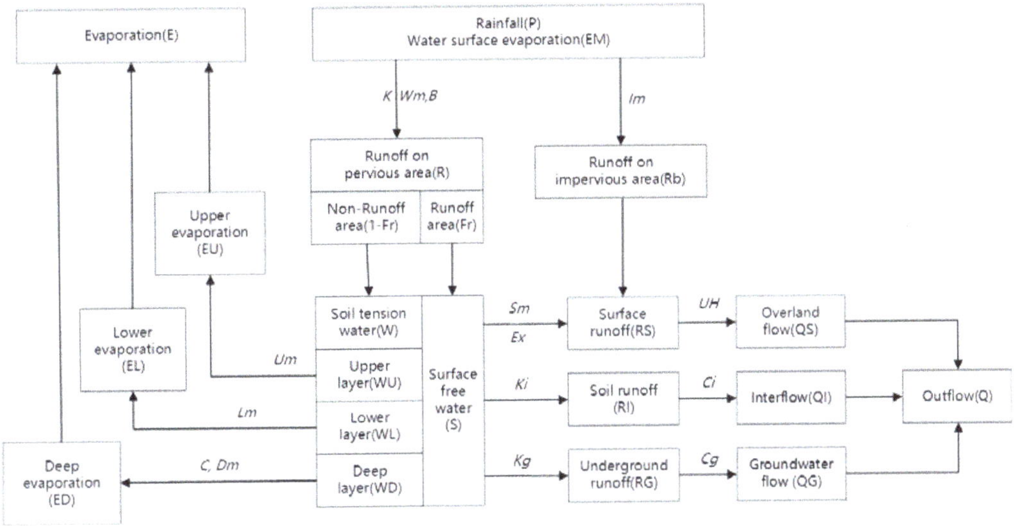


**Figure 2.** Computation flow of the Xinanjiang model.

Table 2 shows the parameters of the Xinanjiang model, all of which need to be determined through parameter calibration.

**Table 2.** Physical meanings and units of model parameters.

| | Parameter | Physical Description | Unit | Param Range |
|---|---|---|---|---|
| 1 | $K$ | Ratio of potential evapotranspiration to pan evaporation | [-] | 0.5–1.2 |
| 2 | $U_m$ | Averaged soil moisture storage capacity of the upper layer | [mm] | 10–40 |
| 3 | $L_m$ | Averaged soil moisture storage capacity of the lower layer | [mm] | 50–90 |
| 4 | $D_m$ | Averaged soil moisture storage capacity of the deep layer | [mm] | 10–80 |
| 5 | $C$ | Coefficient of the deep layer that depends on the proportion of the basin area covered by vegetation with deep roots | [-] | 0.1–0.3 |
| 6 | $B$ | Exponential parameter with a single parabolic curve, which represents the non-uniformity of the spatial distribution of the soil moisture storage capacity over the catchment | [-] | 0.1–0.9 |
| 7 | $I_m$ | Percentage of impervious and saturated areas in the catchment | [-] | 0.0–1.0 |
| 8 | $S_m$ | Areal mean free water capacity of the surface soil layer, which represents the maximum possible deficit of free water storage | [mm] | 10–80 |
| 9 | $E_x$ | Exponent of the free water capacity curve influencing the development of the saturated area | [-] | 0.1–2.0 |
| 10 | $K_g$ | Outflow coefficients of the free water storage to groundwater relationships | [-] | 0.1–0.5 |
| 11 | $K_i$ | Outflow coefficients of the free water storage to interflow relationships | [-] | 0.1–0.5 |
| 12 | $C_i$ | Recession constants of the lower interflow storage | [-] | 0.1–0.99 |
| 13 | $C_g$ | Recession constants of the groundwater storage | [-] | 0.5–0.999 |

The geomorphic unit hydrograph model is adopted for overland flow concentration computation, which is based on the results of DEM data analysis (Figure 3).

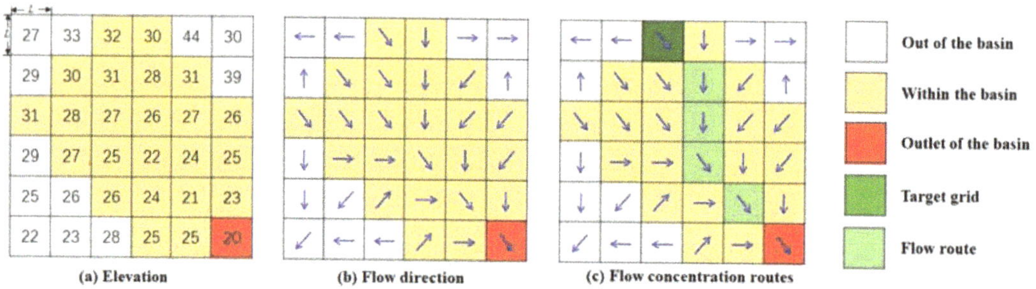


**Figure 3.** Flow direction and flow concentration routes.

The flow direction of each grid is analyzed according to the D8 algorithm [16], and the probability density distribution function of concentration time is determined by computing the time of each water particle falling on the surface of the basin reaching the outlet, so as to further determine the geomorphic unit hydrograph [17]. Based on the principle of energy conversion, this improves the formula of flow velocity and unifies the formula of slope velocity and river velocity [18], as shown in Formula (1).

$$v = \sqrt{\frac{2\mu' g \sum_{k=1}^{n} sin\frac{\theta_k}{2} n_k \Delta h_k}{n}} \quad (1)$$

where $\mu'$ is the energy residual coefficient and its range is [0,1], $\theta$ is the slope angle of the grid outflow direction, $n$ is the total number of grids in the basin upstream of the target grid (including the target grid), $g$ is the gravity acceleration, $\Delta h$ is the elevation difference between the target grid and the outflow grid, $N$ is the number of inflow grids of the target grid, $n_k$ and $v_k$ are the number of upstream grids and the average flow velocity of the kth inflow grid, respectively.

The Muskingum model is used for river network flow concentration [19,20]. Continuous flood routing is realized by segment-by-segment estimation of the model parameters [21].

### 2.3.2. Evaluation Criteria

To evaluate the suitability of the proposed model for the studied Basin, the Nash–Sutcliffe Coefficient of Efficiency (NSCE) is chosen to analyze the degree of goodness of fit [22], which is defined as:

$$NSCE = 1 - \frac{\sum_{i=1}^{N}(Q_s(i) - Q_0(i))^2}{\sum_{i=1}^{N}(Q_0(i) - \overline{Q})^2} \quad (2)$$

where $Q_o(i)$ and $Q_s(i)$ are the observed and simulated flow, respectively, $N$ is the number of data points, and $\overline{Q}$ is the mean value of the observed flow. According to national criteria for flood forecasting in China [23], the scheme is excellent when the average NSCE reaches 0.9. When the average NSCE is greater than 0.7 and less than 0.9, the effect of this scheme is better. This scheme is for reference only; if the average NSCE is greater than 0.5 but less than 0.7, it may not be accurate. Otherwise, the results of the performances of parameter calibration are unsatisfactory for online flood forecasting.

### 2.3.3. Parameter Optimization Method

The shuffled complex evolution (SCE-UA) method is used to optimize the model parameters. The SCE-UA algorithm is a nonlinear hybrid algorithm which combines the advantages of the genetic algorithm and the simplex algorithm, and is based on information exchange and biological evolution laws. It can effectively solve the problems of multi-peak, multi-noise, discontinuity, high-dimension and non-linearity in parameter optimization. Figure 4 shows the calculation flow of the SCE-UA algorithm. This method can efficiently and quickly search for the global optimal solution of model parameters [24,25]. There are 14 parameters that need to be optimized in this study, including 13 parameters of the Xinanjiang model (see Table 1) and 1 parameter of the geomorphic unit hydrograph ($\mu'$). $\mu'$ is the energy residual coefficient and its range is [0,1].

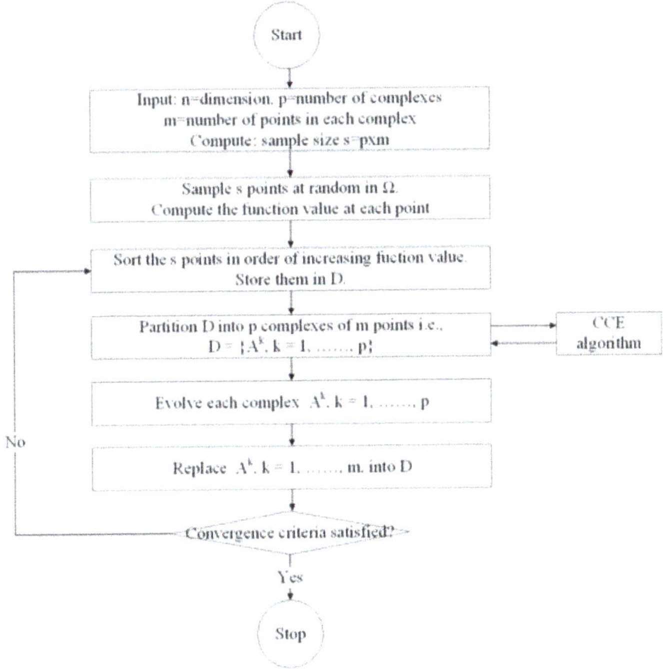


**Figure 4.** Flow of the shuffled complex evolution (SCE-UA) method.

According to the evaluation criteria, the larger the NSCE, the better the simulation effect. Therefore, this study aimed to find the highest mean value of NSCE. Since the goal of the SCE-UA algorithm is to find the minimum, Equation (3) is used as the objective function.

$$F = 1 - \frac{\sum_{i=1}^{t} NSCE_i}{t} \tag{3}$$

where $F$ is the value of objective function, $t$ is the number of floods.

2.3.4. Parameter Regionalization Scheme

The Shortest Distance, Attribute Similarity, Support Vector Regression, Generative Adversarial Networks, Classification and Regression Tree and Random Forest method are used to determine the parameter regionalization scheme, and the final scheme is determined by comparing the simulation results of different methods. For readability, Table 3 lists the abbreviations representing the different methods.

**Table 3.** Abbreviation of parameter regionalization methods.

| Abbreviation | Method Name | Abbreviation | Method Name |
|---|---|---|---|
| SD | shortest distance | GAN | generative adversarial networks |
| AS | attribute similarity | CART | classification and regression tree |
| SVR | support vector regression | RF | random forest |

(1) Shortest Distance (SD)

The nearest basin is determined by computing the spatial distance between the centroid coordinates of the study basin and other basins, and the model parameters of the nearest basin are directly applied to the distributed model of the study basin.

$$D = 2R\sin^{-1}\sqrt{sin\left(\frac{Lat1 - Lat2}{2}\right)^2 + cos(Lat1)cos(Lat2)sin\left(\frac{Lon1 - Lon2}{2}\right)^2} \quad (4)$$

where $D$ is the distance, $R$ is the radius of the earth, about 6,378,137 m, and $Lon1$, $lat1$, $lon2$ and $Lat2$ are the centroid coordinates of the two basins.

(2) Attributes Similarity (AS)

The area ($A$, km$^2$), average slope ($P$), average elevation ($E$, m), average elevation drop ($H$, $m$), shape coefficient ($L$), soil type S = $\{s_1, s_2, s_3\}$ ($s_1$, $s_2$ and $s_3$ are the percentages of clay, silt and sand, %) and land use U = $\{u_1, u_2, u_3, u_4\}$ ($u_1$, $u_2$, $u_3$, $u_4$ are the percentages of forest, grass, cultivated land and other, %) were selected for similarity analysis. The components of U and S range in value from 0 to 1, so no additional processing is required. However, for other attributes, the maximum value method is used for normalization, as follows: collect the maximum values $MaxA$, $MaxP$, $MaxE$, $MaxH$ and $MaxL$ of attribute $A$, $P$, $E$, $H$ and $L$ in 33 catchments, and then let C = $\{A/MaxA, P/MaxP, E/MaxE, H/MaxH, L/MaxL\}$, then C is the normalized result. The similarity index of catchment $x$ and catchment $y$ was defined as Formula (5):

$$T = \frac{\cos(S_x, S_y) + \cos(U_x, U_y) + 1 - D(C_x, C_y)}{3} \quad (5)$$

where $D(a,b)$ and $cos(a,b)$ are Euclidean distances and cosines value of two vectors $a$ and $b$, respectively.

$$D(a,b) = \sqrt{\sum_{i=1}^{n}(a_i - b_i)^2} \quad (6)$$

$$\cos(a, b) = \frac{\sum_{i=1}^{n} a_i b_i}{\sqrt{\sum_{i=1}^{n} a_i^2}\sqrt{\sum_{i=1}^{n} b_i^2}} \quad (7)$$

where $T$ is the similarity index, and its range is [0,1]. The larger the $T$ value, the greater the similarity between the two catchments. Select the basin most similar to the study basin and transplant its parameters.

(3) Support Vector Regression (SVR)

The essence of a support vector machine (SVM) is to map the non-linear function relationship to the linear problem of high-dimensional space, and then find the optimal regression hyperplane in this high-dimensional space, so that all samples are the minimum distance from the optimal hyperplane [26]. Support Vector Regression (SVR) is a method based on a support vector machine to deal with regression problems. It is used to study the relationship between input variables and numerical output variables, and to predict the output value of new variables. It retains the advantages of a support vector machine and is mainly used in the case of a limited or small number of samples [27].

(4) Generative Adversarial Network (GAN)

The generative adversarial network (GAN) is an unsupervised learning model consisting of a discriminator and a generator [28]. The generator automatically generates data, learns the distribution of real samples, and generates pseudo samples that are close to real samples. The discriminator has to distinguish between real samples obtained from the data and fake samples generated by the generator. The two models are iteratively optimized through continuous confrontation training, so that the data distribution generated by the generator is as close as possible to the real data distribution. When the probability of each output of the discriminator is basically 1/2, it indicates that the model has reached the optimal state.

(5) Classification And Regression Tree (CART)

The CART (Classification and Regression Tree) algorithm is a decision tree classification method. It uses a dichotomy recursive segmentation technique to divide the current sample set into two sub sample sets, so that each non leaf node generated has two branches. The decision tree is a weak learning algorithm [29]. The improvement of classification accuracy depends on the reasonable construction and pruning of the tree structure. The CART algorithm generates a decision tree based on the training dataset, and the generated decision tree should be as large as possible. The validation dataset is used to prune the generated tree and select the optimal subtree. At this time, the minimum loss function is used as the pruning standard.

(6) Random Forest (RF)

Random forest model generates multiple different datasets from the original dataset by sampling with put back [30]. The CART tree is used as a weak classifier, and each sub-dataset corresponds to a classifier. Each decision tree selects the attribute with the strongest classification ability for node splitting, without pruning to maximize growth. All final generated decision trees form a random forest. The model can be used for classification or regression prediction, the result of which is determined by the classifier voting.

Based on the above, Figure 5 shows the flow of parameter regionalization. When SVR, GAN, CART and RF are selected for parameter transplantation. The analysis steps are as follows:

(1) For each calibrated catchment A, use the model parameters of any catchment B to compute the average Nash–Sutcliffe coefficient $NSCE_{a\text{-}b}$. Collect all catchment A attributes, catchment B attributes, $NSCE_{a\text{-}b}$ as training dataset for model training. In this study, the sample size of the training set is $25 \times 25$.
(2) For each verified catchment C and calibrated catchment D, use the trained model to take the attributes of C and D as input to predict the mean NSCE, and the parameter group with the highest predictive value is used as the model parameter of C.

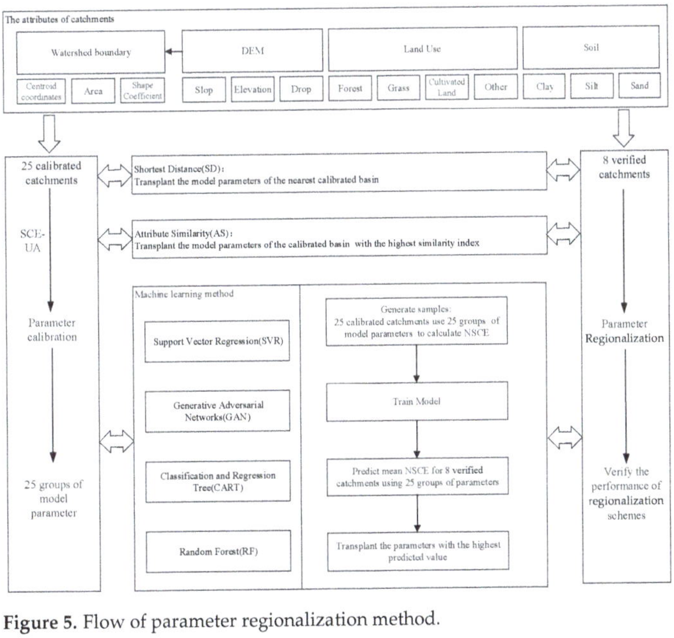


Figure 5. Flow of parameter regionalization method.

## 3. Results

### 3.1. Model Parameter Optimization

The SCE-UA algorithm is used to automatically optimize the model parameters of 25 hydrological stations, and the objective function is to obtain the highest average Nash–Sutcliffe coefficient. The results of parameter calibration are shown in Table 4.

**Table 4.** Simulation results of calibration.

| Station | NSCE | Station | NSCE | Station | NSCE |
|---|---|---|---|---|---|
| NingXiang | 0.78 | HengBanQiao | 0.80 | MaoPing | 0.77 |
| ShiBaZi | 0.76 | HuaQiao | 0.61 | SheBu | 0.83 |
| FeiXian | 0.79 | MaXiPu | 0.72 | HongYanXi | 0.83 |
| FenShi | 0.84 | ShanXiQiao | 0.79 | DaLuPu | 0.82 |
| ZhaiQian | 0.77 | LianYuan | 0.86 | HaoFu | 0.80 |
| JingTouJiang | 0.87 | LouDi | 0.78 | LanShan | 0.67 |
| ShenShanTou | 0.83 | HongYan | 0.74 | JiaYi | 0.86 |
| CaoLongTan | 0.85 | HuangQiao | 0.74 | LiangShuiKou | 0.87 |
| HeTan | 0.79 | | | | |

It can be seen that there are 23 hydrological stations with an average NSCE between 0.7 and 0.9, and 2 between 0.5 and 0.7. According to national criteria for flood forecasting in China, most calibration parameters meet the requirements of online flood forecasting. The distributed model based on the Xinanjiang model and geomorphic unit hydrograph is stable and suitable for most areas of Hunan Province.

The calibration parameters were fed into the distributed model to simulate 426 floods in 25 catchments. Taking LianYuan Station as an example, the calibration result is shown in Figure 6.

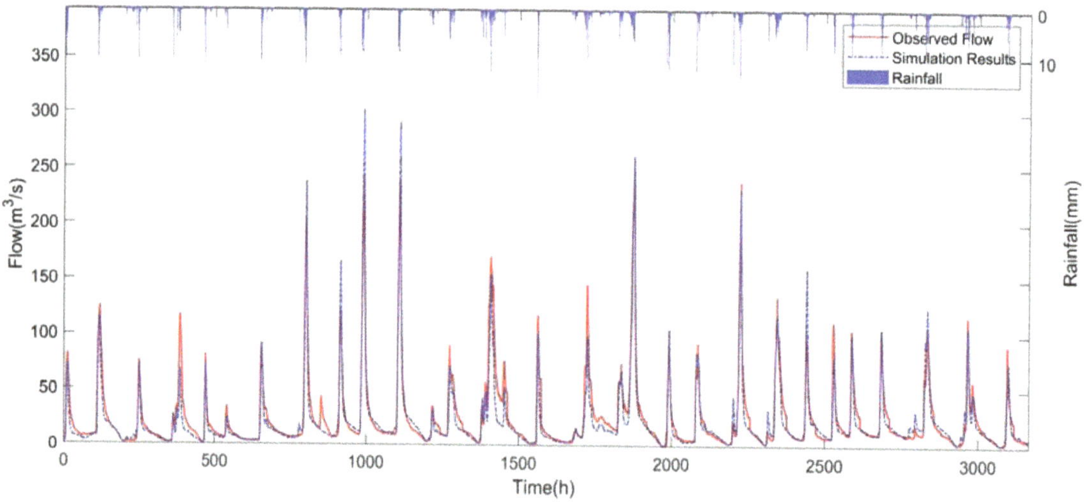


**Figure 6.** Comparison of observed and simulated hydrograph of LianYuan station.

### 3.2. Regionalization Schemes

The shortest distance, attribute similarity, support vector regression, generative adversarial networks, classification and regression tree, and random forest models are selected to construct and verify the parameter regionalization scheme.

According to the catchment attributes, the results of SD and AS can be directly calculated. The centroid coordinates and basic attributes of the 33 catchments are shown in

Table 5, including east longitude (*lon*, ∘), north latitude (*lat*, ∘), area (*A*, $km^2$), average slope (*P*), average elevation (*E*, m), average elevation drop (*H*, m), shape coefficient (*L*), and the percentages of forest ($u_1$), grass ($u_2$), cultivated land ($u_3$), other ($u_3$), clay ($s_1$), silt ($s_2$) and sand ($s_3$). These attributes were extracted during sub-watershed division.

**Table 5.** Information of typical watershed characteristics.

| Station Name | Centroid Coordinates | | Basic Attributes | | | | | Land Use (%) | | | | Soil Type (%) | | |
|---|---|---|---|---|---|---|---|---|---|---|---|---|---|---|
| | *lon* | *lat* | *A* ($km^2$) | *P* | *E* (m) | *H* (m) | *L* | $u_1$ | $u_2$ | $u_3$ | $u_4$ | $s_1$ | $s_2$ | $s_3$ |
| NingXiang | 112.2487 | 28.0972 | 2250 | 0.160 | 167.36 | 145.36 | 0.208 | 49.8 | 1.9 | 42.4 | 5.9 | 44.3 | 55.5 | 0.2 |
| ShiBaZi | 112.3625 | 28.0153 | 564 | 0.125 | 113.57 | 77.57 | 0.289 | 45.2 | 1.4 | 49.6 | 3.8 | 42.3 | 57.7 | 0.0 |
| FeiXian | 112.2963 | 25.6634 | 3559 | 0.213 | 395.16 | 261.16 | 0.244 | 49.8 | 5.5 | 38.3 | 6.4 | 18.3 | 74.9 | 6.8 |
| FenShi | 112.5636 | 25.2776 | 923 | 0.241 | 500.12 | 287.12 | 0.380 | 59.4 | 5.0 | 29.0 | 6.6 | 31.5 | 68.2 | 0.3 |
| ZhaiQian | 113.9350 | 26.0667 | 392 | 0.345 | 1140.59 | 428.59 | 0.629 | 87.8 | 5.2 | 2.7 | 4.3 | 16.8 | 83.2 | 0.0 |
| JingTouJiang | 112.0796 | 26.9399 | 173 | 0.181 | 227.96 | 118.96 | 1.113 | 57.7 | 1.7 | 38.9 | 1.7 | 64.1 | 35.9 | 0.0 |
| ShenShanTou | 112.2196 | 27.0911 | 2930 | 0.176 | 175.06 | 134.06 | 0.131 | 49.0 | 1.6 | 45.9 | 3.5 | 47.0 | 53.0 | 0.0 |
| CaoLongTan | 110.4173 | 28.8383 | 350 | 0.457 | 533.01 | 434.01 | 0.093 | 95.3 | 0.2 | 3.6 | 0.9 | 31.9 | 68.1 | 0.0 |
| HeTan | 109.1290 | 27.1854 | 445 | 0.372 | 616.32 | 274.32 | 0.391 | 84.4 | 4.0 | 10.1 | 1.5 | 90.1 | 9.9 | 0.0 |
| HengBanQiao | 110.5267 | 27.3562 | 31 | 0.323 | 757.67 | 274.67 | 0.424 | 89.5 | 1.2 | 8.5 | 0.8 | 61.1 | 38.9 | 0.0 |
| HuaQiao | 110.2038 | 27.6833 | 81 | 0.279 | 500.44 | 314.44 | 2.249 | 86.6 | 0.8 | 11.7 | 0.9 | 86.5 | 13.5 | 0.0 |
| MaXiPu | 110.4492 | 28.2416 | 342 | 0.383 | 386.66 | 297.66 | 0.119 | 88.6 | 0.8 | 9.3 | 1.3 | 87.7 | 12.2 | 0.1 |
| ShanXiQiao | 110.5982 | 27.5358 | 1211 | 0.342 | 803.06 | 651.06 | 0.173 | 86.6 | 3.6 | 8.3 | 1.5 | 64.0 | 36.0 | 0.0 |
| LianYuan | 111.6015 | 27.6335 | 154 | 0.229 | 248.27 | 128.27 | 0.508 | 54.1 | 2.7 | 36.8 | 6.4 | 59.0 | 41.0 | 0.0 |
| LouDi | 111.7627 | 27.8257 | 1556 | 0.236 | 312.35 | 239.35 | 0.220 | 52.6 | 4.6 | 35.3 | 7.5 | 61.9 | 36.0 | 2.1 |
| HongYan | 110.3664 | 26.8192 | 711 | 0.294 | 634.08 | 303.08 | 0.302 | 78.7 | 1.1 | 18.5 | 1.7 | 66.4 | 33.6 | 0.0 |
| HuangQiao | 110.5639 | 26.7795 | 2689 | 0.226 | 515.43 | 274.43 | 0.268 | 58.3 | 1.4 | 37.1 | 3.2 | 30.5 | 69.3 | 0.2 |
| SheBu | 111.6256 | 27.1677 | 2114 | 0.155 | 322.56 | 137.56 | 0.233 | 38.1 | 3.9 | 51.3 | 6.7 | 42.9 | 57.1 | 0.0 |
| MaoPing | 112.5982 | 27.4498 | 1434 | 0.164 | 154.98 | 122.98 | 0.298 | 58.9 | 0.8 | 38.0 | 2.3 | 23.1 | 76.9 | 0.0 |
| HongYanXi | 109.5954 | 29.3404 | 190 | 0.367 | 689.85 | 322.85 | 0.393 | 84.2 | 1.4 | 13.9 | 0.5 | 42.8 | 57.2 | 0.0 |
| DaLuPu | 111.5408 | 24.8728 | 635 | 0.232 | 487.37 | 269.37 | 0.223 | 54.0 | 5.3 | 35.6 | 5.1 | 18.8 | 80.4 | 0.8 |
| HaoFu | 111.4058 | 25.7095 | 440 | 0.328 | 565.82 | 365.82 | 0.528 | 77.5 | 1.7 | 19.1 | 1.7 | 58.4 | 41.6 | 0.0 |
| LanShan | 112.1479 | 25.2640 | 305 | 0.340 | 675.70 | 428.70 | 0.352 | 82.7 | 1.2 | 12.0 | 4.1 | 54.8 | 45.2 | 0.0 |
| JiaYi | 113.9674 | 28.7773 | 1475 | 0.274 | 316.37 | 248.37 | 0.276 | 77.5 | 2.5 | 17.0 | 3.0 | 44.6 | 55.4 | 0.0 |
| LiangShuiKou | 110.0255 | 29.6915 | 865 | 0.471 | 778.04 | 496.04 | 0.338 | 94.2 | 0.2 | 5.4 | 0.2 | 32.3 | 67.7 | 0.0 |
| LuoLingQiao | 113.3727 | 28.5341 | 340 | 0.170 | 113.07 | 77.07 | 0.652 | 65.2 | 1.2 | 29.9 | 3.7 | 70.0 | 30.0 | 0.0 |
| JiaHe | 112.2656 | 25.3685 | 1501 | 0.263 | 511.87 | 337.87 | 0.276 | 66.2 | 2.7 | 26.1 | 5.0 | 36.8 | 54.3 | 8.9 |
| CaoJiTan | 113.1137 | 26.3820 | 387 | 0.197 | 179.50 | 90.50 | 0.352 | 63.2 | 1.5 | 32.5 | 2.8 | 50.7 | 44.6 | 4.7 |
| ShuangFeng | 112.0508 | 27.4016 | 1552 | 0.164 | 175.11 | 118.11 | 0.301 | 40.4 | 2.9 | 51.6 | 5.1 | 20.5 | 79.5 | 0.0 |
| DongKou | 110.4491 | 27.1576 | 928 | 0.362 | 756.60 | 458.60 | 0.454 | 93.5 | 0.8 | 4.9 | 0.8 | 56.5 | 43.5 | 0.0 |
| JiShou | 109.5497 | 28.3203 | 788 | 0.356 | 621.50 | 453.50 | 0.208 | 81.5 | 3.0 | 14.0 | 1.5 | 58.3 | 41.7 | 0.0 |
| ZhuXiPo | 111.6941 | 28.1490 | 699 | 0.353 | 422.86 | 310.86 | 0.450 | 81.3 | 2.1 | 14.4 | 2.2 | 90.0 | 8.3 | 1.7 |
| ShuangFengTan | 110.5954 | 29.3742 | 444 | 0.360 | 591.95 | 446.95 | 0.257 | 87.0 | 0.6 | 11.2 | 1.2 | 66.1 | 32.5 | 1.4 |

According to the coordinates of the center of the basin, the centroid distance between the verification basin and the calibration basin is calculated by Formula (3), and the calibration basin with the closest distance is selected, and its model parameters are used directly. Normalize the basin properties, calculate the similarity index between the verification basin and the calibration basin using Formula (4), and transfer the model parameters with the highest similarity. Table 6 shows the transplant results of the SD and AS methods.

For SVR, GAN, CART and RF methods, we need to collect samples and train the model first. This required cross-validation of the model parameters for 25 catchments. We apply 25 groups of parameters to the flood simulations of 25 catchments and calculate the mean NSCE.

Figure 7 shows the 25 × 25 cross-validation results for the samples. Among the 625 samples, there are 121 samples with a Nash–Sutcliffe coefficient greater than 0.7, accounting for 19.4% of the total number of samples; 138 samples with a Nash–Sutcliffe coefficient between 0.6 and 0.7, accounting for 22.1%; 112 samples with a Nash–Sutcliffe coefficient between 0.6 and 0.7, accounting for 17.9%; and 254 samples with a Nash–Sutcliffe coefficient less than 0.5, accounting for 40.1%. These samples are used as input to train four models of SVR, GAN, CART and RF, and the results of different parameter groups are used to predict eight verification basins, and the optimal results are selected for parameter transplantation. The Nash–Sutcliffe coefficients of the simulation results are shown in Table 7.

**Table 6.** Transplant results of SD and AS methods.

| Station Name | SD | | AS | |
|---|---|---|---|---|
| | Transplant Station | NSCE | Transplant Station | NSCE |
| LuoLingQiao | JiaYi | 0.66 | LianYuan | 0.86 |
| JiaHe | LanShan | 0.69 | JiaYi | 0.31 |
| CaoJiTan | ZhaiQian | 0.27 | LianYuan | 0.61 |
| ShuangFeng | ShenShanTou | 0.75 | MaoPing | 0.74 |
| DongKou | HengBanQiao | 0.29 | HaoFu | 0.72 |
| JiShou | MaXiPu | 0.71 | ShanXiQiao | 0.46 |
| ZhuXiPo | LouDi | 0.60 | HeTan | 0.72 |
| ShuangFengTan | CaoLongTan | 0.64 | HongYan | 0.69 |
| Average Value | | 0.58 | | 0.64 |

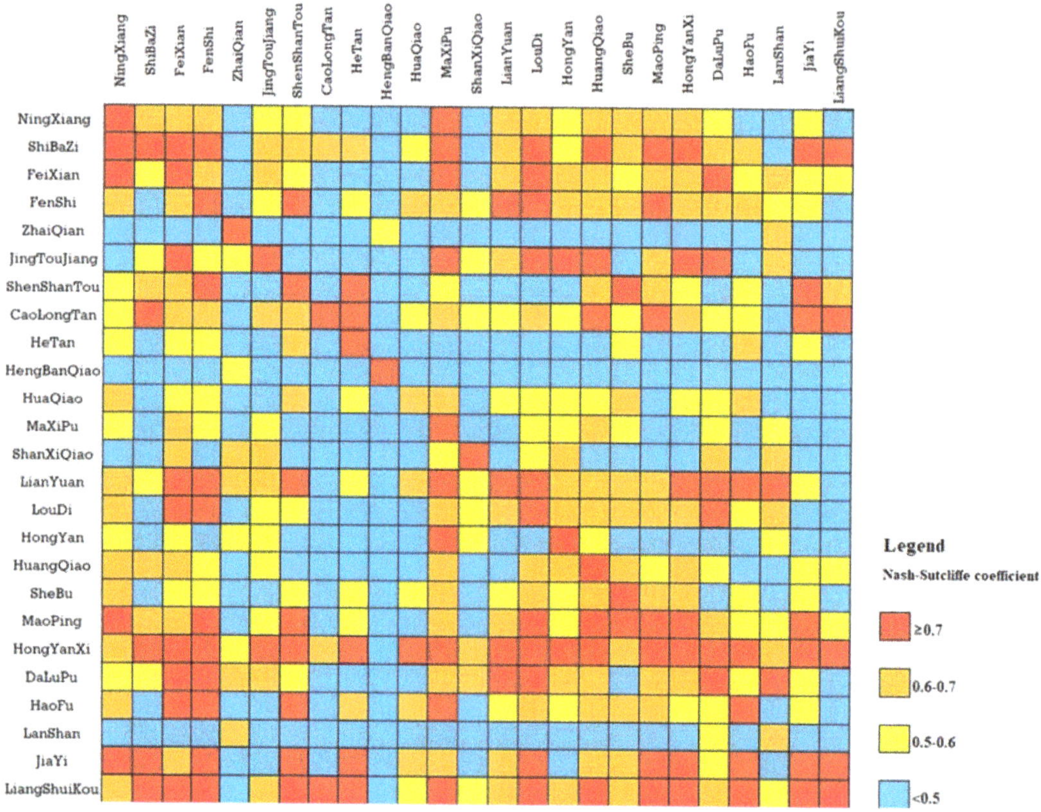


**Figure 7.** Cross validation results of parameter transplantation.

**Table 7.** Transplant results of machine learning methods.

| Station Name | NSCE | | | |
|---|---|---|---|---|
| | SVR | GAN | CART | RF |
| LuoLingQiao | 0.75 | 0.71 | 0.65 | 0.78 |
| JiaHe | 0.55 | 0.69 | 0.51 | 0.65 |
| CaoJiTan | 0.42 | 0.43 | 0.61 | 0.64 |
| ShuangFeng | 0.25 | 0.57 | 0.35 | 0.64 |
| DongKou | 0.80 | 0.78 | 0.80 | 0.65 |
| JiShou | 0.75 | 0.71 | 0.71 | 0.75 |
| ZhuXiPo | 0.52 | 0.64 | 0.57 | 0.59 |
| ShuangFengTan | 0.79 | 0.78 | 0.66 | 0.72 |
| Average Value | 0.60 | 0.66 | 0.61 | 0.68 |

## 4. Discussion

It can be seen from Table 6 that two groups, CaojiTan-ZhaiQian and DongKou-HengBanQiao, performed poorly when using the transplantation parameters of the SD method, with average NSCE of 0.27 and 0.29, respectively. When the AS method was used for transplant parameters, two groups had poor results, namely JiaHe-JiaYi and JiShou-ShanXiQiao, with average NSCEs of 0.31 and 0.46, respectively. Table 8 shows the attributes of these groups of catchments.

**Table 8.** Information on basin attributes.

| Station Name | | CaoJiTan | ZhaiQian | DongKou | HengBanQiao | JiaHe | JiaYi | JiShou | ShanXiQiao |
|---|---|---|---|---|---|---|---|---|---|
| Basin Attributes | Area (km$^2$) | 387 | 392 | 928 | 31 | 1501 | 1475 | 788 | 1211 |
| | Average Slope | 0.197 | 0.345 | 0.362 | 0.323 | 0.263 | 0.274 | 0.356 | 0.342 |
| | Average Elevation (m) | 179.5 | 1140.59 | 756.6 | 757.67 | 511.87 | 316.4 | 621.5 | 803.1 |
| | Average Elevation Drop (m) | 90.5 | 428.59 | 458.6 | 274.67 | 337.87 | 248.4 | 453.5 | 651.1 |
| | Shape Coefficient | 0.352 | 0.629 | 0.454 | 0.424 | 0.276 | 0.276 | 0.208 | 0.173 |
| Land Use (%) | Forest | 63.2 | 87.8 | 93.5 | 89.5 | 66.2 | 77.5 | 81.5 | 86.6 |
| | Grass | 1.5 | 5.2 | 0.8 | 1.2 | 2.7 | 2.5 | 3 | 3.6 |
| | Cultivated Land | 32.5 | 2.7 | 4.9 | 8.5 | 26.1 | 17 | 14 | 8.3 |
| | Other | 2.8 | 4.3 | 0.8 | 0.8 | 5 | 3 | 1.5 | 1.5 |
| Soil (%) | Clay | 50.7 | 16.8 | 56.5 | 61.1 | 36.8 | 44.6 | 58.3 | 64 |
| | Silt | 44.6 | 83.2 | 43.5 | 38.9 | 54.3 | 55.4 | 41.7 | 36 |
| | Sand | 4.7 | 0 | 0 | 0 | 8.9 | 0 | 0 | 0 |

It can be seen from Table 8 that DongKou and HengBanQiao are not only close, but also most of the attributes are similar except for the area and average drop. The area of DongKou is 931 km$^2$, and the area of HengBanQiao is 31 km$^2$. Their average drops are 458.6 m and 274.67 m, respectively. It is obvious that the different areas will lead to large differences in concentration time, and the average drop may significantly affect the concentration speed, which is the most critical factor affecting the geomorphic unit hydrograph [18]. Similarly, compared with JiaHe and JiaYi, their attributes are very similar, except for average elevation and drop. Therefore, we can infer that if the attributes of two catchments are very close, but their average drop difference is significant, this is likely to cause a failed transplantation. The opposite conclusion cannot be established. Table 6 shows an example with the best results (LuoLingQiao-LianYuan). The average NSCE of transplantation can reach 0.86, which is excellent according to the evaluation criteria. However, the attributes of the two catchments, including the average drop, differed significantly (as shown in Table 5).

From the above cases, it can be seen that the applicable conditions and scope of parameter transplantation are relatively complex, and a single factor cannot be considered in isolation. When multiple attributes are considered for parameter transplantation, the results may not be satisfactory for catchments with similar attributes sometimes, so precisely defining the similarity index is a challenge.

In contrast, machine learning methods can discover more hidden rules in data. However, the methods of machine learning cannot all achieve satisfactory results. Comparing only the average NSCE, the results of SVR and CART were even worse than the AS method. In order to better compare the performance of different methods, Table 9 shows the optimal value, worst value and average value obtained using different methods. Figure 8 shows the average NSCEs for the different methods.

**Table 9.** Comparison of parameter regionalization schemes.

| Items | NSCE | | | | | |
|---|---|---|---|---|---|---|
| | SD | AR | SVR | GAN | CART | RF |
| Best | 0.75 | 0.86 | 0.80 | 0.78 | 0.80 | 0.78 |
| Worst | 0.27 | 0.31 | 0.25 | 0.43 | 0.35 | 0.59 |
| Average | 0.58 | 0.64 | 0.60 | 0.66 | 0.61 | 0.68 |

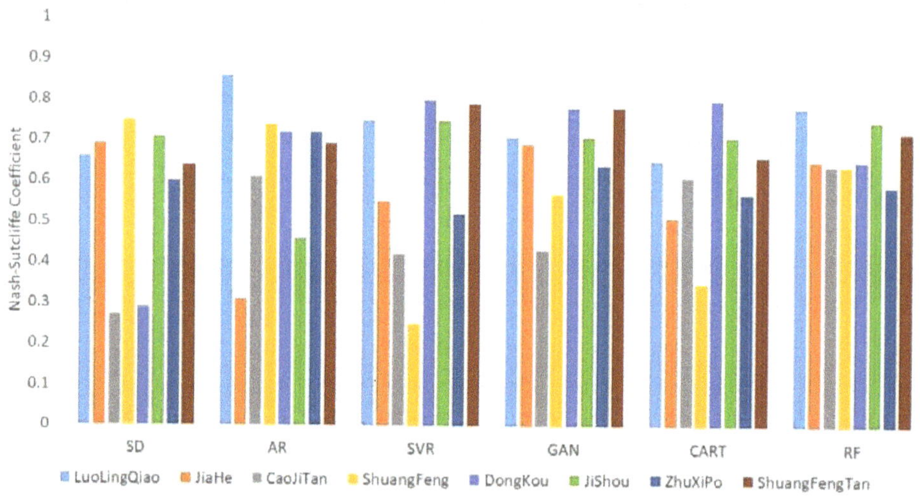


**Figure 8.** Validation results of regionalization schemes.

It can be seen from Table 9 that the average and worst Nash–Sutcliffe coefficients of the simulation results using the random forest model are the highest. Among the best NSCE results in Table 9, AR > SVR ≥ CART > RF ≥ GAN > SD, with AR performing best and SD performing worst. The worst result of NSCE is RF > GAN > CART > AR > SD > SVR; RF is the best and SVR is the worst. According to the NSCE average results, RF > GAN > AR > CART> SVR > SD; RF performed the best and SD performed the worst.

Table 10 summarizes the validation results of the different methods and shows the percentage of catchments with an average NSCE greater than 0.9, greater than 0.7 and less than 0.9, greater than 0.5 and less than 0.7, and less than 0.5.

**Table 10.** NSCE statistical results.

| NSCE | SD | AR | SVR | GAN | CART | RF |
|---|---|---|---|---|---|---|
| ≥0.9 | 0 | 0 | 0 | 0 | 0 | 0 |
| 0.7–0.9 | 25% | 50% | 50% | 50% | 25% | 37.5% |
| 0.5–0.7 | 50% | 25% | 25% | 37.5% | 62.5% | 62.5% |
| <0.5 | 25% | 25% | 25% | 12.5% | 12.5% | 0 |

It can be seen from Table 10 that all of the NSCE results of RF are greater than 0.5, which is not achieved by all of the other methods. According to national criteria for flood forecasting in China, if the average NSCE is less than 0.5, the simulation result is unsatisfactory for online flood forecasting. Therefore, the RF model has better performance than the other methods.

Figure 9 lists the importance of each attribute in the RF model. The most important attribute for prediction using the RF model is the percentage of cultivated land area within the transplanted catchment, followed by the area and average elevation of the calibration catchment. It is well known that slope is a significant impact on hydrological models. However, from the parameter importance of the RF model, the influence of slope is smaller than that of cultivated land, which may be another issue that needs further research.

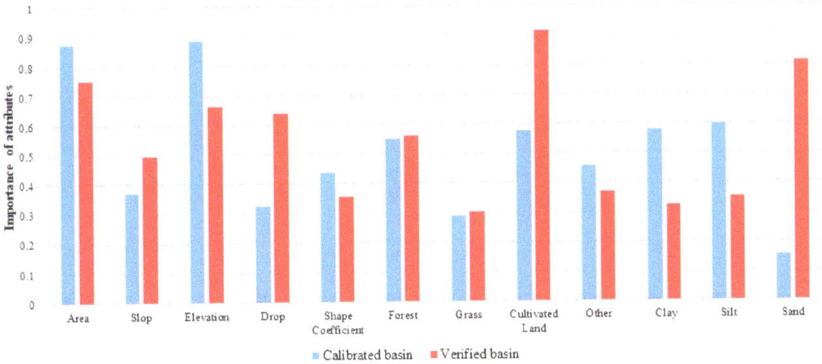


**Figure 9.** Importance of attributes in RF model.

## 5. Conclusions

In this study, the distribution hydrological models of 33 small and medium-sized catchments in Hunan Province were constructed. The model parameters of 25 catchments were calibrated by using the SCE-UA algorithm. The parameter regionalization scheme including Shortest Distance (SD), Attribute Similarity (AS), Support Vector Regression (SVR), Generative Adversarial Networks (GAN), Classification and Regression Tree (CART) and Random Forest (RF) were validated using data from eight catchments. The main conclusions are as follows:

(1) A total of 426 floods of 25 catchments were selected to calibrate the model parameters. Among the simulation results of these 25 catchments, there are 23 catchments with an average NSCE greater than 0.7, and 2 between 0.5 and 0.7. According to national criteria for flood forecasting in China, most calibration parameters meet the requirements of online flood forecasting. The distributed model based on the Xinanjiang model and geomorphic unit hydrograph is suitable for most areas of Hunan Province.

(2) Based on the watershed attributes and cross validation results of model parameters, six parameter regionalization schemes including SD, AR, SVR, GAN, CART and RF were generated, and 136 floods of 8 catchments were used for verification. The average values of the Nash–Sutcliffe coefficients of each scheme were 0.58, 0.64, 0.60, 0.66, 0.61 and 0.68, and the worst values were 0.27, 0.31, 0.25, 0.43, 0.35 and 0.59. The

Nash–Sutcliffe coefficients of the RF model are all greater than 0.5, which cannot be achieved by other methods. The RF model is the most stable solution and significantly outperforms other methods. Using the random forest model to regionalize parameters can improve the accuracy of flood simulation in ungauged areas, which is of great significance for flash flood forecasting and early warning.

(3) The applicable conditions and scope of parameter transplantation are relatively complex, and a single factor cannot be considered in isolation, and the attributes of adjacent catchments may also vary greatly. The result of the attribute similarity method is not very stable, and transplantation can fail when most of the attributes of two catchments are similar, but if the attributes are very different, sometimes good results will be achieved. According to the parameter importance analyzed by the RF model, the slope is not so important, while the cultivated land area is the key to decision making. This result goes against common sense and deserves further research.

There are many factors that affect the accuracy of parameter transplantation. In practice, continuous data collection is required to improve the quality of the underlying dataset. With the accumulation of data and the continuous improvement of the regionalization model, the accuracy of parameter transplantation can be improved.

**Author Contributions:** Conceptualization, W.W. and Y.T.; methodology, Q.M. and Y.T.; software, Y.Z.; validation, R.D.; project administration, C.L. All authors have read and agreed to the published version of the manuscript.

**Funding:** Special project for collaborative innovation of science and technology in 2021 (No: 202121206).

**Institutional Review Board Statement:** Not applicable.

**Informed Consent Statement:** Not applicable.

**Data Availability Statement:** Not applicable.

**Conflicts of Interest:** The authors declare no conflict of interest.

## References

1. Gou, J.J.; Miao, C.Y.; Duan, Q.Y. Progress in parameter sensitivity analysis-optimization-regionalization methods for hydrological models. *Prog. Geogr.* **2022**, *41*, 1338–1348. [CrossRef]
2. Sivapalan, M. Prediction in ungauged basins: A grand challenge for theoretical hydrology. *Hydrol. Process.* **2003**, *17*, 3163–3170. [CrossRef]
3. Young, A.R. Stream flow simulation within UK ungauged catchments using a daily rainfall-runoff model. *J. Hydrol.* **2006**, *320*, 155–172. [CrossRef]
4. Parajka, J.; Bloschl, G.; Merz, R. Regional calibration of catchment models: Potential for ungauged catchments. *Water Resour. Res.* **2007**, *43*. [CrossRef]
5. Li, H.X.; Zhang, Y.Q.; Chiew, F.H.S.; Xu, S. Predicting runoff in ungauged catchments by using Xinanjiang model with MODIS leaf area index. *J. Hydrol.* **2009**, *370*, 155–162. [CrossRef]
6. Yokoo, Y.; Kazama, S.; Sawamoto, M.; Nishimura, H. Regionalization of lumped water balance model parameters based on multiple regression. *J. Hydrol.* **2001**, *246*, 209–222. [CrossRef]
7. Cheng, X.; Ma, X.X.; Wang, W.S.; Liu, X.X.; Wang, Q.L.; Xiao, Y. Applicability Research of HEC-HMS Model Parameter Regionalization in Small Basin of Henan Province. *J. China Hydrol.* **2022**, *42*, 40–46, 102. [CrossRef]
8. Yao, C.; Qiu, Z.Y.; Li, Z.J.; Hu, W.D.; Xu, J. Parameter regionalization study and application of API model and Xin'anjiang model. *J. China Inst. Water Resour. Hydropower Res.* **2019**, *47*, 189–194. [CrossRef]
9. Sun, Z.L.; Liu, Y.L.; Chen, X.; Shu, Z.K.; Wu, H.F.; Wang, J.; Bao, Z.X.; Wang, G.Q. Review of Hydrological Model Parameter Regionalization Method [OL]. *J. China Hydrol.* **2022**, 1–9. [CrossRef]
10. Yi, X.; Zhou, F.; Wang, X.Y.; Yang, Y.H.; Guo, H.C. Classification and runoff simulation of data-scarce basins based on self-organizing maps. *Prog. Geogr.* **2014**, *33*, 1109–1116.
11. Ragettli, S.; Zhou, J.; Wang, H.; Liu, C.; Guo, L. Modeling flash floods in ungauged mountain catchments of China: A decision tree learning approach for parameter regionalization. *J. Hydrol.* **2017**, *555*, 330–346. [CrossRef]
12. Liu, C.J.; Zhou, J.; Wen, L.; Ma, Q.; Guo, L.; Ding, L.Q.; Sun, D.Y. Research on spatio temporally-mixed runoff model and parameter regionalization for small and medium-sized catchments. *J. China Inst. Water Resour. Hydropower Res.* **2021**, *19*, 99–114. [CrossRef]
13. Zhao, R.J. The Xinanjiang model applied in China. *J. Hydrol.* **1992**, *135*, 371–381. [CrossRef]

14. Zhao, R.J.; Zhang, Y.L.; Fang, L.R. The Xinanjiang model. In *Hydrological Forecasting, Proceeding Oxford Symposium, Oxford, UK, 15–18 April 1980*; IASH: Washington, DC, USA, 1980; pp. 351–356.
15. Wang, W.C.; Cheng, C.T.; Chau, K.W.; Xu, D.M. Calibration of Xinanjiang model parameters using hybrid genetic algorithm based fuzzy optimal model. *J. Hydroinform.* **2012**, *14*, 784–799. [CrossRef]
16. Fairfield, J.; Leymarie, P. Drainage networks from grid digital elevation models. *Water Resour. Res.* **1991**, *27*, 709–717. [CrossRef]
17. Rui, X.; Yu, M.; Liu, F.; Gong, X. Calculation of watershed flow concentration based on the grid drop concept. *Water Sci. Eng.* **2008**, *1*, 1–9. [CrossRef]
18. Wang, W.-C.; Zhao, Y.-W.; Chau, K.-W.; Xu, D.-M.; Liu, C.-J. Improved flood forecasting using geomorphic unit hydrograph based on spatially distributed velocity field. *J. Hydroinform.* **2021**, *23*, 724–729. [CrossRef]
19. Tewolde, M.; Smithers, J. Flood routing in ungauged catchments using Muskingum methods. *Water SA* **2006**, *32*, 379–388. [CrossRef]
20. Todini, E. A mass conservative and water storage consistent variable parameter Muskingum-Cunge approach. *Hydrol. Earth Syst. Sci.* **2007**, *11*, 1645–1659. [CrossRef]
21. Song, X.-M.; Kong, F.-Z.; Zhu, Z.-X. Application of Muskingum routing method with variable parameters in ungauged basin. *Water Sci. Eng.* **2011**, *4*, 1–12. [CrossRef]
22. Nash, J.E.; Sutcliffe, J.V. River flow forecasting through conceptual models part I—A discussion of principles. *J. Hydrol.* **1970**, *10*, 282–290. [CrossRef]
23. National Center of Hydrological Information. *The National Criteria for Hydrological Forecasting*; Hydroelectric Press: Beijing, China, 2000.
24. Duan, Q.Y.; Gupta, V.K.; Sorooshian, S. Shuffled complex evolution approach for effective and efficient global minimization. *J. Optim. Theory Appl.* **1993**, *76*, 501–521. [CrossRef]
25. Sorooshian, S.; Duan, Q.Y.; Gupta, V.K. Optimal use of the SCE-UA global optimization method for calibrating watershed models. *J. Hydrol.* **1994**, *158*, 265–284. [CrossRef]
26. Okkan, U.; Serbes, Z.A. Rainfall-runoff modeling using least squares support vector machines. *Environmetrics* **2012**, *23*, 549–564. [CrossRef]
27. Panahi, M.; Sadhasivam, N.; Pourghasemi, H.R.; Rezaie, F.; Lee, S. Spatial prediction of groundwater potential mapping based on convolutional neural network (CNN) and support vector regression (SVR). *J. Hydrol.* **2020**, *588*, 125033. [CrossRef]
28. Wang, M.Q.; Yuan, W.W.; Zhang, J.Y. Overview of research on Generative Adversarial Network GAN. *Comput. Eng. Des.* **2021**, *42*, 3389–3395. [CrossRef]
29. Li, X.N.; Zhang, Y.J.; She, Y.J.; Chen, L.W.; Chen, J.X. Estimation of impervious surface percentage of river network regions using an ensemble leaning of CART analysis. *Remote Sens. Land Resour.* **2013**, *25*, 174–179. [CrossRef]
30. Breiman, L. Random forests. *Mach. Learn.* **2001**, *45*, 5–32.

*water*

MDPI

*Article*

# Prediction of the Periglacial Debris Flow in Southeast Tibet Based on Imbalanced Small Sample Data

**Jun Du [1,2,*], Hong-ya Zhang [1,2], Kai-heng Hu [3], Lin Wang [4] and Lin-yao Dong [1,2]**

1 Department of Soil and Water Conservation, Changjiang River Scientific Research Institute (CRSRI), Wuhan 430010, China
2 Research Center on Mountain Torrent and Geologic Disaster Prevention, Ministry of Water Resources, Wuhan 430010, China
3 Institute of Mountain Hazards and Environment, Chinese Academy of Sciences (CAS), Chengdu 610041, China
4 Chuo Kaihatsu Corporation, Tokyo 169-8612, Japan
* Correspondence: dxjx2006@126.com

**Abstract:** Using data sourced from 15 periglacial debris flow gullies in the Parlung Zangbo Basin of southeast Tibet, the importance of 26 potential indicators to the development of debris flows was analyzed quantitatively. Three machine learning approaches combined with the borderline resampling technique were introduced for predicting debris flow occurrences, and several scenarios were tested and compared. The results indicated that temperature and precipitation, as well as vegetation coverage, were closely related to the development of periglacial debris flow in the study area. Based on seven selected indicators, the Random Forest-based model, with its weighted recall rate and Area Under the ROC Curve (AUC) greater than 0.76 and 0.77, respectively, performed the best in predicting debris flow events. Scenario tests indicated that the resampling was necessary to the improvement of model performance in the context of data scarcity. The new understandings obtained may enrich existing knowledge of the effects of main factors on periglacial debris flow development, and the modeling method could be promoted as a prediction scheme of regional precipitation-related debris flow for further research.

**Keywords:** periglacial debris flow; southeast Tibet; small sample imbalanced data; prediction model; random forest

**Citation:** Du, J.; Zhang, H.-y.; Hu, K.-h.; Wang, L.; Dong, L.-y. Prediction of the Periglacial Debris Flow in Southeast Tibet Based on Imbalanced Small Sample Data. *Water* **2023**, *15*, 310. https://doi.org/10.3390/w15020310

Academic Editors: Cristiana Di Cristo and Marcel Hürlimann

Received: 25 November 2022
Revised: 26 December 2022
Accepted: 9 January 2023
Published: 11 January 2023



## 1. Introduction

The term 'periglacial debris flow' here refers to a special torrent containing a large amount of sediment and rocks formed by the mechanisms such as ice avalanches, rock avalanches, snowmelt, permafrost degradation, and glacier lake outbursts in the margin of modern glaciers and snow-covered areas. As an important type of mountain disaster in high-latitude and high-altitude areas, the periglacial debris flow presents great destructive power because of its huge scale, sudden outbreak, and rapid process, and it is widely distributed in, e.g., Switzerland, France, Italy, Russia, Canada, the USA and China. In recent decades, with the increase in extreme weather events caused by global warming, periglacial debris flow events have become more and more frequent, which has attracted extensive attention from academia and society [1–4].

As one of the most sensitive areas to global climate change on the earth [5], periglacial debris flow events in southeast Tibet have increased significantly in recent decades [6]. Simultaneously, with the increase in human activities, there are more and more reports of debris flow blocking traffic, inundating houses, and causing casualties, which has brought great trouble to the production and life of local people [3,7,8]. As a result, relevant prevention work is becoming a focus of public attention.

It is difficult to implement engineering measures to control periglacial debris flows because of the harsh environment at high-altitude areas. Monitoring and early warning have become the main means of corresponding disaster prevention and mitigation. Therefore, it is crucial to clarify the initial conditions and formation mechanism of the debris flow process. Previous studies have shown that the formation and development of periglacial debris flow not only is related to general precipitation and topographic factors but also is significantly affected by temperature changes [1,6]. Because of the phenomenon of ice segregation, the ice-filled frozen rock joint is widely developed in the glacier area [9]. When the temperature rises and the ice and snow melt, the originally stable rock mass may fail and collapse with a massive magnitude caused by the mechanisms of unloading and brittle fracture [10]. For the permafrost, the so-called thaw consolidation effect caused by the temperature rise will increase the pore water pressure of the soil and then trigger landslides and debris flows [11].

The loose materials of periglacial debris flows mainly come from ice avalanche, rock avalanche and freeze–thaw erosion in high-altitude areas. In many cases, these materials have high potential energy and large volume [2–4]. Under the erosion and infiltration of snowmelt water, they are easy to roll off the hillside, causing vibration liquefaction and saturation liquefaction [12]. Therefore, unlike commonly reported rainfall-related debris flow events [13–15], periglacial debris flows may erupt even without significant precipitation [16]. Apparently, the compound action of temperature and precipitation is a catalyst for stimulating the transformation of these loose materials to debris flows. Even worse, greater uncertainty in the occurrence and propagation of debris flows has resulted from the effects of material accumulation and scale upgrades as materials roll downslope, in which the destruction and formation of cascade barrier dams continues [3].

Accordingly, it is evident that the formation mechanism of periglacial debris flows is more complicated than the common type of debris flows triggered by a rainfall–runoff process. Although previous case studies have provided some qualitative knowledge and techniques, the results from mechanism analysis based on energy conservation and mechanical balance are not universal and remain hard to popularize. In addition, due to the difficulty of observation in high-latitude and high-altitude areas, the empirical relationships between periglacial debris flows and potential influencing factors are also insufficient. Therefore, how to quantitatively identify the main impact factors of regional debris flow development under the constraints of poor mechanism knowledge and insufficient data and then establish a feasible event-based prediction model is not only the need of disaster prevention and mitigation but also of great scientific research significance.

Machine learning, here refers to a method of data analysis whereby rule-structured classifiers for predicting the classes of newly sampled cases are obtained from a "training set" of pre-classified cases [17], has become a research hotspot in academia in recent years. As a kind of general theory and technology in big data analysis, machine learning has performed outstandingly in many fields. In the field of mountain disasters, Costache [18] compared the application performance of different machine learning models and their hybrid algorithms in a risk assessment of mountain torrents in typical watersheds in Romania, and the results showed that the model with hybrid algorithms had the best accuracy (AUC > 0.87). Zhou et al. [19] divided the step-like landslide displacement in the Three Gorges Reservoir Area into trend term and periodic term, and they believed that using the particle swarm optimization–support vector machine (PSO–SVM) model better represented the relationship between reservoir water-level fluctuations and periodic displacement than the traditional method. Liang et al. [13] used Bayesian network (BN), SVM, and artificial neural network to assess the hazard pattern of debris flow in mainland China and found that the BN method provided the best detection accuracy. Focusing on the post-fire debris flow event in California (USA), some scholars built multi-element index systems from the aspects of rainfall, terrain, vegetation, soil erodibility, and disaster degree, and used machine learning algorithms, such as logistic regression, decision tree, and naive Bayes, to establish classification models to estimate the occurrence rate of debris flow [14,15,20]. The AUC of relevant models could reach more than 0.7.

It is true that the machine learning algorithms have achieved many successes in the above areas. However, most of the cases come from the risk or hazard assessment of mountain disasters, whereas real-time prediction of disaster-prone events is rarely involved. In most cases, the applied algorithm requires a large-scale data set to obtain more reliable results, such as lower risk of overfitting and higher AUC [21,22]. In addition, for the prediction of debris flow with strong uncertainty, the imbalance between positive and negative samples, i.e., the debris flow events occurred or not, is quite common. Although the influence can be reduced by resampling or weight adjustment, in theory [23–25], few relevant successful cases have been reported.

Accordingly, based on the highly available remote sensing data and the real-time temperature and precipitation information, this study took the imbalanced small sample data of typical periglacial debris flow gullies in the Parlung Zangbo Basin of southeast Tibet as an example. Then, we (1) established a comprehensive index system using analysis methods such as field survey, geostatistical analysis, and geographic information system (GIS)-based map algebraic calculation; (2) analyzed the main indicators of regional debris flow development quantitatively utilizing correlation analysis, recursive feature elimination based on support vector machine (SVM-RFE) and GainRatio to select the monitoring and early warning indicators; (3) integrated the intelligent resampling technique to optimize the data samples; and finally (4) built the classification models for comparative tests and prediction scheme optimization.

## 2. Materials and Methods

### 2.1. Study Area

The Parlung Zangbo River is located northeast of the Yarlung Zangbo River between the Eastern Himalayas and the Nyainqentanglha Mountains. It is a tributary of the Yarlung Zangbo River with the largest discharge. There are steep slopes, deeply incised valleys, and developed faults in this area. The Parlong Zangbo River originates from the Azar Glacier in the south of Lake Ranwu, which is about 4900 m.a.s.l. It flows from southeast to northwest and then enters the Yarlung Zangbo River at an altitude of about 1540 m. The relative elevation difference in the basin is between 1500 and 4000 m.

Affected by the warm moist flow from the Bay of Bengal, the basin has abundant precipitation and contrasting seasons of rain and drought. The average annual precipitation is nearly 900 mm, mainly from May to October [26]. The annual average temperature is between 10 and 12 °C, and the distribution of vertical temperature gradient is obvious. Generally, the climate in the low-altitude valley is warm and humid, where lush forests, agriculture and animal husbandry economy have developed. In contrast, the high-altitude area of the basin is occupied by a typical frigid alpine climate, where the marine glaciers and their heritage (i.e., ice lakes) are widely developed [27]. Specially, this high-altitude area is characterized by severe cold and freezing weathering, and consequently broken surface materials, collapse, landslide, and other gravity processes are distributed widely. In short, it is a typical monsoon-type periglacial debris flow disaster-prone area in China, and it is reported that 67 debris flow gullies have been identified in the Ranwu–Peilong section only before 1999 [28].

According to literature investigation, field investigation and observation, a certain number of debris flow records have accumulated in the Parlung Zangbo basin since the 1950s [29]. However, considering that global warming has caused significant changes in the local debris flow development environment [6], applicability of the early part of these data is limited. Thus, we selected 23 debris flow events that have occurred in 15 debris flow gullies in the Ranwu–Peilong section between 2012 and 2020 as the research objects (Figure 1, Table 1). The gullies are characterized by a total area of 1–28 km$^2$, with their outlets elevated at 2400–3800 m and having an over 30° average slope. The bedrock of these gullies is composed of granite gneiss, slate and quartzite, and the low-altitude part of the gullies is basically a forest covered area [7]. These established the general geographical features of the regional periglacial debris flow gullies.

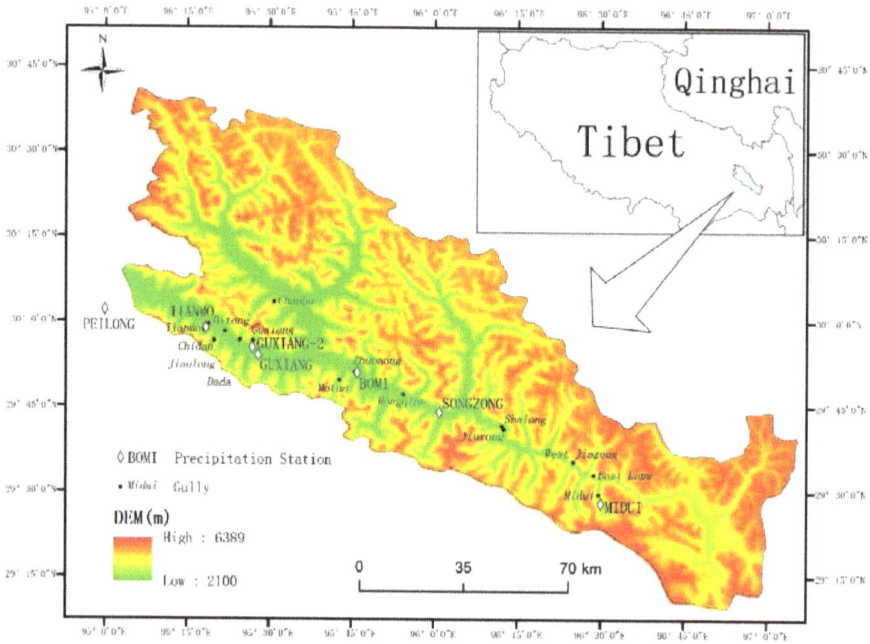


**Figure 1.** Debris flow gullies and precipitation stations in the Ranwu–Peilong section of the Parlung Zangbo Basin.

**Table 1.** Debris flow events recorded in the study area from 2012 to 2020.

| Gully No. | Gully | Area (km$^2$) | Event No. | Date | Time | Possible Causes | Monitoring Station |
|---|---|---|---|---|---|---|---|
| 1 | Bitong | 24 | DF1 | 2016-09-05 | 06:30 | Snowmelt, Precipitation | PEILONG |
| | | | DF2 | 2018-07-11 | 03:00 | Snowmelt, Precipitation | GUXIANG |
| 2 | Chaobu | 16 | DF3 | 2017-08-03 | 17:00 | Ice and Snow melt | GUXIANG |
| 3 | Chidan | 28 | DF4 | 2016-09-05 | 11:00 | Snowmelt, Precipitation | PEILONG |
| 4 | Dada | 3 | DF5 | 2020-07-10 | 16:10 | Snowmelt, Precipitation | TIANMO |
| 5 | East Lapu | 4 | DF6 * | 2013-07-05 | 21:00 | Snowmelt, Precipitation | MIDUI/ SONGZONG |
| | | | DF7 * | 2014-07-24 | 20:00 | Snowmelt, Precipitation | |
| | | | DF8 | 2018-05-22 | 17:00 | Snowmelt, Precipitation | |
| 6 | Guxiang | 25 | DF9 | 2020-07-09 | 21:00 | Ice and Snow melt | GUXIANG-2 |
| 7 | Jiaolong | 22 | DF10 | 2016-09-05 | — | Snowmelt, Precipitation | PEILONG |
| 8 | Jiurong | 7 | DF11 * | 2014-08.18 | 23:00 | Snowmelt, Precipitation | SONGZONG |
| | | | DF12 * | 2014-08-23 | 02:00 | Snowmelt, Precipitation | |
| | | | DF13 * | 2015-08-04 | 22:30 | Snowmelt, Precipitation | |
| | | | DF14 * | 2015-08-20 | 07:40 | Snowmelt, Precipitation | |

**Table 1.** *Cont.*

| Gully No. | Gully | Area ($km^2$) | Event No. | Date | Time | Possible Causes | Monitoring Station |
|---|---|---|---|---|---|---|---|
| 9 | Midui | 1 | DF15 * | 2015-08-19 | 23:30 | Snowmelt, Precipitation | MIDUI/ SONGZONG |
| 10 | Motuo | 3 | DF16 * | 2015-08-19 | 22:30 | Snowmelt, Precipitation | BOMI/ SONGZONG |
| 11 | Rongqian | 5 | DF17 * | 2015-08-19 | 18:30 | Snowmelt, Precipitation | BOMI/ SONGZONG |
| 12 | Shalong | 15 | DF18 * | 2015-08-19 | 21:00 | Snowmelt, Precipitation | SONGZONG |
| 13 | Tianmo | 18 | DF19 | 2018-07-11 | 03:00 | Snowmelt, Precipitation | GUXIANG |
| 14 | West Jiazong | 2 | DF20 * | 2012-09-22 | 09:00 | Snowmelt, Precipitation | MIDUI |
| | | | DF21 * | 2013-07-05 | 21:00 | Snowmelt, Precipitation | |
| | | | DF22 * | 2013-07-31 | 18:00 | Snowmelt, Precipitation | |
| 15 | Zhuonong | 5 | DF23 * | 2015-08-19 | 19:20 | Snowmelt, Precipitation | BOMI/ SONGZONG |

Note: Data from Deng et al. [7], Zeng et al. [30], Li et al. [31], and field observation; * refers to the events used for training and validation.

### 2.2. Method

The quantitative analysis methods used in this paper mainly include Pearson correlation analysis, Borderline Synthetic Minority Over-Sampling Technique (Borderline SMOTE), Random Forest (RF), and Support Vector Machine (SVM). Because these methods have many separate applications reported [13–15,32], in this study, we provided only the necessary introductions to the main machine learning algorithms.

#### 2.2.1. Borderline SMOTE

In this study, we summarized the prediction of debris flow occurrence into a bivariate classification problem, that is, the events with or without debris flow. Data from 23 debris flow events in 15 debris flow gullies were collected in the study area. Excluding the 9 events used for testing after 2015, only 14 events relevant to 8 gullies could be used for training and validation (Table 1). According to the monitoring results of the other gullies during a debris flow event period, we expanded the training data to 165 samples, of which 16 were positive samples of debris flow (12 events plus two events used twice by using two different but adjacent precipitation stations), and 149 were negative samples of non-debris flow (Figure 2). The ratio of positive and negative samples was close to 1:10. Hence, it was a typical imbalanced small sample data set. If these data were trained directly without special processing, it could be found that even if all positive samples were judged to be negative, the accuracy of the model, i.e., the ratio of correctly judged samples to the whole sample set here, still reached 91%, which was not the desired result. Therefore, we introduced a resampling technique for such data sets.

Commonly used resampling techniques include over-sampling and under-sampling. They can achieve the purpose of balancing data by directly copying minority class samples or subtracting redundant information from some of the majority class samples. Obviously, the over-sampling makes the decision boundary more specific and easily leads to over-fitting, whereas the under-sampling directly reduces the potentially useful information. To eliminate these adverse effects, Chawla [23] proposed the Synthetic Minority Over-Sampling Technique (SMOTE); the fundamental idea of this technique is to extract several of the nearest congener samples b around the selected minority class center sample a and then interpolate any points on the ab line as the new synthetic samples.

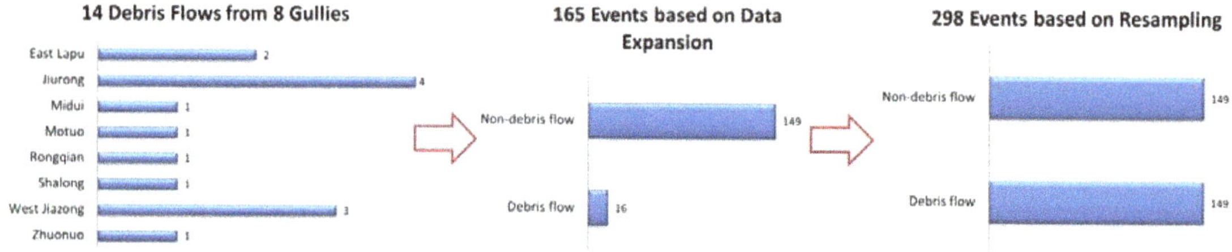


**Figure 2.** The preparation process of the data set for training and validation.

Technically, SMOTE can alleviate the overfitting phenomenon caused by ordinary oversampling to a certain extent, but treating all minority samples equally also causes problems, such as high variance (overgeneralization) or data overlap of the new data sets. In addition, the classification idea of SVM tells us that the samples located at the decision boundary often play a decisive role in correct classification [33], and the expansion of these samples is usually more valuable. Accordingly, Hui [34] proposed an improved version of SMOTE (i.e., Borderline SMOTE), which can determine the decision boundary based on the proximity relationship between each minority sample and the surrounding majority samples. Then, only the minority samples near the boundary (i.e., the danger zone) can be interpolated linearly, such that the new data set can be more prominent in the decision boundary.

The method can be divided into two types (i.e., Borderline SMOTE-1 and Borderline SMOTE-2). The difference between these two types is that the Borderline SMOTE-1 performs ordinary SMOTE interpolation on the minority samples of danger zone after the decision boundary has been determined, whereas the Borderline SMOTE-2's minority center samples will be connected with the neighboring majority samples for interpolating new samples. Theoretically, Borderline SMOTE-1 is more helpful to highlight the decision boundary. In this study, we used Borderline SMOTE-1 to interpolate the 16 positive samples from the 165 samples of the training set, to make the ratio of positive and negative samples 1:1. Then, a training set of 298 samples was obtained (Figure 2). The specific algorithm was implemented in Python.

#### 2.2.2. Random Forest (RF)

The RF, a representative method of ensemble learning [35], was selected as one of the basic classifiers to avoid possible overfitting phenomenon. In essence, it is a collection of weak classifiers containing multiple decision tree structures. For classification, the algorithm generates a result through the "voting" of each decision tree [36]. RF can be regarded as an application extension of the information entropy theory. The construction of a decision tree (i.e., the basic unit of RF) follows the basic principle of rapidly reducing system entropy [37].

For a sample set, the Decision Tree model can take the following steps: First, the total value of information entropy of each feature will be calculated according to the probability that the attributes (such as high, medium, or low) of different input features (such as temperature, precipitation intensity, or slope) appear in each output category (e.g., the event occurs/does not occur). Second, based on the principle of rapidly reducing information entropy, the input feature with the most reduced information entropy is selected as the root node to classify the samples according to attribute differences of different features. If the output category corresponding to these samples is pure or single (e.g., debris flows occur in all samples with precipitation intensity exceeding 15 mm/h), the sample set under this attribute will not be subdivided downward as a leaf node. On the contrary, for the samples in non-leaf nodes, repeat these two steps until all the leaf nodes are generated.

Because the samples are inevitably mixed with noise data, the overfitting phenomenon may occur after the samples are completely subdivided into the final leaf nodes, which affects the generalization performance of the model [38]. Therefore, in the modeling process, the splitting scale of a decision tree generally can be controlled within a certain range based on specific principles or standards, such as setting a GainRatio rate or Gini coefficient. The RF resamples the number and feature of samples before building the decision tree, which means that each time, only part of the data in the total set is taken to build the tree [35]. In this way, many decision trees can be obtained according to the different samples collected. Because only part of the data in the total set is extracted each time, the interference of noisy data can be effectively reduced. Therefore, the probability of overfitting in the RF model is theoretically lower [35,39].

We used the RF plug-in in WEKA software to train, validate, and test the debris flow prediction model. Generally, it is believed that the two most important hyperparameters of RF modeling are the number of decision trees and the maximum number of features selected per sampling [40,41], i.e., numiterations and numfeatures in WEKA, respectively. The Gridsearch method was used to adjust the hyperparameters, firstly. However, it was found that for some algorithms, there was still the possibility of overfitting, or the given results consumed unnecessary computing power. In view of these situations, the hyperparameters were adjusted manually based on the results of Gridsearch so as to reduce calculation consumption while ensuring the accuracy. Finally, we determined that the numiterations was 70 and the numfeatures was 3. In addition, considering that the GainRatio can be used to reflect the relative importance of each feature, the GainRatio method was used to investigate the contribution of each indicator to the occurrence of debris flow in the preliminary indicator selection phase.

#### 2.2.3. Support Vector Machine (SVM)

SVM is originally formulated as a binary classifier that obtains the optimal hyperplane by setting the maximum interval between the two classes of samples [33]. When using some specific modes, such as decomposing the multiclass datasets into multiple binary subsets, SVM can also be used to deal with multiclass problems. Because the setting of maximum interval requires only a small number of samples (i.e., the support vectors) that are closest to the hyperplane, this method is relatively friendly to small sample data sets [21]. The maximum interval can be classified into two types, i.e., the soft interval and the hard interval. The hyperplanes of the hard margin strictly distinguish all samples, which is suitable for linearly separable data sets. The soft margin means that a slack variable is added to select the optimal hyperplane on the premise of pursuing the maximum interval width. It is suitable for a data set that has a few outliers but is still linearly separable in general. When the data set is linearly non-separable, SVM introduces the kernel functions to map the samples into high-dimensional space, transform it into a linearly separable data set, and then finish the classification work.

The objective function of SVM not only has a loss function that reflects the accuracy of the fitting but also includes a regularization term that expresses the complexity of the model [25,33]. As a result, the optimization goal is not only to reduce the empirical risk but also to consider the structural risk of the model to avoid generating complex models that produce the overfitting phenomena due to the pursuit of local optima [42]. Thus, the generalization ability of the SVM model can be improved. The hyperparameters that need to be paid attention to in the actual operation of SVM are the C (penalty-factor) in the loss function term and the $\sigma^2$ in the kernel function (usually a Gaussian kernel function). The higher the C value is set, the greater the weight of the loss function term in the objective function, which will seek to further reduce the empirical risk, and then the model will be more prone to overfitting. Otherwise, the weight of the regularization term in the objective function is increased in a disguised way, and the generalization ability of the model will be strengthened, but there may be underfitting for the model classification. The higher the $\sigma^2$ value in the kernel function, the smoother the generated function, the more samples can become the support vectors, and the more likely the final model is to be underfitted.

This study used the LibSVM plug-in in WEKA to train, validate, and test the debris flow prediction model. According to the results of Gridsearch and manual tests, we decided to use the SVM model with radial basis kernel function for prediction, where gamma (corresponding to $1/2\sigma^2$) = 0.0001, and cost (corresponding to C) = 1. Using the weights generated by SVM for each feature in the process of solving the maximum interval, the relative importance of each feature could be calculated, i.e., the so-called SVM–RFE algorithm. This algorithm was employed to investigate the contribution of each indicator to the occurrence of debris flow. The linear SVM classifier (LibLINEAR plug-in in WEKA) was also introduced into the modeling to compare the performance of linear and nonlinear classifiers intuitively.

### 2.3. Data and Indicators

#### 2.3.1. Data Sources

In this study, remote sensing and meteorological data with high availability were collected as the main information sources, mainly involving precipitation, temperature, landform, geology, vegetation, and snow cover (Table 2).

**Table 2.** Data types and sources.

| Type | Content | Sources |
|---|---|---|
| Precipitation | Hourly data of the stations of PEIlONG, BOMI, GUXIANG, SONGZONG, MIDUI on the day and the adjacent days of debris flow events in 2012–2018, and daily data of the 10 days before the event day; hourly data of GUXIANG-2 and TIANMO stations on the day and the adjacent days of debris flow events in 2020, and daily data of the 7 days before the event day | Data from 2012 to 2018 collected from the Bomi Geologic Hazard Observation Station of the Institute of Mountain Hazards and Environment, CAS; data of 2020 sourced from the field observation |
| Temperature | Daily maximum, minimum, and average data of Bomi station in 2012–2020 and Tianmo station in 2020 on the event day and the 15 days before the event day | Data of the Bomi station collected from the China Meteorological Data Service Centre; data of the Tianmo station sourced from the field observation |
| Landform | DEM of the Parlong Zangbo Basin | SRTM 90 m DEM and ASTER 30 m GDEMv2 |
| Geology | National 1:2.5 million geological map of China | National Geological Archives Data Center, China |
| Vegetation Cover | NDVI of the study area in typical months from 2012 to 2020 | MODIS 500 m monthly synthetic product |
| Snow Cover | Snow cover products from 2012 to 2015 | Maximum_Snow_Extent MOD_Grid_Snow_500 m products of MOD10A2 |

#### 2.3.2. Basic Indicators

- Meteorology

In view of the natural relationships between the development of periglacial debris flow and the changes in precipitation and temperature, many scholars have explored the indicative factors of debris flow events according to these two factors. The precipitation indicators included antecedent precipitation, precipitation intensity, and precipitation duration, and the temperature indicators included accumulated temperature and average temperature. Based on outcomes from previous studies and the convenience of data collection [29,43–45], we selected 14 potential indicators, including accumulated effective precipitation over 3/5/10 days before the event ($A_3/A_5/A_{10}$), maximum and average precipitation intensity of the event day (before the event, of course, $I_a/I_m$), precipitation duration ($D$), temperature rise rate before the flood season (March to July normally, $T_r$), coefficient of variation of the daily temperature range over 5/10/15 days before the event ($C_v{}^5/C_v{}^{10}/C_v{}^{15}$), antecedent three-day effective accumulated temperature (greater than 0 °C, $T_a$), and the average daily temperature over 5/10/15 days before the event ($T_5/T_{10}/T_{15}$).

The antecedent accumulated effective precipitation ($A_3/A_5/A_{10}$) was calculated based on the Antecedent Precipitation Index (API, formula 1, Deng et al. [7]).

$$r_0 = kr_1 + k^2r_2 + k^3r_3 + \ldots + k^n r_n \quad (1)$$

where $r_0$ is the antecedent accumulated effective precipitation (mm) of the event day (day 0); $k$ is a constant reflecting surface runoff, which is related to soil texture and infiltration capacity; and rn is the maximum regional precipitation (mm) on the nth day before day 0.

It has been reported that the moraine soil in the study area was coarse in general [46], which was consistent with results from the field investigation of this research (Figure 3). Accordingly, it was considered that the gully soil of the study area had good permeability, and the two parameters (i.e., $n$ and $k$) of the formula were set as 3 and 0.5 ($A_3$), respectively. In addition, the scenarios of $n = 5/k = 0.6$ ($A_5$) and $n = 10/k = 0.7$ ($A_{10}$) were set to understand the influence of different parameter settings on the prediction of debris flow events (Table 3).

**Table 3.** Precipitation indicators data of related stations in training/validation phase (2012–2015).

| No. | Precipitation Station | Year | Date | Event No. * | $D$ (h) | $I_a$ (mm/h) | $I_m$ (mm/h) | $A_3$ (mm) | $A_5$ (mm) | $A_{10}$ (mm) |
|---|---|---|---|---|---|---|---|---|---|---|
| 1 | BOMI | 2012 | 09-21 | 1 | 2 | 0.8 | 1.0 | 10.0 | 14.2 | 21.3 |
| 2 | | 2012 | 09-22 | 1 | 1 | 1.0 | 1.0 | 5.8 | 9.4 | 16.2 |
| 3 | | 2012 | 09-22 | 2 | 3 | 2.0 | 5.0 | 5.8 | 9.4 | 16.2 |
| 4 | | 2012 | 09-22 | 3 | 13 | 1.4 | 4.0 | 5.8 | 9.4 | 16.2 |
| 5 | | 2013 | 07-05 | 1 | 2 | 0.8 | 1.0 | 0.1 | 0.3 | 2.8 |
| 6 | | 2013 | 07-05 | 2 | 9 | 2.1 | 4.5 | 0.1 | 0.3 | 2.8 |
| 7 | | 2014 | 07-23 | 1 | 4 | 1.9 | 4.5 | 2.9 | 3.7 | 6.9 |
| 8 | | 2014 | 07-24 | 1 | 1 | 1.5 | 1.5 | 5.7 | 7.2 | 10.7 |
| 9 | | 2014 | 07-24 | 2 | 4 | 1.0 | 2.0 | 5.7 | 7.2 | 10.7 |
| 10 | | 2014 | 07-24 | 3 | 4 | 1.8 | 3.5 | 5.7 | 7.2 | 10.7 |
| 11 | | 2014 | 07-25 | 1 | 3 | 0.8 | 1.5 | 6.5 | 8.8 | 12.6 |
| 12 | | 2014 | 08-18 | 1 | 1 | 1.0 | 1.0 | 4.4 | 8.5 | 14.7 |
| 13 | | 2014 | 08-22 | 1 | 3 | 0.8 | 1.0 | 4.5 | 5.8 | 10.4 |
| 14 | | 2014 | 08-22 | 2 | 1 | 1.0 | 1.0 | 4.5 | 5.8 | 10.4 |
| 15 | | 2014 | 08-22 | 3 | 3 | 0.7 | 1.0 | 4.5 | 5.8 | 10.4 |
| 16 | | 2014 | 08-22 | 4 | 2 | 1.3 | 1.5 | 4.5 | 5.8 | 10.4 |
| 17 | | 2014 | 08-22 | 5 | 2 | 1.8 | 2.0 | 4.5 | 5.8 | 10.4 |
| 18 | | 2014 | 08-23 | 1 | 14 | 1.0 | 3.5 | 7.3 | 9.4 | 14.1 |
| 19 | SONGZONG | 2012 | 09-21 | 1 | 3 | 1.2 | 1.5 | 10.8 | 15.5 | 22.5 |
| 20 | | 2012 | 09-21 | 2 | 4 | 1.9 | 2.5 | 10.8 | 15.5 | 22.5 |
| 21 | | 2012 | 09-22 | 1 | 11 | 2.9 | 7.0 | 12.5 | 17.7 | 25.8 |
| 22 | | 2012 | 09-23 | 1 | 10 | 1.6 | 2.5 | 28.4 | 36.7 | 49.0 |
| 23 | | 2013 | 07-05 | 1 | 5 | 1.3 | 3.0 | 0.8 | 0.9 | 2.8 |
| 24 | | 2013 | 07-05 | 2 | 1 | 1.5 | 1.5 | 0.8 | 0.9 | 2.8 |
| 25 | | 2013 | 07-06 | 1 | 4 | 1.8 | 3.0 | 6.4 | 7.7 | 10.3 |
| 26 | | 2014 | 07-23 | 1 | 1 | 1.5 | 1.5 | 0.4 | 0.6 | 2.4 |
| 27 | | 2014 | 07-24 | 1 | 3 | 1.0 | 1.5 | 0.9 | 1.2 | 2.7 |
| 28 | | 2014 | 07-24 | 2 | 2 | 3.0 | 2.5 | 0.9 | 1.2 | 2.7 |
| 29 | | 2014 | 08-18 | 1 | 2 | 5.0 | 2.5 | 2.7 | 6.4 | 11.7 |
| 30 | | 2014 | 08-22 | 1 | 4 | 0.6 | 2.5 | 3.9 | 5.3 | 9.4 |

Table 3. *Cont.*

| No. | Precipitation Station | Year | Date | Event No. * | $D$ (h) | $I_a$ (mm/h) | $I_m$ (mm/h) | $A_3$ (mm) | $A_5$ (mm) | $A_{10}$ (mm) |
|---|---|---|---|---|---|---|---|---|---|---|
| 31 | | 2014 | 08-23 | 1 | 9 | 2.4 | 5.0 | 8.6 | 11.5 | 16.2 |
| 32 | | 2015 | 08-03 | 1 | 2 | 4.0 | 2.5 | 0.0 | 0.0 | 0.2 |
| 33 | | 2015 | 08-04 | 1 | 3 | 3.2 | 6.0 | 6.5 | 7.8 | 9.1 |
| 34 | | 2015 | 08-04 | 2 | 2 | 3.8 | 5.0 | 6.5 | 7.8 | 9.1 |
| 35 | | 2015 | 08-17 | 1 | 2 | 1.0 | 1.0 | 1.0 | 1.3 | 2.1 |
| 36 | | 2015 | 08-17 | 2 | 3 | 1.2 | 1.5 | 1.0 | 1.3 | 2.1 |
| 37 | | 2015 | 08-17 | 3 | 4 | 0.6 | 1.0 | 1.0 | 1.3 | 2.1 |
| 38 | | 2015 | 08-18 | 1 | 1 | 1.0 | 1.0 | 7.0 | 8.6 | 10.3 |
| 39 | | 2015 | 08-19 | 1 | 9 | 1.1 | 2.0 | 11.4 | 14.7 | 18.3 |
| 40 | | 2015 | 08-19 | 2 | 10 | 2.5 | 6.0 | 11.4 | 14.7 | 18.3 |
| 41 | MIDUI | 2012 | 09-22 | 1 | 17 | 1.3 | 3.0 | 3.9 | 5.9 | 10.3 |
| 42 | | 2012 | 09-22 | 2 | 13 | 1.2 | 3.0 | 3.9 | 5.9 | 10.3 |
| 43 | | 2012 | 09-23 | 1 | 1 | 1.5 | 1.5 | 15.6 | 19.9 | 26.3 |
| 44 | | 2012 | 09-23 | 2 | 9 | 1.5 | 5.5 | 15.6 | 19.9 | 26.3 |
| 45 | | 2013 | 07-04 | 1 | 4 | 1.8 | 2.5 | 0.0 | 0.0 | 1.4 |
| 46 | | 2013 | 07-05 | 1 | 2 | 1.5 | 1.5 | 2.8 | 3.3 | 4.8 |
| 47 | | 2014 | 07-23 | 1 | 1 | 1.0 | 1.0 | 1.6 | 2.3 | 3.7 |
| 48 | | 2014 | 07-23 | 2 | 2 | 1.0 | 1.0 | 1.6 | 2.3 | 3.7 |
| 49 | | 2014 | 07-23 | 3 | 2 | 1.3 | 1.5 | 1.6 | 2.3 | 3.7 |
| 50 | | 2014 | 07-24 | 1 | 6 | 2.1 | 3.5 | 3.4 | 4.7 | 6.4 |
| 51 | | 2014 | 08-22 | 1 | 4 | 0.8 | 1.0 | 0.9 | 2.1 | 5.4 |
| 52 | | 2014 | 08-23 | 1 | 9 | 1.7 | 2.5 | 3.1 | 4.5 | 7.3 |
| 53 | | 2014 | 08-24 | 1 | 8 | 1.6 | 3.5 | 8.6 | 11.2 | 15.1 |

Note: * means event number on that day.

Considering that the debris flow events in the study area mainly occurred from July to September, the event may have been related to the amount of ice and snow melting from March or April every year. Therefore, we selected the daily average temperature and days from March to July of each year to construct the scatter diagram and used the slope of the linear trend equation of the scatter diagram to reflect the temperature rise rate in this period (Figure 4) and to obtain the indicator of the temperature rise rate before the flood season ($T_r$).

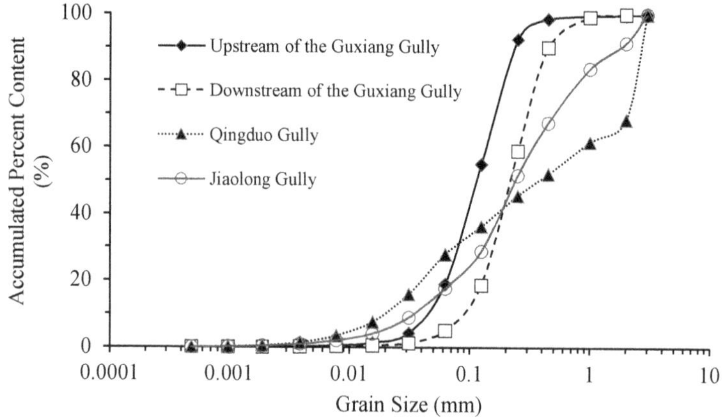


**Figure 3.** Grain size characteristics of samples from typical debris flow gullies in field investigation.

Assuming that the more severe the temperature change before the event, the more conducive it would be to freeze–thaw erosion and loose material accumulation, we added the coefficient of variation of the daily temperature range over 5/10/15 days before the event ($C_v{}^5/C_v{}^{10}/C_v{}^{15}$), which were calculated by the coefficient of variation of the difference between daily maximum temperature and minimum temperature over 5/10/15 days (Table 4). To further investigate the influence of short-term temperature change on the development of periglacial debris flow, the indicators of antecedent three-day effective accumulated temperature (greater than 0 °C, $T_a$) and the average of daily temperature over 5/10/15 days before the event ($T_5/T_{10}/T_{15}$) were also calculated.

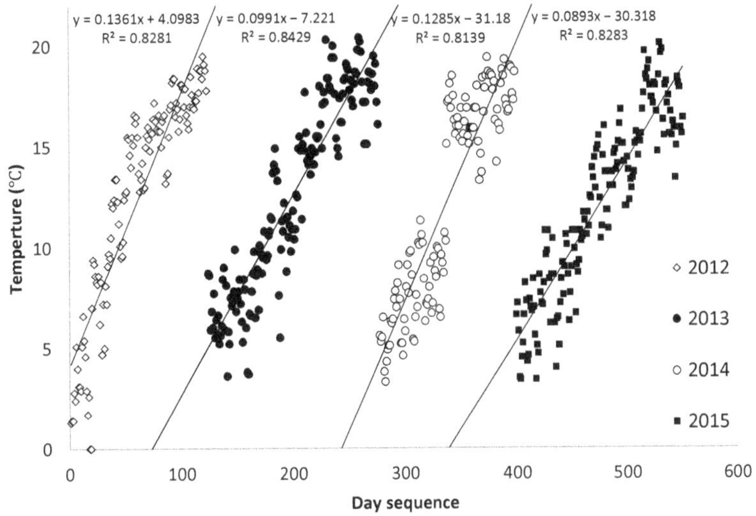


**Figure 4.** Scatter plot between daily average temperature and day sequence from March to July of each year (missing data for April 2012 and May 2014).

**Table 4.** Temperature indicators data of BOMI station in training and validation phase in 2012–2015 (unit: °C).

| Year | Date | $C_v{}^5$ | $C_v{}^{10}$ | $C_v{}^{15}$ | $T_5$ | $T_{10}$ | $T_{15}$ | $T_a$ |
|---|---|---|---|---|---|---|---|---|
| 2012 | 09-22 | 0.29 | 0.37 | 0.37 | 14.1 | 13.7 | 14.4 | 43.3 |
| 2012 | 09-21 | 0.30 | 0.38 | 0.38 | 13.8 | 13.8 | 14.6 | 44.0 |
| 2012 | 09-23 | 0.40 | 0.44 | 0.41 | 13.9 | 13.4 | 14.1 | 39.6 |
| 2013 | 07-06 | 0.11 | 0.26 | 0.29 | 18.7 | 17.5 | 17.5 | 56.8 |
| 2013 | 07-04 | 0.14 | 0.33 | 0.30 | 17.9 | 16.8 | 17.5 | 55.8 |
| 2013 | 07-05 | 0.11 | 0.29 | 0.30 | 18.5 | 17.2 | 17.6 | 56.4 |

**Table 4.** *Cont.*

| Year | Date | $C_v^5$ | $C_v^{10}$ | $C_v^{15}$ | $T_5$ | $T_{10}$ | $T_{15}$ | $T_a$ |
|---|---|---|---|---|---|---|---|---|
| 2013 | 07-31 | 0.26 | 0.30 | 0.33 | 18.5 | 18.0 | 17.8 | 54.3 |
| 2014 | 07-23 | 0.23 | 0.34 | 0.29 | 18.1 | 17.2 | 17.6 | 56.6 |
| 2014 | 07-25 | 0.23 | 0.24 | 0.30 | 18.2 | 17.2 | 17.4 | 53.6 |
| 2014 | 07-24 | 0.20 | 0.36 | 0.30 | 18.4 | 17.1 | 17.5 | 55.1 |
| 2014 | 08-18 | 0.55 | 0.39 | 0.32 | 14.6 | 16.3 | 17.2 | 42.0 |
| 2014 | 08-22 | 0.26 | 0.42 | 0.35 | 16.3 | 15.6 | 16.5 | 48.9 |
| 2014 | 08-23 | 0.30 | 0.41 | 0.35 | 16.0 | 15.3 | 16.2 | 45.8 |
| 2014 | 08-24 | 0.35 | 0.42 | 0.36 | 15.6 | 15.0 | 16.0 | 44.2 |
| 2015 | 08-04 | 0.15 | 0.19 | 0.25 | 17.3 | 17.2 | 16.6 | 54.6 |
| 2015 | 08-03 | 0.14 | 0.19 | 0.25 | 17.0 | 16.8 | 16.5 | 53.4 |
| 2015 | 08-17 | 0.31 | 0.37 | 0.31 | 18.8 | 17.9 | 17.9 | 54.2 |
| 2015 | 08-18 | 0.46 | 0.41 | 0.38 | 18.0 | 18.0 | 17.7 | 50.6 |
| 2015 | 08-19 | 0.46 | 0.49 | 0.43 | 17.0 | 17.9 | 17.6 | 47.8 |
| 2015 | 08-20 | 0.38 | 0.53 | 0.48 | 15.9 | 17.5 | 17.3 | 44.4 |

- Topography

Topographic relief has had a great influence on runoff generation and flow confluence as well as on the storage and release of gravitational potential energy from loose materials. Hence, it is one of the preconditions for the development of periglacial debris flow. In this study, we selected five indicators, including watershed height difference ($H_d$), gully gradient of mainstream ($G_g$), concave or convex characteristic from longitudinal profile ($P$), watershed average slope ($S_a$), and slope aspect ($A_s$), to reflect the influences of topographic factors (Table 5).

In practice, the longitudinal profile of the gully mainstream was extracted first using the three-dimensional analyst function in GIS. Then, we obtained $H_d$ by calculating the difference between the outlet elevation and maximum elevation; calculated $G_g$ from the ratio of Hd to the projection length of the longitudinal profile by arctangent function transformation; characterized $P$ by a quantitative method (Formular 2); and obtained $S_a$ by calculating the average slope using the zonal statistical function of GIS. The $A_s$ was summarized based on several main slope aspects of each watershed.

**Table 5.** Topographic and geological characteristics of the eight debris flow gullies in the training set.

| Gully | $H_d$ (m) | $S_a$ (°) | $G_g$ (°) | $A_s$ | $P$ (%) | $F$ | $L$ |
|---|---|---|---|---|---|---|---|
| Zhuonong | 1832.62 | 32 | 11.4 | East–Northwest | −65.5 | 0.33 | Extremely hard |
| West Jiazong | 1858.96 | 37 | 34.0 | South–Southwest | −12.6 | 0.00 | Extremely hard |
| Shalong | 1508.94 | 32 | 13.4 | Southwest–West | −3.8 | 0.31 | Secondary hard |
| Rongqian | 1993.82 | 32 | 29.4 | South–Southwest | −4.9 | 0.10 | Extremely hard |
| Motuo | 1510.14 | 27 | 28.6 | East–Southeast | −2.2 | 0.25 | Extremely hard |
| Midui | 1262.14 | 34 | 31.8 | East–Northeast | 4.2 | 0.33 | Secondary hard |
| Jiurong | 1751.91 | 36 | 23.5 | West–Northwest | −28.0 | 0.21 | Secondary hard |
| East Lapu | 1563.89 | 36 | 30.1 | South–Southwest | 3.1 | 0.10 | Extremely hard |

$$P = (a_1 - a_2) / a_2 \quad (2)$$

where $P$ is the concave or convex characteristic of a longitudinal profile (%), and a positive value means that the profile is mainly convex, whereas a negative value is opposite; $a_1$ is the area of the profile; and $a_2$ is the triangle area enclosed by the three vertices of the profile.

- Geology

Geological indicators included lithologic hardness ($L$) and fault buffer ($F$, Table 5). The indicator of $L$, mainly hard and sub-hard, was qualitatively classified based on field investigation and the 1:2.5 million national geological map. The study area had many faults according to the map, among which the most famous one named Jiali-Ranwu fault was distributed in the south of the area, and the rest were mostly unproved faults. According to the fault data provided by the map, we scored five buffers (1 km, 2 km, 3 km, 5 km, and 10 km) according to the distance to the fault (Figure 5), and then, the average value of each gully watershed was extracted and normalized using the zonal statistics function of GIS to obtain the $F$ value.

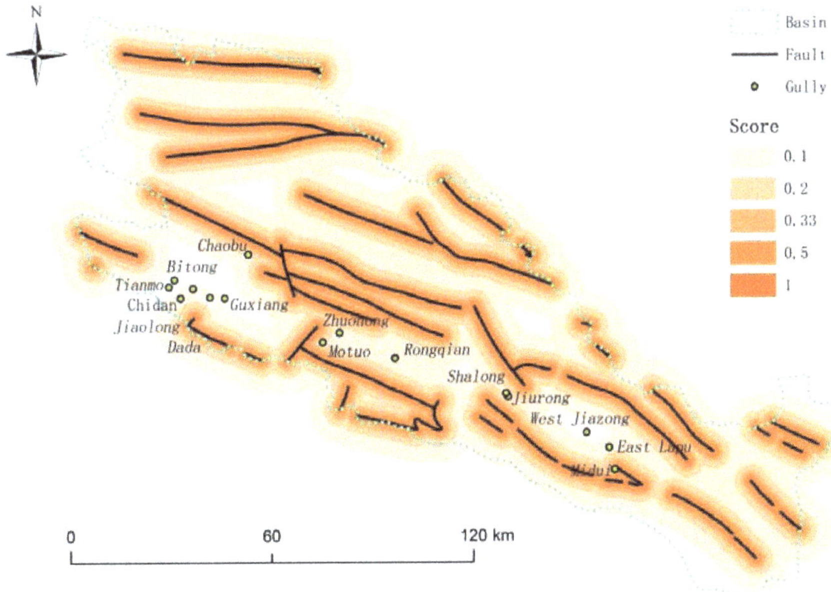


**Figure 5.** Distribution of faults and their buffers in the study area.

- Underlying surface

Snow melting is one of the basic reasons for the formation of periglacial debris flow [1,10]; thus, the change of snow cover is an important consideration in the underlying surface conditions. In contrast, the influence of vegetation on periglacial debris flow development is more complicated. On the one hand, vegetation can enhance the strength of soil by root system [47], and the hydrological effect of vegetation, such as intercepting rainfall and extracting soil moisture, is also conducive to the stability of a slope [48]. On the other hand, the surcharge load caused by plant weight and the transmission of drag force from wind reduce the slope stability [49,50]. Therefore, vegetation can promote or inhibit the formation of debris flows by increasing or reducing the supply of loose materials to a certain extent.

For the study area, vegetation can hardly affect the initiation of loose materials in the glacier area. However, the forests in the low-altitude valley may play an important role on the development of periglacial debris flow. Hence, the effects of vegetation and snow cover would be focused on this part, including four indicators: antecedent spatial scale Normalized Difference Vegetation Index ($NDVI_s$), antecedent temporal scale NDVI ($NDVI_t$), average maximum snow cover rate ($MSC_a$), and antecedent snow cover decrease rate ($ASC_d$). Among them, we calculated the antecedent NDVI indicators, basically, the vegetation coverage of each gully watershed in the month before the debris flow events in 2012–2020, based on the monthly synthetic product of MODIS 500 m NDVI (Table 6). Considering that both interannual and regional differences on vegetation coverage were significant, the $NDVI_t$ and $NDVI_s$ were obtained by normalizing the annual and regional sequences, respectively, to reflect the impact of this factor comprehensively.

**Table 6.** NDVI changes of 15 debris flow gullies in the study area from 2012 to 2020.

| Gully | NDVI | | | | | | | | |
|---|---|---|---|---|---|---|---|---|---|
| | Training/Validation Phase | | | | Testing Phase | | | | |
| | 2012-08 | 2013-06 | 2014-07 | 2015-07 | 2016-08 | 2017-07 | 2018-04 | 2018-06 | 2020-07 |
| Tianmo | 0.12 | 0.18 | 0.20 | 0.70 | | | | 0.40 | 0.51 |
| Chidan | 0.23 | 0.37 | 0.31 | 0.75 | 0.69 | | | | |
| Jiaolong | 0.12 | 0.16 | 0.17 | 0.51 | 0.37 | | | | |
| Chaobu | 0.13 | 0.20 | 0.52 | 0.83 | | 0.64 | | | |
| Bitong | 0.12 | 0.18 | 0.20 | 0.51 | 0.44 | | | 0.29 | 0.39 |
| Dada | 0.16 | 0.19 | 0.43 | 0.61 | | | | | 0.39 |
| Guxiang | 0.12 | 0.17 | 0.19 | 0.42 | | 0.30 | | 0.28 | |
| Zhuonong | 0.24 | 0.48 | 0.31 | 0.69 | | | | | |
| West Jiazong | 0.16 | 0.42 | 0.28 | 0.43 | | | | | |
| Shalong | 0.18 | 0.27 | 0.23 | 0.38 | | | | | |
| Rongqian | 0.24 | 0.34 | 0.22 | 0.51 | | | | | |
| Motuo | 0.29 | 0.47 | 0.38 | 0.62 | | | | | |
| Midui | 0.06 | 0.11 | 0.08 | 0.16 | | | | | |
| Jiurong | 0.34 | 0.41 | 0.42 | 0.57 | | | 0.22 | | |
| East Lapu | 0.26 | 0.30 | 0.23 | 0.32 | | | 0.04 | | |

Based on the products of MOD10A2 from 2012 to 2015, we calculated the proportion of snow and cloud cover area of each watershed at training and validation phase, respectively. The layers with more than 5% cloud were removed in this process, and the proportion of the largest snow coverage area before the debris flow events was selected as the maximum snow coverage rate of the current year. According to the difference between this maximum rate and the latest snow coverage rate before the event, the $ASC_d$ could be calculated, and the $MSC_a$ could also be obtained based on the average statistics of this maximum rate from 2012 to 2015 (Table 7).

**Table 7.** Snow cover conditions of the eight debris flow gullies at training and validation phase in 2012–2015 (unit: %).

| Gully | $ASC_d$ | | | | $MSC_a$ |
|---|---|---|---|---|---|
| | **2012** | **2013** | **2014** | **2015** | |
| Zhuonong | 89.2 | 93.6 | 86.0 | 93.0 | 86.5 |
| West Jiazong | 84.3 | 91.6 | 100.0 | 80.0 | 98.5 |
| Shalong | 43.8 | 100.0 | 80.1 | 94.5 | 93.1 |
| Rongqian | 93.7 | 100.0 | 83.9 | 79.5 | 77.5 |
| Motuo | 90.9 | 100.0 | 86.3 | 100.0 | 100.0 |
| Midui | 80.0 | 90.0 | 80.0 | 100.0 | 100.0 |
| Jiurong | 100.0 | 100.0 | 82.3 | 100.0 | 92.4 |
| East Lapu | 91.4 | 91.4 | 100.0 | 96.0 | 90.9 |

Based on the above-mentioned 25 indicators and the watershed area ($A$) of each gully, a total of 26 preliminary indicators were selected and normalized for subsequent analysis.

## 3. Results and Discussion

### *3.1. Selection of Prediction Indicators*

To avoid the possible interference from over-resampling, 165 events data without resampling were used here for further indicator selection. Based on the principle of reducing the indicators as much as possible and high operability, different methods such as Pearson correlation and SVM-RFE were employed to determine the final indicators to be included in the classification model.

First, we used Pearson correlation analysis to eliminate the variables with obvious false correlation; that is, the correlation that cannot reflect the true relationship between the two variables, even if it is statistically significant. The results indicated that 13 indicators related to vegetation, temperature, and precipitation were significant with the development of periglacial debris flow in the study area (Table 8 left). Most of them presented a positive correlation except for the temperature rise rate before the flood season ($T_r$). As mentioned before, periglacial debris flows usually occur with the increase in temperature [6]. However, the relationship between $T_r$ and debris flow events seemed to contradict this cognition: that is, the faster the warming, the fewer debris flow events. Further analysis showed that $T_r$ was mainly related to the initial temperature in March of each year. The higher the initial temperature level, the more limited the space for subsequent warming, and the slower the warming rate (Figure 6), whereas a higher initial temperature may indeed lead to the increase in snow melting rate and the occurrence of periglacial debris flow. Therefore, for the indicator of $T_r$, its correlation with debris flow development cannot reflect the actual relationship between temperature rise rate and regional debris flow occurrence; thus, the indicator needed to be eliminated.

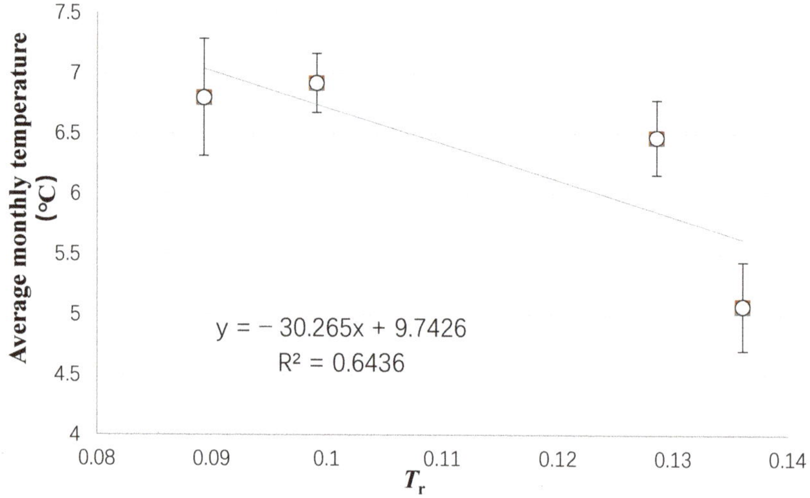


**Figure 6.** Relationship between $T_r$ and the average temperature of March from 2012 to 2015 (with the Standard Error of Mean on daily average temperature of each March).

Furthermore, we employed SVM-RFE and the GainRatio method to investigate the importance of the remaining 25 indicators in debris flow development, in which the 10-fold cross-validation was used in SVM-RFE. The results showed that although the importance or correlation ranking of each indicator from the three methods was quite different, the top indicators were highly similar, mainly the temporal sequences related to vegetation, precipitation, and temperature. Conversely, the indicators reflecting the spatial heterogeneity of landform, geology, and watershed characteristics ranked lower in general, which might have been related to the range of selected sample data.

Statistically speaking, the method of sample expansion in this study limits the performance of indicators reflecting spatial heterogeneity to a certain extent. The data from 165 events collected here included 16 events with debris flow and 149 events without debris flow. The 16 events with debris flow were obtained based on eight gullies monitored by the stations of Bomi, Songzong, and Midui from 2012 to 2015 (Table 1), while the 149 events without debris flow were expanded from the same eight gullies on different dates. Taking the debris flow event on 23 August 2014 in Jiurong gully as an example, on the event day, if no debris flow was recorded in the other seven gullies, the seven gullies on 23 August 2014 would be registered as non-debris flow events. In addition, if no debris flow was recorded in the other seven gullies on the day of 22 August 2014 or 24 August 2014, the seven gullies on that day would also be registered as non-debris flow events. In this way, many gullies have experienced almost the same frequency of debris flow and non-debris flow events, and the difference of watershed indicators, such as the mainstream slope, among these eight debris flow gullies is much smaller than that between debris flow gullies and non-debris flow gullies [51,52]. Thus, it is difficult to find a significant correlation between the indicators reflecting spatial heterogeneity and debris flow events in such a data set.

According to the results from the three assessment methods (Table 8), indicators closely related to the development of periglacial debris flow were $NDVI_t$, $D$, $I_m$, $I_a$, $C_v^{15}$, $C_v^{10}$, $C_v^{5}$, $T_{15}$, $T_{10}$, $A_{10}$, $A_5$, and $A_3$. Because the indicators in $I_m/I_a$, $C_v^{15}/C_v^{10}/C_v^{5}$, $T_{15}/T_{10}$, and $A_{10}/A_5/A_3$ presented repeatability to some extent, we selected the indicator with the highest comprehensive ranking of each category as an input of the classification model. Considering that the comprehensive ranking of Sa was the highest among all spatial related indicators, the final seven indicators were $D$, $I_m$, $C_v^{15}$, $T_{10}$, $A_3$, $NDVI_t$, and $S_a$.

**Table 8.** Assessment methods of indicator selection and related ranking.

| Rank | Correlation Analysis | | SVM-RFE | | GainRatio | |
|---|---|---|---|---|---|---|
| | Coefficient | Indicator | Average Merit | Indicator | Score | Indicator |
| 1 | 0.42 ** | $NDVI_t$ | 23 ± 2.7 | $D$ | 0.47 | $C_v^{10}$ |
| 2 | −0.36 ** | $T_r$ | 22.5 ± 3.0 | $T_{10}$ | 0.47 | $Cv^{15}$ |
| 3 | 0.34 ** | $D$ | 21.4 ± 2.6 | $A_5$ | 0.26 | $C_v^5$ |
| 4 | 0.30 ** | $C_v^{15}$ | 21.2 ± 1.5 | $T_{15}$ | 0.24 | $A_3$ |
| 5 | 0.27 ** | $T_{10}$ | 20.3 ± 4.1 | $Cv^{15}$ | 0.24 | $I_m$ |
| 6 | 0.26 ** | $I_m$ | 20.3 ± 1.1 | $A_{10}$ | 0.16 | $NDVI_t$ |
| 7 | 0.25 ** | $A_3$ | 19.8 ± 1.9 | $I_m$ | 0.11 | $T_{10}$ |
| 8 | 0.24 ** | $C_v^5$ | 19.3 ± 2.1 | $C_v^{10}$ | 0.11 | $T_{15}$ |
| 9 | 0.24 ** | $C_v^{10}$ | 18.4 ± 1.4 | $NDVI_t$ | 0.11 | $D$ |
| 10 | 0.23 ** | $A_5$ | 16.9 ± 5.2 | $I_a$ | 0.06 | $I_a$ |
| 11 | 0.22 ** | $I_a$ | 15.1 ± 2.7 | $A_3$ | 0 | $A_{10}$ |
| 12 | 0.21 ** | $T_{15}$ | 14.1 ± 0.3 | $S_a$ | 0 | $T_5$ |
| 13 | 0.19 * | $A_{10}$ | 12.4 ± 1.0 | $H_d$ | 0 | $A_5$ |
| 14 | 0.11 | $S_a$ | 8.8 ± 3.2 | $ASC_d$ | 0 | $MSC_a$ |
| 15 | 0.08 | $T_5$ | 8.8 ± 4.3 | $C_v^5$ | 0 | $ASC_d$ |
| 16 | 0.06 | $ASC_d$ | 8.3 ± 2.8 | $T_5$ | 0 | $F$ |
| 17 | −0.06 | $L$ | 8.1 ± 4.7 | $MSC_a$ | 0 | $P$ |
| 18 | 0.039 | $A$ | 6.8 ± 3.6 | $G_g$ | 0 | $A_s$ |
| 19 | 0.03 | $A_s$ | 6.7 ± 1.9 | $A_s$ | 0 | $NDVI_s$ |
| 20 | −0.03 | $F$ | 6.4 ± 2.65 | $P$ | 0 | $G_g$ |
| 21 | −0.02 | $T_a$ | 6.3 ± 3.55 | $NDVI_s$ | 0 | $A$ |
| 22 | −0.02 | $H_d$ | 6.2 ± 3.25 | $T_a$ | 0 | $H_d$ |
| 23 | 0.02 | $P$ | 5.7 ± 3.44 | $L$ | 0 | $L$ |
| 24 | 0.01 | $MSC_a$ | 4.1 ± 2.74 | $F$ | 0 | $S_a$ |
| 25 | 0.01 | $G_g$ | 4.1 ± 3.05 | $A$ | 0 | $T_a$ |
| 26 | 0.01 | $NDVI_s$ | | | | |

Note: ** and * are the significance of 0.01 and 0.05, respectively.

## 3.2. Modeling and Comparative Analysis

### 3.2.1. Training and Validation

As mentioned in Section 2.2.1, after the indicator selection work, Borderline SMOTE-1 was used to obtained a balanced training set, and the data were imported into the RF, SVM and Linear SVM (LSVM) models as the inputs, respectively. We employed the 10-fold cross-validation to train and validate the binary classification process on whether the debris flow occurred. Simultaneously, the precision rate, recall rate, Matthews Correlation Coefficient (MCC), Receiver Operating Characteristic Curve (ROC) and other model assessment indices were employed to evaluate the performance of each classification model. The calculation formulas of the main assessment indices are as follows:

$$\text{TPR or Recall} = TP/(TP + FN) \quad (3)$$

$$\text{FPR} = FP/(FP + TN) \quad (4)$$

$$\text{Precision} = TP/(TP + FP) \quad (5)$$

$$\text{F-Measure} = 2 \cdot \text{Precision} \cdot \text{Recall}/(\text{Precision} + \text{Recall}) \quad (6)$$

$$\text{MCC} = (TP \cdot TN - FP \cdot FN)/\text{sqrt}\,[(TP + FP)(TP + FN)(TN + FP)(TN + FN) \quad (7)$$

All these indices are based on a fundamental concept named confusion matrix (Table 9), which is a form of contingency table showing the differences between the true and predicted classes for a set of labeled examples [53]. In the confusion matrix, *TP* and *TN* are the number of true positives and true negatives, respectively, indicating that the predicted values are the same as the true values, and the predicted values are positive and negative examples, respectively; *FN* and *FP* are the number of false negatives and false positives, respectively, indicating that the predicted values are opposite to the true values, and the predicted values are negative and positive, respectively.

The assessment indices can evaluate models from different aspects. The true positive rate (TPR) or recall rate reflects the ratio of correctly classified positive cases to the total number of positive cases [54]; the false positive rate (FPR) represents the ratio of misclassified negative cases to the total number of negative cases, i.e., the false alarm rate in early warning system. Precision refers to the ratio of true positive cases to all cases classified as positive. The F-Measure is the harmonic weighted average of precision rate and recall rate, reflecting the comprehensive situation of the two indices [54]. The MCC describes the correlation between the prediction set and the observation set, with a value range of −1 to 1, and it is regarded as one of the most adaptable comprehensive evaluation indices, which was applicable to imbalanced data sets [55]; and the AUC is the area under the ROC curve [53], reflecting the probability that a positive sample is correctly classified. A value of 1 meant that all positive samples can be correctly classified, and a value of 0.5 indicates that the performance of the classifier was no different from random classification. Except for the FPR, the higher the value of the assessment indices, the better the model classification performance.

**Table 9.** A typical confusion matrix.

| True Class | Predicted Class | |
|---|---|---|
| | + | − |
| + | *TP* | *FN* |
| − | *FP* | *TN* |

Note: + and − are positive and negative cases, respectively.

The results showed that the RF-based model outperformed the SVM and LSVM-based model (Table 10). The values of recall rate, F-Measure, MCC, AUC, and other major assessment indices of the RF model in the training and validation phase all exceeded 0.95, and the prediction performance on positive and negative events was equivalent. Although the values of main assessment indices of the SVM model were also above 0.92, the recall rate of the negative events was 8% lower than that of positive events, the weighted FPR of the SVM model was more than three times higher than that of the RF model, and the MCC was also 10.3% lower than that of the RF model. As a typical linear classifier, LSVM lagged behind the SVM and RF-based model in almost all the assessment indices, reflecting the complexity of periglacial debris flow prediction.

Further analysis of the confusion matrix showed that the SVM model had 17 false predictions of negative events, which was much higher than that of the RF model, and the number of false predictions of positive events was also higher than that of the RF model (Table 11). In short, although the overall prediction accuracy of the two models in the training and validation phase could reach a high level, the RF model was the preferred solution.

**Table 10.** Comparison of the results from different models on periglacial debris flow prediction.

| Phase | Model | Class | TPR | FPR | Precision | F-Measure | MCC | AUC |
|---|---|---|---|---|---|---|---|---|
| Training and Validation | RF | No | 0.966 | 0.013 | 0.986 | 0.976 | 0.953 | 0.997 |
| | | Yes | 0.987 | 0.034 | 0.967 | 0.977 | | |
| | | WA | 0.977 | 0.023 | 0.977 | 0.977 | | |
| | SVM | No | 0.886 | 0.034 | 0.964 | 0.923 | 0.855 | 0.926 |
| | | Yes | 0.966 | 0.114 | 0.894 | 0.929 | | |
| | | WA | 0.926 | 0.074 | 0.929 | 0.926 | | |
| | LSVM | No | 0.765 | 0.342 | 0.691 | 0.726 | 0.425 | 0.711 |
| | | Yes | 0.658 | 0.235 | 0.737 | 0.695 | | |
| | | WA | 0.711 | 0.289 | 0.714 | 0.711 | | |
| Testing | RF | No | 0.750 | 0.222 | 0.750 | 0.750 | 0.528 | 0.778 |
| | | Yes | 0.778 | 0.250 | 0.778 | 0.778 | | |
| | | WA | 0.765 | 0.237 | 0.765 | 0.765 | | |
| | SVM | No | 0.500 | 0.222 | 0.667 | 0.571 | 0.290 | 0.639 |
| | | Yes | 0.778 | 0.500 | 0.636 | 0.700 | | |
| | | WA | 0.647 | 0.369 | 0.651 | 0.639 | | |
| | LSVM | No | 0.375 | 0.111 | 0.750 | 0.500 | 0.311 | 0.632 |
| | | Yes | 0.889 | 0.625 | 0.615 | 0.727 | | |
| | | WA | 0.647 | 0.383 | 0.679 | 0.620 | | |

Note: "Yes" for debris flow events and "No" for non-debris flow events; WA represents weighted average.

Table 11. Comparison of results from different models on the confusion matrix.

| Phase | Model | RF | | SVM | | LSVM | |
|---|---|---|---|---|---|---|---|
| | Class | No | Yes | No | Yes | No | Yes |
| Training and Validation | No | 144 | 5 | 132 | 17 | 114 | 35 |
| | Yes | 2 | 147 | 5 | 144 | 51 | 98 |
| Testing | No | 6 | 2 | 4 | 4 | 3 | 5 |
| | Yes | 2 | 7 | 2 | 7 | 1 | 8 |

3.2.2. Testing

According to the RF, SVM and LSVM models established by the training and validation results, 17 events in 2016–2020 were imported into the model for testing (Table 12). The results showed that the MCC and AUC of the RF model were 82% and 22% higher than those of the SVM model, respectively (Table 10). Combined with the confusion matrix of the test set (Table 11), it can be found that the prediction performance of both models on debris flow events is good, and the recall rate can reach more than 0.77. The disadvantage of SVM is reflected in the prediction of non-debris flow events: the recall rate is only 0.5, which is much lower than that of RF (0.75).

Compared with the performance in the training and validation phase, the LSVM-based model performed significantly better in this phase, and some assessment indices were even higher than that from the RF or SVM model. However, this is mainly because the model predicted 13 of the 17 events as debris flow, which improved the recall rate of debris flow events (0.889) but also enlarged the false alarm rate (0.625) simultaneously (Table 10). The LSVM has no kernel function processing procedure, so it is only applicable to the linearly separable datasets [21,33]. For the complex data set in this case, the classifier is not competent for this prediction task.

If we focus on the prediction results of each event (Table 12), it can be found that the outputs of the RF and SVM models are more sensitive to the precipitation indicators, especially $A_3$ (0.79 ** for RF and 0.573 * for SVM on Pearson correlation coefficient, the same below) and $I_m$ (0.587 * for RF and 0.726 ** for SVM). In addition, the response of SVM to temperature indicator is also outstanding, especially for $C_v^{15}$ (0.787 **). Taking the outputs as the dependent variable and the seven indicators as the independent variable for stepwise regression, the results showed that the goodness of fit from the SVM model was higher than that from the RF model (Table 13). Although the processing mechanism of nonparametric models (such as SVM and RF) does not depend on the correlation between the input and output, the prediction result of SVM in this case seems to be closer to the linear relationships between the indicators and the events, even with the support of kernel functions.

**Table 12.** Prediction results of different models on the 17 events of the testing set.

| No. | Gully | Date | Station | $A_3$ | $D$ | $I_m$ | $NDVI_t$ | $S_a$ | $C_v^{15}$ | $T_{10}$ | Debris Flow | RF | SVM |
|---|---|---|---|---|---|---|---|---|---|---|---|---|---|
| 1 | Chidan | 2016-09-05 | PEILONG | 21.75 | 15.00 | 4.70 | 0.69 | 30.70 | 0.43 | 16.80 | Yes | √ | √ |
| 2 | Jiaolong | 2016-09-05 | PEILONG | 21.75 | 15.00 | 4.70 | 0.37 | 36.20 | 0.43 | 16.80 | Yes | √ | √ |
| 3 | Bitong | 2016-09-05 | PEILONG | 21.75 | 15.00 | 4.70 | 0.44 | 34.50 | 0.43 | 16.80 | Yes | √ | √ |
| 4 | Chaobu | 2017-08-03 | GUXIANG | 3.10 | 1.00 | 0.20 | 0.64 | 35.00 | 0.17 | 18.30 | Yes | √ | × |
| 5 | Chaobu | 2017-08-02 | GUXIANG | 0.09 | 6.00 | 1.60 | 0.64 | 35.00 | 0.17 | 18.40 | No | √ | √ |
| 6 | Guxiang | 2017-08-03 | GUXIANG | 3.10 | 1.00 | 0.20 | 0.30 | 35.60 | 0.17 | 18.30 | No | √ | √ |
| 7 | East Lapu | 2018-05-22 | SONGZONG | 0.00 | 0.00 | 0.00 | 0.04 | 36.00 | 0.21 | 14.00 | Yes | × | × |
| 8 | East Lapu | 2018-05-21 | SONGZONG | 0.00 | 0.00 | 0.00 | 0.01 | 36.00 | 0.21 | 13.10 | No | √ | √ |
| 9 | Jiurong | 2018-05-22 | SONGZONG | 0.00 | 0.00 | 0.00 | 0.22 | 36.00 | 0.21 | 14.00 | No | √ | √ |
| 10 | Tianmo | 2018-07-11 | GUXIANG | 12.94 | 10.00 | 3.10 | 0.40 | 37.70 | 0.24 | 18.20 | Yes | √ | √ |
| 11 | Bitong | 2018-07-11 | GUXIANG | 12.94 | 10.00 | 3.10 | 0.29 | 34.50 | 0.24 | 18.20 | Yes | √ | √ |
| 12 | Guxiang | 2018-07-11 | GUXIANG | 12.94 | 10.00 | 3.10 | 0.28 | 35.60 | 0.24 | 18.20 | No | × | × |
| 13 | Dada | 2020-07-09 | TIANMO | 0.65 | 2.35 | 6.00 | 0.39 | 33.00 | 0.37 | 16.26 | No | √ | × |
| 14 | Dada | 2020-07-10 | TIANMO | 6.50 | 3.00 | 7.60 | 0.39 | 33.00 | 0.35 | 16.19 | Yes | √ | √ |
| 15 | Tianmo | 2020-07-10 | TIANMO | 6.50 | 3.00 | 7.60 | 0.51 | 37.70 | 0.35 | 16.19 | No | × | × |
| 16 | Guxiang | 2020-07-08 | GUXIANG-2 | 0.20 | 0.00 | 0.00 | 0.39 | 35.60 | 0.37 | 15.98 | No | √ | × |
| 17 | Guxiang | 2020-07-09 | GUXIANG-2 | 0.35 | 3.80 | 2.33 | 0.39 | 35.60 | 0.37 | 16.26 | Yes | × | √ |

**Table 13.** Main model parameters of the stepwise regression analysis based on test set.

| Dependent Variable | Independent Variable | Standardized Coefficients | t-Value | $p$-Value | F-Value | Adjusted $R^2$ |
|---|---|---|---|---|---|---|
| RF outputs | $A_3$ | 0.790 | 4.986 | 0.000 | 24.862 | 0.599 |
| SVM outputs | $C_v^{15}$<br>$T_{10}$ | 0.818<br>0.338 | 5.891<br>2.433 | 0.000<br>0.029 | 19.167 | 0.694 |

The key of success classification for SVM is to find the support vectors. Borderline SMOTE highlights the decision boundary through interpolation, so it is more conducive to the classification of the SVM model, theoretically [19,56]. Actually, the optimal hyperparameter value of gamma given by Gridsearch was 0.001. Under this condition, the weighted recall rate, MCC, and AUC of SVM were 0.977, 0.953, and 0.977, respectively, which was basically as good as those of the RF model. However, the model classified 16 of the 17 test events into non-debris flow events, resulting in the recall rate of debris flow events being only 0.11, i.e., the overfitting occurred. This indicates that the SVM algorithm is still prone to overfitting even if the Borderline SMOTE is applied, whereas the RF shows better robustness at this point, partly because it is relatively less sensitive to the hyperparameter setting [57].

Some scholars believed that the ensemble learning-based RF model was insensitive to the lack of input data [58]. In other words, even if there was no process of indicator selection, theoretically, the model established by all indicators had a higher probability of making an effective prediction, even in the absence of some input indicators in the testing phase. In this study, we also tested the RF model built by 26 full indicators. The results showed that the performance of the full indicator model was basically the same as that of the simplified model in the phases of training, validation, and testing, presenting no advantage in prediction accuracy. Moreover, full indicator modeling would have increased the amount of calculation and affected the speed of model construction and response. Considering the performance of the model and the cost of data collection and processing, the early warning efficiency of the simplified model was obviously higher.

Another thing worth noting is that although the RF model shows acceptable accuracy in this phase, there are still some shortcomings. First, the prediction effect for debris flow events without precipitation is still poor, such as the seventh event in Table 12. Second, the model takes the indicators reflecting spatial heterogeneity into account, but the influences of these indicators are limited in prediction. For example, the Dada gully and adjacent Tianmo gully share the same precipitation station; thus, most monitoring data are identical except the $NDVI_t$ and $S_a$. However, for the events on 10 July 2020, the model cannot distinguish whether debris flow occurred in these two gullies only based on the data difference of this degree (Table 12). Therefore, the model proposed in this paper is still a regional prediction model, and its spatial resolution cannot be accurate to specific gullies at present.

### 3.2.3. Comparison

#### I-D Model and I-D-A Model

The rainfall indicator method is currently the most widely used debris flow monitoring and early warning method in the world. Its representative method is the power law equation (*I-D* method) proposed by Caine [59]. This method assumes that the critical rainfall intensity for debris flow is nonlinear with the rainfall duration. As the rainfall duration increases, the critical rainfall intensity decreases accordingly. The traditional *I-D* method does not consider the impact of antecedent rainfall, so its prediction accuracy may be poor for watersheds with better vegetation coverage and greater water storage potential. Fortunately, the so-called *I-A* method based on the relationships between antecedent effective rainfall and rain intensity can make up for this shortcoming [7]. However, some scholars have pointed out that the warning effect of the *I-D* method is better than that of the *I-A* method for sandy soil areas with strong infiltration capacity [60].

The proposed RF model, i.e., the original model here, is essentially an event-based prediction model for periglacial debris flow. By introducing the antecedent effective cumulative precipitation ($A_3$) of each event, the model can simultaneously consider the effects of antecedent precipitation ($A_3$), precipitation duration ($D$), and precipitation intensity ($I_m$) on the development of debris flow. Therefore, it is actually an *I-D-A* model. Herein, we tested and compared the performance of the original model and the RF model without $A_3$, i.e., the *I-D* model in this paper. The results showed that the advantage of the *I-D-A* model in the training and validation phase was not obvious, and the difference between them was less than 3% in MCC, which was the assessment index with the largest difference. On the contrary, in the testing phase, the classification accuracy of the *I-D-A* model was significantly higher than that of the *I-D* model, and the difference between them in MCC was more than two-fold. Without the support of $A_3$, the false alarm rate for both positive and negative samples increased (Table 14). Considering that this result is obtained even in the area with good infiltration condition, which is reported as more suitable for the *I-D* model [60], it is believed that the antecedent precipitation index is of great significance to improve the performance of debris flow prediction model, at least in this case.

Resampling Model and Non-Resampling Model

Previous studies have found that because of the modeling idea and model structure, some machine learning algorithms are inherently capable of dealing with the imbalanced data sets to some extent [19,59]. For this type of algorithm, such as the RF, even if the original data are not resampled, a satisfactory classification result may be obtained. Herein, we tested the RF model with the original non-resampled data. The results showed that the resampling processing in this case was essential to improve the performance of the model. In the training and validation phase, although the recall rate of the debris flow event was only 0.625, the weighted recall rate exceeded 0.94 because of the larger weights on the non-debris flow event. This resulted in the model predictions being more biased toward non-debris flow events during the testing phase (Table 15). Compared with the original model that had been resampled, the MCC of the non-resampled model had dropped by 32% in the training and validation phase, and the drop in the testing phase had reached 83%. The negative effect brought by the imbalanced data set was very significant.

SMOTE Model and Borderline SMOTE Model

The SMOTE is an intelligent oversampling algorithm commonly used in the current academic community, and the Borderline SMOTE, as an improved version of SMOTE, theoretically has a higher interpolation efficiency at the classification boundary. Using the SMOTE from WEKA to expand the original 16 debris flow events by eight times, the ratio of debris flow events to non-debris flow events was close to 1:1. We tested the RF model built by this data set and compared it with the original model using Borderline SMOTE. The results showed that in the training and validation phase, the SMOTE-based model had almost the same performance as the original model (Table 14). In the testing phase, however, the SMOTE-based model has a significantly lower recall rate for debris flow events (Table 15), and its MCC was only 38% of the original model. The use of Borderline SMOTE modeling is of practical significance for the improvement of prediction accuracy in this case.

Based on the results of the three tests, it can be found that the situation with/without the $A_3$ test and the SMOTE/Borderline SMOTE test is similar; that is, the original model presents obvious advantages only in the testing phase. This is partly due to the lack of data in the testing set. However, it should also be noted that in each test, the assessment indices of the original model were ahead of the other models in all phases (Table 14), indicating that these two measures did improve the model prediction performance. On the contrary, the result of whether to resample the test is quite different from that of the other two tests. Regardless of the training and validation phase or the testing phase, the performance of the model using the resampling technique (whether ordinary SMOTE or Borderline SMOTE) is much better than that of the model without resampling, which indicates that the technique is necessary for preprocessing the original small and imbalanced data set.

**Table 14.** Comparison of prediction performance of RF models in different conditions.

| Phase | Model | Class | TPR | FPR | MCC | ROC |
|---|---|---|---|---|---|---|
| Training and Validation | Original | No | 0.966 | 0.013 | 0.953 | 0.997 |
| | | Yes | 0.987 | 0.034 | | |
| | | WA | 0.977 | 0.023 | | |
| | Without $A_3$ | No | 0.946 | 0.020 | 0.927 | 0.995 |
| | | Yes | 0.980 | 0.054 | | |
| | | WA | 0.963 | 0.037 | | |
| | Non-resample | No | 0.980 | 0.375 | 0.664 | 0.947 |
| | | Yes | 0.625 | 0.020 | | |
| | | WA | 0.945 | 0.341 | | |
| | SMOTE | No | 0.966 | 0.028 | 0.939 | 0.995 |
| | | Yes | 0.972 | 0.034 | | |
| | | WA | 0.969 | 0.031 | | |
| Testing | Original | No | 0.750 | 0.222 | 0.528 | 0.778 |
| | | Yes | 0.778 | 0.250 | | |
| | | WA | 0.765 | 0.237 | | |
| | Without $A_3$ | No | 0.500 | 0.333 | 0.169 | 0.736 |
| | | Yes | 0.667 | 0.500 | | |
| | | WA | 0.588 | 0.422 | | |
| | Non-resample | No | 0.750 | 0.667 | 0.091 | 0.729 |
| | | Yes | 0.333 | 0.250 | | |
| | | WA | 0.529 | 0.446 | | |
| | SMOTE | No | 0.750 | 0.556 | 0.203 | 0.750 |
| | | Yes | 0.444 | 0.250 | | |
| | | WA | 0.588 | 0.394 | | |

Note: "Yes" for debris flow events and "No" for non-debris flow events; WA represents weighted average.

**Table 15.** Confusion matrix of prediction results of RF models in different conditions.

| Model | Phase | Training & Validation | | Testing | |
|---|---|---|---|---|---|
| | Class | No | Yes | No | Yes |
| Original | No | 144 | 5 | 6 | 2 |
| | Yes | 2 | 147 | 2 | 7 |
| Without $A_3$ | No | 141 | 8 | 4 | 4 |
| | Yes | 3 | 146 | 3 | 6 |
| Non-resample | No | 146 | 3 | 6 | 2 |
| | Yes | 6 | 10 | 6 | 3 |
| SMOTE | No | 144 | 5 | 6 | 2 |
| | Yes | 4 | 140 | 5 | 4 |

## 4. Conclusions

Taking 15 typical periglacial debris flow gullies in the Parlung Zangbo Basin of southeast Tibet as an example, this study preliminarily explored how to establish a periglacial debris flow prediction model with acceptable accuracy and certain promotion value under the constraints of insufficient mechanism knowledge and imbalanced small sample data. Through field investigation and geostatistical analysis, a comprehensive index system covering 26 indicators from vegetation, meteorology, geology, and geomorphology was constructed firstly, and then seven of the indicators were selected as the recommended monitoring and early warning indicators, which were used as the inputs of the classification model after resampling. In addition, several comparative tests were also carried out for understanding the performance of the established model, and the following conclusions were drawn.

For the periglacial debris flow gully in the Peilong–Ranwu section of the Parlung Zangbo Basin, various temperature and precipitation indicators, as well as vegetation coverage, were closely related to the development of debris flow. Seven indicators, i.e., (1) precipitation duration ($D$), (2) maximum precipitation intensity ($I_m$), (3) coefficient of variation of the daily temperature range ($C_v{}^{15}$), (4) antecedent temperature ($T_{10}$), (5) antecedent accumulated precipitation ($A_3$), (6) antecedent NDVI ($NDVI_t$), and (7) average slope ($S_a$), were recommended as the monitoring and early warning indicators.

Based on these seven indicators, we constructed the periglacial debris flow prediction model using the Borderline SMOTE and RF classification algorithm. The main assessment indices, such as the weighted recall ratio, MCC, and AUC, exceeded 0.95 in the training and validation phase, and most of them exceeded 0.7 in the testing phase, showing acceptable accuracy under the limited sample data. Therefore, it is considered that the proposed workflow and model, as a prediction scheme of regional precipitation-related periglacial debris flow, have a certain promotion value for further research.

The performance of the SVM-based prediction scheme in this case is also acceptable. However, compared with the RF scheme, the SVM scheme has a lower recall rate on non-debris flow events, which means that this scheme may lead to a higher false alarm rate. In addition, the SVM-based classification model may still have overfitting phenomenon under the interpolation of Borderline SMOTE, while the RF based on ensemble learning idea shows better robustness; that is, the model presents more stable performance under the condition of imbalanced small sample data interpolation.

For imbalanced small sample data sets, resampling is necessary to improve the classification accuracy of the model. In this case, the prediction results from the RF model without resampling were more biased toward non-debris flow events, because the non-debris flow events accounted for the majority of the data set. Compared with the RF model with resampling, the prediction accuracy of the model without resampling had dropped significantly, which could reach 83%.

The test results indicated that considering the antecedent accumulated precipitation in a precipitation event helped to improve the model prediction accuracy, even for an area with strong permeable sandy soil. Similarly, as an improved version of SMOTE, the application of Borderline SMOTE was also of positive significance to the improvement of model performance.

**Author Contributions:** Conceptualization, J.D.; methodology, J.D. and H.-y.Z.; software, J.D. and H.-y.Z.; validation, J.D.; formal analysis, J.D.; investigation, J.D. and K.-h.H.; resources, K.-h.H.; data curation, J.D. and K.-h.H.; writing—original draft preparation, J.D.; writing—review and editing, K.-h.H. and L.W.; supervision, L.-y.D.; project administration, J.D. All authors have read and agreed to the published version of the manuscript.

**Funding:** This research was funded by the National Key R&D Program of China (Grant No. 2021YFE0111900, 2018YFC1505201), the Special Fund of Chinese Central Government for Basic Scientific Research Operations in Commonweal Research Institutes (No. CKSF2021435/TB, CKSF2021744/TB), and the National Natural Science Foundation of China (No. 41977171).

**Data Availability Statement:** Experimental data from this study will be available upon request. Other event data can be found in relevant literature [7,30,31].

**Acknowledgments:** Special thanks to Ming-feng Deng from the Bomi Geologic Hazard Observation Station of the Institute of Mountain Hazards and Environment, Chinese Academy of Sciences, for debris flow monitoring data supporting.

**Conflicts of Interest:** The authors declare no conflict of interest.

## References

1. Decaulne, A.; Saemundsson, P.; Petursson, O. Debris flow triggered by rapid snowmelt: A case study in the glei.arhjalli area, northwestern Iceland. *Geogr. Ann. Ser. A Phys. Geogr.* **2005**, *87*, 487–500. [CrossRef]
2. Legg, N.T.; Meigs, A.J.; Grant, G.E.; Kennard, P. Debris flow initiation in proglacial gullies on Mount Rainier, Washington. *Geomorphology* **2014**, *226*, 249–260. [CrossRef]
3. Wei, R.; Zeng, Q.; Davies, T.; Yuan, G.; Wang, K.; Xue, X.; Yin, Q. Geohazard cascade and mechanism of large debris flows in Tianmo gully, SE Tibetan Plateau and implications to hazard monitoring. *Eng. Geol.* **2018**, *233*, 172–182. [CrossRef]

4. Kumar, A.; Bhambri, R.; Tiwari, S.K.; Verma, A.; Gupta, A.K.; Kawishwar, P. Evolution of debris flow and moraine failure in the Gangotri Glacier region, Garhwal Himalaya: Hydro-geomorphological aspects. *Geomorphology* **2019**, *333*, 152–166. [CrossRef]
5. Liu, X.; Chen, B. Climatic warming in the Tibetan Plateau during recent decades. *Int. J. Climatol.* **2015**, *20*, 1729–1742. [CrossRef]
6. Yu, G.-A.; Yao, W.; Huang, H.Q.; Liu, Z. Debris flows originating in the mountain cryosphere under a changing climate: A review. *Prog. Phys. Geogr. Earth Environ.* **2020**, *45*, 339–374. [CrossRef]
7. Deng, M.; Chen, N.; Ding, H. Rainfall characteristics and thresholds for periglacial debris flows in the Parlung Zangbo Basin, southeast Tibetan Plateau. *J. Earth Syst. Sci.* **2018**, *127*, 11. [CrossRef]
8. Gao, B.; Zhang, J.J.; Wang, J.C.; Chen, L.; Yang, D.X. Formation mechanism and disaster characteristics of debris flow in the Tianmo gully in Tibet. *Hydrogeol. Eng. Geol.* **2019**, *46*, 144–153.
9. Gruber, S.; Haeberli, W. Permafrost in steep bedrock slopes and its temperature-related destabilization following climate change. *J. Geophys. Res. Atmos.* **2007**, *112*, F2. [CrossRef]
10. Krautblatter, M.; Funk, D.; Günzel, F.K. Why permafrost rocks become unstable: A rock-ice-mechanical model in time and space. *Earth Surf. Process. Landforms* **2013**, *38*, 876–887. [CrossRef]
11. Harris, C.; Davies, M.C.R.; Etzelmüller, B. The assessment of potential geotechnical hazards associated with mountain permafrost in a warming global climate. *Permafr. Periglac. Process.* **2001**, *12*, 145–156. [CrossRef]
12. Iverson, R.; Vallance, J. New views of granular mass flows. *Geology* **2001**, *29*, 115–118. [CrossRef]
13. Liang, W.-J.; Zhuang, D.-F.; Jiang, D.; Pan, J.-J.; Ren, H.-Y. Assessment of debris flow hazards using a Bayesian Network. *Geomorphology* **2012**, *171–172*, 94–100. [CrossRef]
14. Staley, D.M.; Negri, J.A.; Kean, J.W.; Laber, J.L.; Tillery, A.C.; Youberg, A.M. Prediction of spatially explicit rainfall intensity–duration thresholds for post-fire debris-flow generation in the western United States. *Geomorphology* **2017**, *278*, 149–162. [CrossRef]
15. Addison, P.; Oommen, T.; Sha, Q. Assessment of post-wildfire debris flow occurrence using classifier tree. *Geomat. Nat. Hazards Risk* **2019**, *10*, 505–518. [CrossRef]
16. Walter, F.; Amann, F.; Kos, A.; Kenner, R.; Phillips, M.; de Preux, A.; Huss, M.; Tognacca, C.; Clinton, J.; Diehl, T.; et al. Direct observations of a three million cubic meter rock-slope collapse with almost immediate initiation of ensuing debris flows. *Geomorphology* **2020**, *351*, 106933. [CrossRef]
17. Michie, D.; Spiegelhalter, D.J.; Taylor, C.C. *Machine Learning, Neural and Statistical Classification*; Citeseer: Princeton, NJ, USA, 1994. [CrossRef]
18. Costache, R. Flash-Flood Potential assessment in the upper and middle sector of Prahova river catchment (Romania). A comparative approach between four hybrid models. *Sci. Total. Environ.* **2018**, *659*, 1115–1134. [CrossRef]
19. Zhou, C.; Yin, K.; Cao, Y.; Ahmed, B. Application of time series analysis and PSO–SVM model in predicting the Bazimen landslide in the Three Gorges Reservoir, China. *Eng. Geol.* **2016**, *204*, 108–120. [CrossRef]
20. Cui, Y.; Cheng, D.; Chan, D. Investigation of Post-Fire Debris Flows in Montecito. *ISPRS Int. J. Geo-Inf.* **2018**, *8*, 5. [CrossRef]
21. Liu, Z.; Wang, L.; Zhang, Y.; Chen, C.P. A SVM controller for the stable walking of biped robots based on small sample sizes. *Appl. Soft Comput.* **2016**, *38*, 738–753. [CrossRef]
22. Squarcina, L.; Villa, F.M.; Nobile, M.; Grisan, E.; Brambilla, P. Deep learning for the prediction of treatment response in depression. *J. Affect. Disord.* **2020**, *281*, 618–622. [CrossRef] [PubMed]
23. Chawla, N.V.; Bowyer, K.W.; Hall, L.O.; Kegelmeyer, W.P. SMOTE: Synthetic Minority Over-sampling Technique. *J. Artif. Intell. Res.* **2002**, *16*, 321–357. [CrossRef]
24. Schlögl, M.; Stuetz, R.; Laaha, G.; Melcher, M. A comparison of statistical learning methods for deriving determining factors of accident occurrence from an imbalanced high resolution dataset. *Accid. Anal. Prev.* **2019**, *127*, 134–149. [CrossRef] [PubMed]
25. Cheng, D.; Wang, J.; Wei, X.; Gong, Y. Training mixture of weighted SVM for object detection using EM algorithm. *Neurocomputing* **2015**, *149*, 473–482. [CrossRef]
26. Cheng, Z.L.; Liang, S.; Liu, J.K.; Liu, D.X. Distribution and change of glacier lakes in the upper Palongzangbu River. *Bull. Soil Water Conserv.* **2012**, *32*, 8–12.
27. Liu, J.; Yao, X.J.; Gao, Y.P.; Qi, M.M.; Duan, H.; Zhang, D. Glacial Lake variation and hazard assessment of glacial lakes outburst in the Parlung Zangbo River Basin. *J. Lake Sci.* **2019**, *31*, 244–255.
28. Lv, R.R.; Tang, B.X.; Zhu, P.Y. *Debris Flow and Environment in Tibet*; Press of Chengdu Science and Technology: Chengdu, China, 1999.
29. Jia, Y. The Impact Mechanism of Climate Warming on Mountain Hazards in the Southeast Tibet. Ph.D. Thesis, University of Chinese Academy of Sciences, Beijing, China, 2018; pp. 118–143.
30. Zeng, X.Y.; Zhang, J.J.; Yang, D.X.; Wang, J.C.; Gao, B.; Li, Y. Characteristics and Geneses of Low Frequency Debris Flow along Parlongzangbo River Zone: Take Chaobulongba Gully as an Example. *Sci. Technol. Eng.* **2019**, *19*, 103–107.
31. Li, Y.L.; Wang, J.C.; Chen, L.; Liu, J.; Yang, D.; Zhang, J. Characteristics and geneses of the group-occurring debris flows along Parlung Zangbo River zone in 2016. *Res. Soil. Water Conserv.* **2018**, *25*, 401–406.
32. Dong, L.-J.; Li, X.-B.; Peng, K. Prediction of rockburst classification using Random Forest. *Trans. Nonferrous Met. Soc. China* **2013**, *23*, 472–477. [CrossRef]
33. Burges, C.J.C. A Tutorial on Support Vector Machines for Pattern Recognition. *Data Min. Knowl. Discov.* **1998**, *2*, 121–167. [CrossRef]
34. Hui, H.; Wang, W.Y.; Mao, B.H. Borderline-SMOTE: A new over-sampling method in imbalanced data sets learning. In *International Conference on Intelligent Computing*; Springer: Berlin/Heidelberg, Germany, 2005; Volume 3644, pp. 878–887. [CrossRef]

35. Breiman, L. Random forests. *Mach. Learn.* **2001**, *45*, 5–32. [CrossRef]
36. Svetnik, V.; Liaw, A.; Tong, C.; Culberson, J.C.; Sheridan, R.P.; Feuston, B.P. Random Forest: A Classification and Regression Tool for Compound Classification and QSAR Modeling. *J. Chem. Inf. Comput. Sci.* **2003**, *43*, 1947–1958. [CrossRef]
37. Quinlan, J.R. Introduction of decision trees. *Mach. Learn.* **1986**, *1*, 81–106. [CrossRef]
38. Mitchell, T.M. *Machine Learning*; McGraw-Hill: Boston, MA, USA, 1997.
39. Dimitriadis, S.I.; Liparas, D.; Dni, A. How random is the random forest? Random forest algorithm on the service of structural imaging biomarkers for Alzheimer's disease: From Alzheimer's disease neuroimaging initiative (ADNI) database. *Neural Regen. Res.* **2018**, *13*, 962–970. [CrossRef]
40. Ming, D.; Zhou, T.; Wang, M.; Tan, T. Land cover classification using random forest with genetic algorithm-based parameter optimization. *J. Appl. Remote. Sens.* **2016**, *10*, 035021. [CrossRef]
41. Zhou, T.N.; Ming, D.P.; Zhao, R. Land cover classification based on algorithm of parameter optimization random forests. *Sci. Surv. Mapp.* **2017**, *42*, 88–94.
42. Vapnik, V. *The Nature of Statistical Learning Theory*; Springer: Berlin/Heidelberg, Germany, 1999.
43. Rebetez, M.; Lugon, R.; Baeriswyl, P.A. Climatic change and debris flows in high mountain regions: The case study of the Ritigraben Torrent (Swiss Alps). In *Climatic Change at High Elevation Sites*; Diaz, H.F., Beniston, M., Bradley, R.S., Eds.; Springer: Dordrecht, The Netherlands, 1997; pp. 139–157.
44. Chleborad, A.F. *Use of Air Temperature Data to Anticipate the Onset of Snowmelt-Season Landslides (USGS Open-File Report 98-0124)*; US Geological Survey: Reston, VA, USA, 1998.
45. Deng, M.; Chen, N.; Liu, M. Meteorological factors driving glacial till variation and the associated periglacial debris flows in Tianmo Valley, south-eastern Tibetan Plateau. *Nat. Hazards Earth Syst. Sci.* **2017**, *17*, 345–356. [CrossRef]
46. Chen, N.S.; Wang, Z.; Tian, S.F.; Zhu, Y.H. Study on debris flow process induced by moraine soil mass failure. *Quat. Sci.* **2019**, *39*, 1235–1245.
47. Greenway, D. Vegetation and Slope Stability. In *Slope Stability: Geotechnical Engineering and Geomorphology*; Anderson, M.G., Richards, K.S., Eds.; John Wiley and Sons Ltd: Hoboken, NJ, USA, 1987.
48. Wilkinson, P.L.; Anderson, M.G.; Lloyd, D.M. An integrated hydrological model for rain-induced landslide prediction. *Earth Surf. Process. Landforms* **2002**, *27*, 1285–1297. [CrossRef]
49. Coppin, N.J.; Richards, I.G. *Use of Vegetation in Civil Engineering*; Construction Industry Research and Information Association (CIRIA): London, UK, 1990.
50. Emadi-Tafti, M.; Ataie-Ashtiani, B.; Hosseini, S.M. Integrated impacts of vegetation and soil type on slope stability: A case study of Kheyrud Forest, Iran. *Ecol. Model.* **2021**, *446*, 109498. [CrossRef]
51. Takahashi, T. *Debris Flow Mechanics, Prediction and Countermeasures*; Taylor and Francis Group: London, UK, 2007; p. 108.
52. Du, J.; Fan, Z.-J.; Xu, W.-T.; Dong, L.-Y. Research Progress of Initial Mechanism on Debris Flow and Related Discrimination Methods: A Review. *Front. Earth Sci.* **2021**, *9*, 629567. [CrossRef]
53. Bradley, P. The use of the area under the ROC curve in the evaluation of machine learning algorithms. *Pattern Recogn.* **1997**, *30*, 1145–1159. [CrossRef]
54. Fawcett, T. An Introduction to ROC analysis. *Pattern Recogn. Lett.* **2006**, *27*, 861–874. [CrossRef]
55. Boughorbel, S.; Jarray, F.; El-Anbari, M. Optimal classifier for imbalanced data using Matthews Correlation Coefficient metric. *PLoS ONE* **2017**, *12*, e0177678. [CrossRef]
56. Lee, T.; Kim, M.; Kim, S.-P. Improvement of P300-Based Brain–Computer Interfaces for Home Appliances Control by Data Balancing Techniques. *Sensors* **2020**, *20*, 5576. [CrossRef]
57. Lei, C.; Deng, J.; Cao, K.; Xiao, Y.; Ma, L.; Wang, W.; Ma, T.; Shu, C. A comparison of random forest and support vector machine approaches to predict coal spontaneous combustion in gob. *Fuel* **2019**, *239*, 297–311. [CrossRef]
58. Tang, Z.; Mei, Z.; Liu, W.; Xia, Y. Identification of the key factors affecting Chinese carbon intensity and their historical trends using random forest algorithm. *J. Geogr. Sci.* **2020**, *30*, 743–756. [CrossRef]
59. Caine, N. The Rainfall Intensity—Duration Control of Shallow Landslides and Debris Flows. *Geogr. Ann. Ser. A Phys. Geogr.* **1980**, *62*, 23–27. [CrossRef]
60. Abancó, C.; Hürlimann, M.; Moya, J.; Berenguer, M. Critical rainfall conditions for the initiation of torrential flows. Results from the Rebaixader catchment (Central Pyrenees). *J. Hydrol.* **2016**, *541*, 218–229. [CrossRef]

*water*

MDPI

*Article*

# Model-Based Mechanism Analysis of "7.20" Flash Flood Disaster in Wangzongdian River Basin

**Sijia Hao [1,2], Wenchuan Wang [1], Qiang Ma [2,*], Changzhi Li [2], Lei Wen [3], Jiyang Tian [2] and Changjun Liu [2]**

[1] College of Water Resources, North China University of Water Resources and Electric Power, Zhengzhou 450046, China
[2] China Institute of Water Resources and Hydropower Research, Beijing 100038, China
[3] College of Hydrology and Water Resources, Hohai University, Nanjing 210098, China
* Correspondence: maqiang@iwhr.com

**Abstract:** With digital information technology based on limited data, disaster simulation review is an important guideline for analyzing disaster mechanisms, planning post-disaster reconstruction, and improving defense capability. Taking the "7.20" flash flood in the Wangzongdian river basin as a research area, a hydrological-hydrodynamic model was established using limited measured data. The results showed that the extreme rainstorm caused flooding in mountainous areas and the collapse of subgrade water damming, and the high-level flood quickly flowed into Wangzongdian Village in a short distance, which was the main cause of this serious disaster. Considering the collapse due to the congestion of the upstream bridge, the simulated flood flow in Wangzongdian Village reached 782 $m^3/s$, which was basically consistent with the post-disaster survey results, with a relative error of only +8%. The modeling strategy proposed in this paper is applicable in the review of sudden heavy rainstorms and flash floods and can provide technical guidance for future flash flood simulation review analysis in other areas.

**Keywords:** hydrological model; hydrodynamic model; disaster review analysis; heavy rainfall in Henan

**Citation:** Hao, S.; Wang, W.; Ma, Q.; Li, C.; Wen, L.; Tian, J.; Liu, C. Model-Based Mechanism Analysis of "7.20" Flash Flood Disaster in Wangzongdian River Basin. *Water* **2023**, *15*, 304. https://doi.org/10.3390/w15020304

Academic Editor: Chang Huang

Received: 29 November 2022
Revised: 6 January 2023
Accepted: 9 January 2023
Published: 11 January 2023



## 1. Introduction

Flash floods in China mainly refer to river floods caused by heavy rainfall in small watersheds (within an area of 200 $km^2$) in hilly areas, which are characterized by small scope, short duration, high suddenness, and high destructive power, and are very likely to cause massive casualties and socioeconomic losses [1]. According to the statistics of the Ministry of Emergency Management of China, a total of 42 heavy rainfall events occurred in China in 2021, with an average precipitation of about 659 mm, an increase in about 6% over previous years, resulting in 59.01 million people being affected and an economic loss of about 245.89 billion yuan [2]. It is of great significance to carry out a review simulation analysis of flash floods in small mountain watersheds to understand the causes of flash floods, to improve the flood prevention and mitigation capacity in mountainous areas of China, to summarize prevention and control experience, and to establish and improve a sound flash flood disaster prevention and control system [3]. The basic monitoring facilities in China's mountainous areas still need to be further improved, which makes it difficult to obtain refined data and more difficult to review flash floods [4]. However, a refined post-disaster review can be achieved on the basis of limited data by reasonably developing a flash flood review strategy based on domestic models.

The improvement of numerical algorithms and the rapid development of computer technology have guaranteed accurate and efficient flood analysis and simulation. Distributed hydrological simulations allow the reenactment of the effects of changes in different hydrological elements on water cycle processes according to the spatial variability of the watershed's underlying surface [5,6]. Shu et al. [7] put forward a distributed hydrological model SHUD, which uses the finite volume method to solve surface-underground coupling and uses the irregular triangular network to construct the basin simulation space to realize the high-resolution numerical simulation of meters-kilometers in space and seconds-hours in time. Zhao et al. [8] established a distributed hydrological model VRGL for hilly regions by considering the role of soil-weathering bedrock interface in runoff generation and carried out the application in Tunxi streams in humid, hilly areas in China. The results showed that VRGL has good accuracy (the relative errors of peak discharge and flood volume are within $\pm$ 20%) and can be used for flood forecasting in hilly areas. Liu et al. [9] proposed a spatio temporally-mixed runoff model SKBY and the modular distributed hydrological model FFMS by analyzing the transformation and combination of small watershed infiltration and storage mechanism in two dimensions of time and space. It was proved that the model improves the simulation accuracy of floods under short-duration and heavy rainfall conditions. Hydrological models have been widely used in flood forecasting and warning in China. However, the output results of the model are the process of runoff generation obtained according to the hydrological conditions of the basin, which cannot achieve the dynamic description of flood routing in the inundated area [10]. Moreover, the simulation accuracy is greatly affected when the natural runoff generation mechanism of the basin changes due to the topographic changes caused by floods. The hydrodynamic model can not only simulate the evolution process of water flow in the river course and floodplain but also the waterlogging in the urban and rural areas. According to hydrodynamic calculation, the spatio-temporal variation of the submerged area can be obtained. The finite volume method is a commonly used numerical calculation method in the current hydrodynamic model. The numerical shallow water simulation using a Godunov-type scheme for the riemann approximate solver has good accuracy and is widely used [11,12]. Zhang et al. [13] established a two-dimensional hydrodynamic model of rainfall-runoff based on the Godunov-type scheme with unstructured grids and verified that with this model, the surface runoff movement could be finely simulated.

Scholars have conducted plenty of research on flash flood simulation review and disaster-causing mechanism analysis. Liu et al. [14] compared flood simulations with three methods: the watershed model method, reasoning formula method, and empirical formula method, and concluded that the results of the watershed model method were more reasonable. Zhang et al. [15] constructed a mountain hydrological model for the Guanshan River basin by introducing the rainstorm interflow mechanism and found that the rainstorm interflow mechanism is one of the important runoff generation mechanisms in small watersheds in humid mountainous areas, and it is considered that the hydrological model of small mountain watershed with this mechanism is more effective in flash flood forecasting and early warning. Sandrine et al. [16] simulated flash floods based on a distributed hydrological model and found that flash flood dynamics showed two phases: the first phase was mainly controlled by soil properties, and the second phase was by rainfall due to soil saturation. In summary, an accurate description of the geometry and underlying surface characteristics of the small watershed and a clear understanding of the characteristics of runoff generation are the keys to simulating the process of and analyzing the causes of flash floods. However, when the natural geomorphological conditions of the region change severely during the flooding process, the simulation results of the hydrological model often deviate greatly from the actual results. Lin et al. [17] analyzed the inundation of flash floods in small mountain watersheds based on one-dimensional and two-dimensional coupled hydrodynamic models, respectively, and verified that one-dimensional and two-dimensional coupled hydrodynamic models can effectively simulate the flood characteristics and evolution process in a small mountain watershed. Zhang et al. [13] simulated and analyzed the flood evolution after dam failure by using the dam failure model based on the Godunov-type scheme. The two-dimensional hydrodynamic model can automatically adapt to the change of flow regime and has unique advantages in the simulation of dam failure, dam break, and other flood types. Segura et al. [18] simulated the 2007 flash flood of the Girona

River in Spain by using the coupled hydrological and hydrodynamic model and found that coupled model simulation considering geomorphological characteristics can help analyze flood causes and provide technical guidance for early warning. El-Saoud et al. [19] found that infrastructure construction aggravated flash flood disasters to some extent by using a hydrological model (HEC-1) with two-dimensional hydraulic modeling (HEC-RAS 5.0.7) software to draw the early warning map of flash flood risk in Mecca, Saudi Arabia. Although the hydrodynamic model can describe flood routing in detail, the amount of data required for modeling is relatively large, which is often limited by time, cost, and local realities in flood review, so it is difficult to be widely used. At present, flash flood disaster simulation reviews are mostly conducted by a single type of model. It is difficult to accurately reproduce the whole process of flood development, whether the flood simulation is realized by calculating the peak discharge through the hydrological model or by calculating the submerged water depth through the hydrodynamic model. Therefore, there is an urgent need for guidelines and strategy for the joint application of multiple models to optimize the whole process simulation of watershed flood routing.

Compared with foreign specialized models of water resources, China has a late-developing advantage in studying flood disasters based on hydrological and hydrodynamic models. However, in practical applications, especially in flash flood simulation, it is still necessary to develop appropriate model simulation strategies according to the extent of post-disaster data collection. From 17 to 24 July 2021, Henan Province was hit by unusually heavy rainfall, which triggered urban waterlogging and flash flood disasters, causing serious loss of life and property. In this paper, a hydrological-hydrodynamic model was used to analyze the "7.20" flash flood in Wangzongdian village based on national flash flood investigation results and postdisaster field survey data. A rapid modeling strategy of flash flood disasters using limited data was put forward based on the simulation results, helping to quickly clarify the sources and causes of floods and providing important technical and decision-making support for flash flood disaster prevention and early warning.

## 2. Research Materials

In July 2021, China's Henan Province was hit by heavy rainfall. The daily rainfall in Xingyang exceeded the historical extreme value, causing a significant impact on people's life and social production (Figure 1). According to statistics, this extraordinary rainstorm left 242,069 population affected in Xingyang City, with an affected area of 21,745.05 hectares of crops and 10,629 houses out of 3260 households collapsed. In addition, it caused varying degrees of damage to 630 communication base stations, 5473.95 km of fiber optic cables, 44 base station houses, 13 small and medium-sized reservoirs, five erosion and torrent control works, five rivers, two channels, and one irrigation station, resulting in direct economic losses of 5.051 billion yuan in Xingyang City [20]. This disaster event was selected for review to figure out the causes and mechanism of the disaster, which has technical reference significance for disaster review analysis in other areas affected by flash floods in China.

### 2.1. Study Area

Wangzongdian Village is located in the central part of Henan Province, a shallow hilly area with complex topography, straddling the two major basins of the Yellow River and the Huai River. Its rainfall varies greatly interannually and is unevenly distributed within the year, with rainfall from June to September accounting for 65% of the annual rainfall. The upstream catchment area of Wangzongdian Village is 21.95 km$^2$, including three branch ditches. The partially bare land in the upper reaches carried a large amount of sediment flowing with flash floods, causing damage to villages and infrastructure along the lower reaches of the river. Wangzongdian Village is surrounded by mountains on all sides, such as Dalong Mountain, Xiaolong Mountain, Xiaoding Mountain, Jianshan Mountain, Matou Mountain, and Jinjinggou Mountain. Such a geographical situation caused the flood to flow down mountains and converge in Wangzongdian Village, which was prone to cause flood disasters. The distribution of the watershed and location of Wangzongdian Village are shown in Figure 2.

(a) (b)

**Figure 1.** Comparison of before and after the "7.20" flash flood disaster in Wangzongdian Village, Henan Province. (a) before, (b) after.

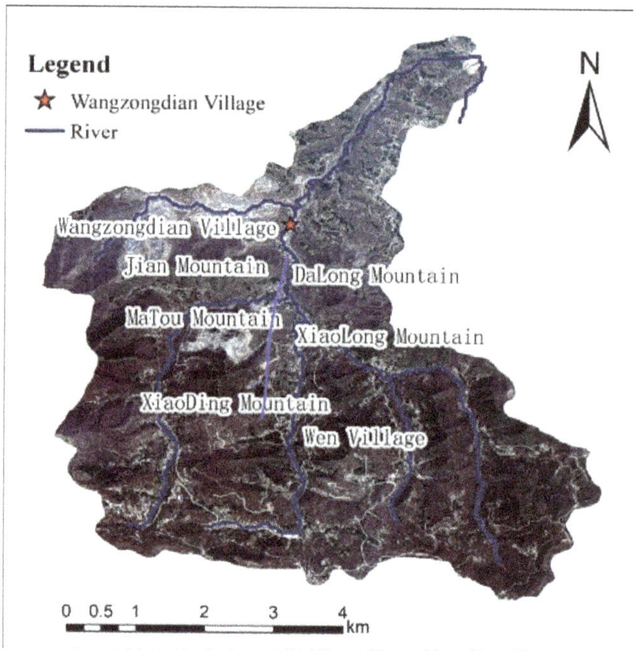


**Figure 2.** Overview of the study area.

## 2.2. Data Collection

The digital data used to build hydrological and hydrodynamic models mainly include terrain elevation data, small watersheds, water systems, nodes, soil texture, and land use. Table 1 records the details and source of the data.

**Table 1.** Basic data information of study area.

| Data Type | Resolution | Time | Data Source |
|---|---|---|---|
| Rainfall data | - | 2021 | Rain gauge data |
| DEM | 30 m | 2021 | Geospatial data cloud web site |
| Remote sensing image | 2.5 m,30 m | 2015 | National flash flood disaster investigation and assessment |
| Watershed | 10~50 km$^2$ | 2015 | |
| Land use | 30 m | 2015 | |
| Soil texture | 30 m | 2015 | |
| Dam break data | - | 2021 | Postdisaster field survey data. |

2.2.1. Underlying Attribute Data

The modeled watershed area of the upper reaches of Wangzongdian was 29.1 km$^2$, which was divided into 24 calculation units with an area of 0.16–1.6 km$^2$. For each calculation unit, 75 items of basic attribute information, including spatial attributes, land use, and soil type of the small watershed, were extracted, and the spatial topological relationship with the small watershed as the unit was established, forming the basic data set of Wangzongdian Village flash flood small watershed. Statistically, the land use type of the watershed is dominated by arable land and forest land, accounting for 63% and 33% of the whole, respectively. The main types of soils are sandy loam and sandy clay, accounting for 66% and 34%, respectively. The distribution of land use, soil texture, and water system in the Wangzongdian watershed is shown in Figure 3. The 30 m DEM data are shown in Figure 4, and the red point is the dam-failure site.

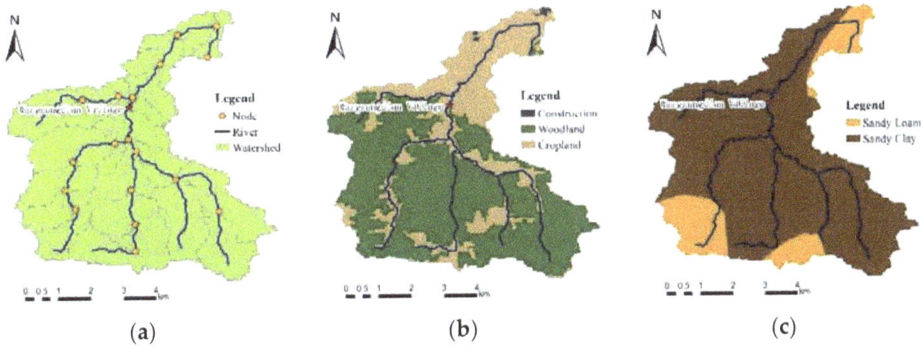


**Figure 3.** (**a**) Calculation unit division. (**b**) land use attribute data. (**c**) soil texture data.

2.2.2. Rainfall Data

By analyzing the measured rainfall data of Wangzongdian Rainfall Station, it was found that the rainfall was wide in scope, large in amount, short in duration, and high in rainfall level. From 8:00 a.m. on July 18 to 8:00 a.m. on 22 July 2021, the total rainfall was 764.5 mm, exceeding its annual average rainfall (608 mm) by 156.5 mm in Wangzongdian Village, of which the maximum daily rainfall was 353 mm on July 20, exceeding half of the annual average rainfall (58.05%), with a rain intensity of 58.5 mm/h from 2:00 a.m. to 3:00 a.m. [21,22]. All broke the extreme historical value of local meteorological observation records. The cumulative and hourly rainfall distribution of Wangzongdian Village is shown in Figure 5 below.

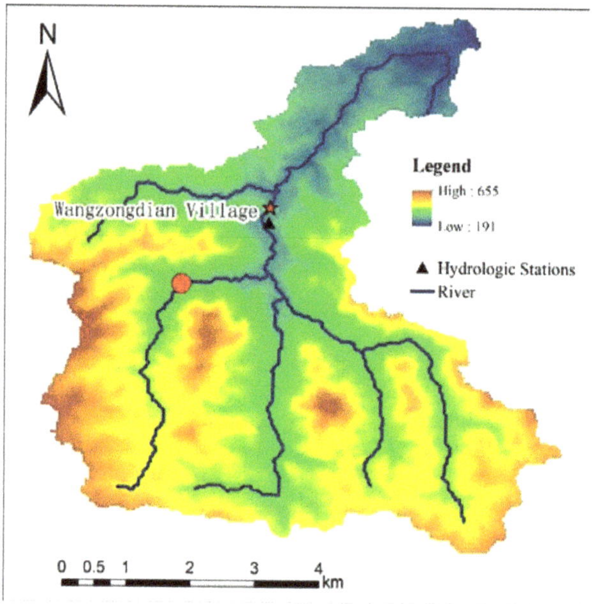


**Figure 4.** Digital Elevation data of Wangzongdian watershed.

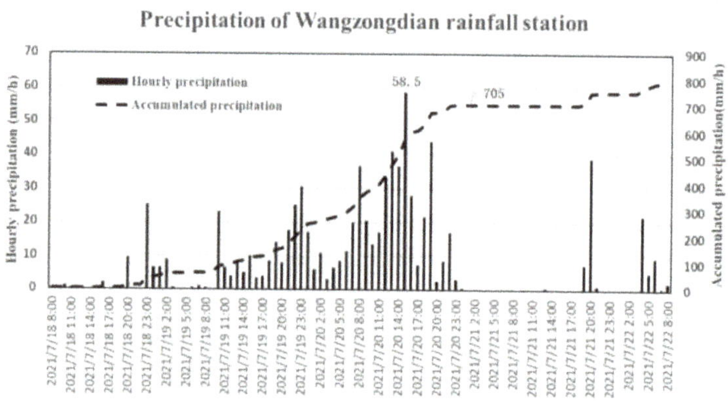


**Figure 5.** Cumulative and hourly rainfall distribution of Wangzongdian Rainfall Station.

### 2.2.3. Post-Disaster Investigation Data

On the afternoon of August 6, 2021, Xingyang Municipal Water Resources Bureau commissioned a five-person survey team from Zhengzhou Hydrology and Water Resources Survey Bureau to investigate the river from upstream downward using instruments and equipment such as total station, RTK, and UAV. The team finally selected a relatively straight 100.4-m section of the channel segment as the survey section, as shown in the following Figure 6 [23], and the red line refers to the measuring line for the cross section.

**Figure 6.** A real shot of the survey section after the flood.

The survey team measured the upper, middle, and lower sections, and it was known that the upper section was 60 m away from the middle section and the middle section 40.4 m from the lower section. According to the on-site survey of the river, the roughness $n$ was determined to be 0.032, the average section area $A$ was calculated to be 144 $m^2$, the average wetted perimeter was 48 m, the hydraulic radius R was 2.99, and the surface slope S was 0.006. By using the mathematical formula $Q = \frac{1}{n}AR^{2/3}S^{1/2}$, the peak discharge was preliminarily estimated to be 724 $m^3/s$, and the average section velocity is 5.03 m/s. In this study, the disaster survey data will be compared with the simulation results as the result of actual measurement to verify the accuracy of the simulation results.

## 3. Simulation Analysis

### 3.1. Modeling Strategy

The crux of flash flood disaster simulation review lies in the collection of basic data and the optimization of simulation methods. Due to the strong destructive power of flash floods, had a great impact on the traffic and topography of the affected areas, which brought great difficulties to post-disaster data collection. In this case, investigators were unable to quickly collect all disaster-related data at one time, making the construction of the model more challenging. In this paper, a two-phase flash flood disaster review strategy was put forward, and different models were adopted to simulate and analyze the disaster according to different phases of data collection:

In Phase I, the distributed flash flood hydrological model was mainly used to simulate and review the disaster by using public data such as satellite remote sensing, collecting information on watershed underlying surface and rainfall data monitored by meteorological and hydrological departments. The main goal of this phase was to analyze the flood magnitude of the target flash flood disaster through simulation and to clarify the flood source.

In Phase II, focusing on the refined hydrodynamic model simulation, based on the simulation results of Phase I, a two-dimensional hydrodynamic model was constructed to simulate and analyze the spatial and temporal changes of the flood routing and inundation area in the more severely affected areas based on the actual measured, refined terrain, to replay the flood occurrence, the collapse of embankments and dams, and the inundation of villages and towns, to truly clarify the causes of flash flood disaster, and to provide guidance for the design of embankments, buildings and roads in some villages and towns in post-disaster reconstruction.

### 3.2. Modeling Tools

#### 3.2.1. Hydrological Model

In this study, a modular small watershed flood analysis system, FFMS (Flash Flood Modelling System), was used to simulate the flooding process. As a new generation of small and medium-sized watershed heavy rains and flooding simulation software developed by the research team of China Institute of Water Resources and Hydropower Research, this software integrated the advantages of computer programming languages such as C++, Fortran, Java, and multi-format data pre- and post-processing technology supports the automatic division of small watersheds and parameter extraction. Moreover, it can realize modular parallel calculation of runoff generation and provide both intelligent parameter calibration schemes and two-dimensional visualization and display [24,25]. The software supports several domestic and international hydrological simulation methods (HEC, PRMS, Xinanjiang model, spatio-temporal variable source mixed runoff model) for automatic modeling and manual drag-and-drop modeling and has an intelligent expert parameter database containing watershed underlying surface runoff generation and river channel parameters under different conditions.

The hydrological simulation of the flash flood disaster in Wangzongdian was conducted based on this model software. The model constructed a distributed hydrological model of a small watershed in terms of planar mixing, vertical mixing, and temporal mixing of excess infiltration/storage [26] by dividing different geomorphic hydrological response units and generalizing the one-to-one correspondence between each small watershed geomorphic hydrological response unit and different runoff generation mechanisms [27]. In particular, for flash flood runoff generation, the model used CARTO one-dimensional unsaturated infiltration simulation method [28,29]. The soil moisture content domain was discretized into an interval with constant moisture content, and the one-dimensional infiltration equation was derived to quickly obtain the vertical displacement of the wetting front. Considering the mutual suction between different moisture content units, the lateral displacement of the wetting front was redistributed. Compared with Green-Ampt and Richard's equations, this method used a numerical scheme without estimating the nonlinear gradient, which was more efficient and accurate [30].

3.2.2. Hydrodynamic Model

IFMS (Integrated Flood Modeling System) is a one- and two-dimensional visual hydrodynamic simulation software developed by the China Institute of Water Resources and Hydropower Research [31,32]. The software includes a 1D river network calculation engine supporting complex hydraulic engineering scheduling simulation, a high-resolution 2D flood analysis calculation engine, and a fast unstructured grid generation tool. The software is based on the GIS platform to complete model pre- and post-processing, which can realize 1D and 2D flood simulation and urban pipe network and 2D model coupling simulation. The software has strong visualization functions and simple operation, and it has now realized the coupled calculation of hydrological and hydrodynamic models with a CPU-GPU parallel computing strategy, which greatly improves computational efficiency [33].

The two-dimensional shallow water equation of water depth average can be simplified as follows:

$$\frac{\partial h}{\partial t}+\frac{\partial hu}{\partial x}+\frac{\partial hv}{\partial y}=0 \tag{1}$$

$$\frac{\partial hu}{\partial t}+\frac{\partial}{\partial x}\left(hu^2+\frac{1}{2}gh^2\right)+\frac{\partial huv}{\partial y}=s_x \tag{2}$$

$$\frac{\partial hu}{\partial t}+\frac{\partial hvu}{\partial x}+\frac{\partial}{\partial y}\left(hv^2+\frac{1}{2}gh^2\right)=s_y \tag{3}$$

where, $h$ is the water depth; $u$ is the velocity in the $x$ direction; $v$ is the velocity in the $y$ direction; $S_x$, $S_y$ are the source item

*3.3. Modeling Process*

The distributed hydrological model of the small watershed above Wangzongdian Village was established by using FFMS software based on Wangzongdian small watershed basic data set layer data and rainfall data of main stations. The inverse distance weighting method was used to obtain the rainfall at the surface of small watersheds, and the spatio-temporal variable source mixed runoff generation simulation method was selected for runoff generation calculation. The triangular generalized unit line method was used for slope confluence, the linear reservoir method for base flow, and the motion wave method for flood routing. This modeling process required a small amount of data and took a short time.

Based on the high-resolution DEM data (Figure 4), the red point in the Figure is the dam-failure site. Applied IFMS software to construct a two-dimensional hydrodynamic model of a small Wangzongdian watershed. The guiding principle was that the dam-failure formula was adopted to calculate the dam-failure site, and the results of the hydrological model were still used as boundary conditions for other tributaries.

The required site survey data and laser point cloud measurement data for the dam failure are shown in Figure 7. The calculation of the peak flow of the breach was based on the empirical formula of the Research Institute of China's Ministry of Railways [34].

$$Q_{max} = 0.27\sqrt{g}(L/B)^{1/10}(B/b)^{1/3}b(h_0 - kh)^{3/2} \quad (4)$$

where $Q_{max}$ is the maximum flow rate of the dam bursting flood ($m^3/s$); $g$ is the acceleration of gravity ($m/s^2$); $g$ is the length of the reservoir area; when $L/B > 5$, take the $L/B = 5$; $b$ is the average width of the breach (m); $h_0$ is the water depth in front of the dam (m); $h$ is the average height of the residual dam body at the breach (m); $k$ is the empirical coefficient, and $k$ and $h$ are taken as 0. The laser point cloud measurement value $L = 600$ in Figure 7; $B = 41.9$(yellow line); $b = 36.9$(red line); $h_0 = 17.1$(dark red line), $Q_{max} = 3013$ $m^3/s$ can be obtained.

Figure 7. Dam failure site and laser point cloud measurement data.

## 4. Simulation Results and Analysis

### 4.1. Simulation Results

The model used default parameters for the distributed hydrological simulation of the hourly flash flood process in the Wangzongdian watershed from 0:00 a.m. on July 20 to 22:00 p.m. on July 2021. The simulated flood process line of the FFMS model is shown in Figure 8. The calculated runoff coefficient was about 0.79. Considering the large rain intensity and short duration of the flash flood in Wangzongdian Village, the simulated runoff coefficient was more reasonable. But the peak flow was 414.25 $m^3/s$, only about half of the disaster survey results (724 $m^3/s$). Accordingly, it was preliminarily judged that there was a special case in a flood event which changed the natural process of runoff generation and flood routing. Based on the proposed flash flood simulation review strategy, the need for a second phase of hydrodynamic simulation analysis was assessed.

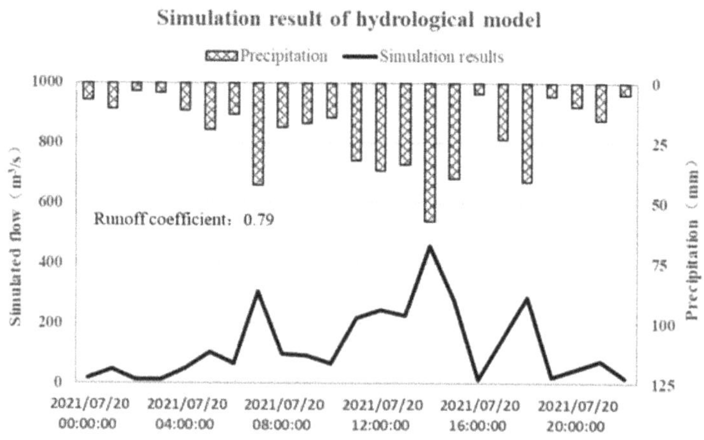


**Figure 8.** Simulation results of Wangzongdian watershed.

Based on the conclusion of the hydrological simulation, field investigations showed that there was an arch bridge dam failure in the upper reaches of the watershed. This dam break process disrupted the natural runoff mechanisms of the watershed, making the distributed hydrologic model unable to better simulate the development of this flood. According to the review strategy, the disaster was simulated in the second step by a coupled hydrological and hydrodynamic simulation.

Based on IFMS software, the flooding process of Wangzongdian Village from 0:00 a.m. to 20:00 p.m. on 20 July 2021 was simulated in two cases: considering dam failure and not considering dam failure. The results are shown in Figure 9. Without considering dam failure, the simulated peak discharge of the hydrodynamic model was about 422 $m^3/s$, close to the hydrological simulation result of 414.25 $m^3/s$. However, the hydrodynamic model simulation result was about 782 $m^3/s$ under the consideration of dam failure, which was basically consistent with the disaster investigation results, with a relative simulation flood error of +8%. The results showed that the hydrodynamic model could review the process of this flash flood and support the analysis of the causes of the flash flood under the consideration of dam failure.

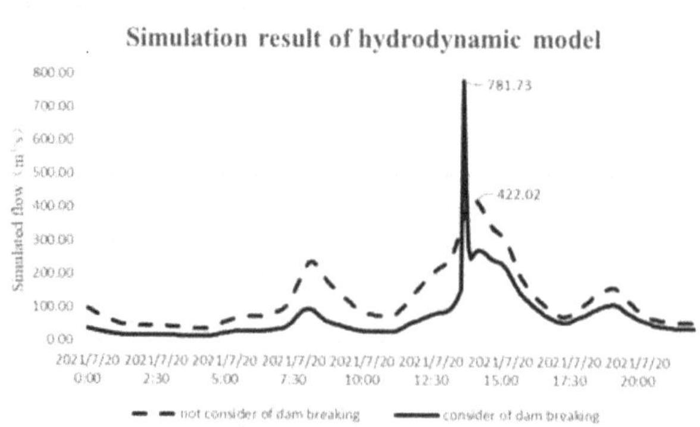


**Figure 9.** Hydrodynamic model simulation results.

According to hydrodynamic calculation, the submerged water depth of the river at different times is shown in Figure 10. According to the rainfall distribution map of Wangzongdian Village (Figure 5), it can be seen that its cumulative precipitation was 162 mm from 0:00 a.m. to 10:00 a.m. on July 20, and the submerged water depth of some areas downstream reached 2–5 m. The rainfall gradually increased after 10:00 a.m., and at 12:00 p.m., some areas downstream of the Village were flooded to a depth of more than 5 m; the rainfall continued to increase from 12:00 p.m. to 14:00 p.m., and the maximum rainfall (57.6 mm) was at 12:00 p.m.; the submerged water depth of most areas in the upper reaches of Wangzongdian Village reaches 2–5 m, while that of most areas near Wangzongdian Village and its downstream exceeded 5 m. Subsequently, with the gradual decrease in rainfall in the watershed, the submerged water level near Wangzongdian Village gradually decreased, and the process of water withdrawal started. Based on the simulation evolution of the flood in the Wangzongdian watershed, it can be concluded that the flood is characterized by high flow and rapid rise.

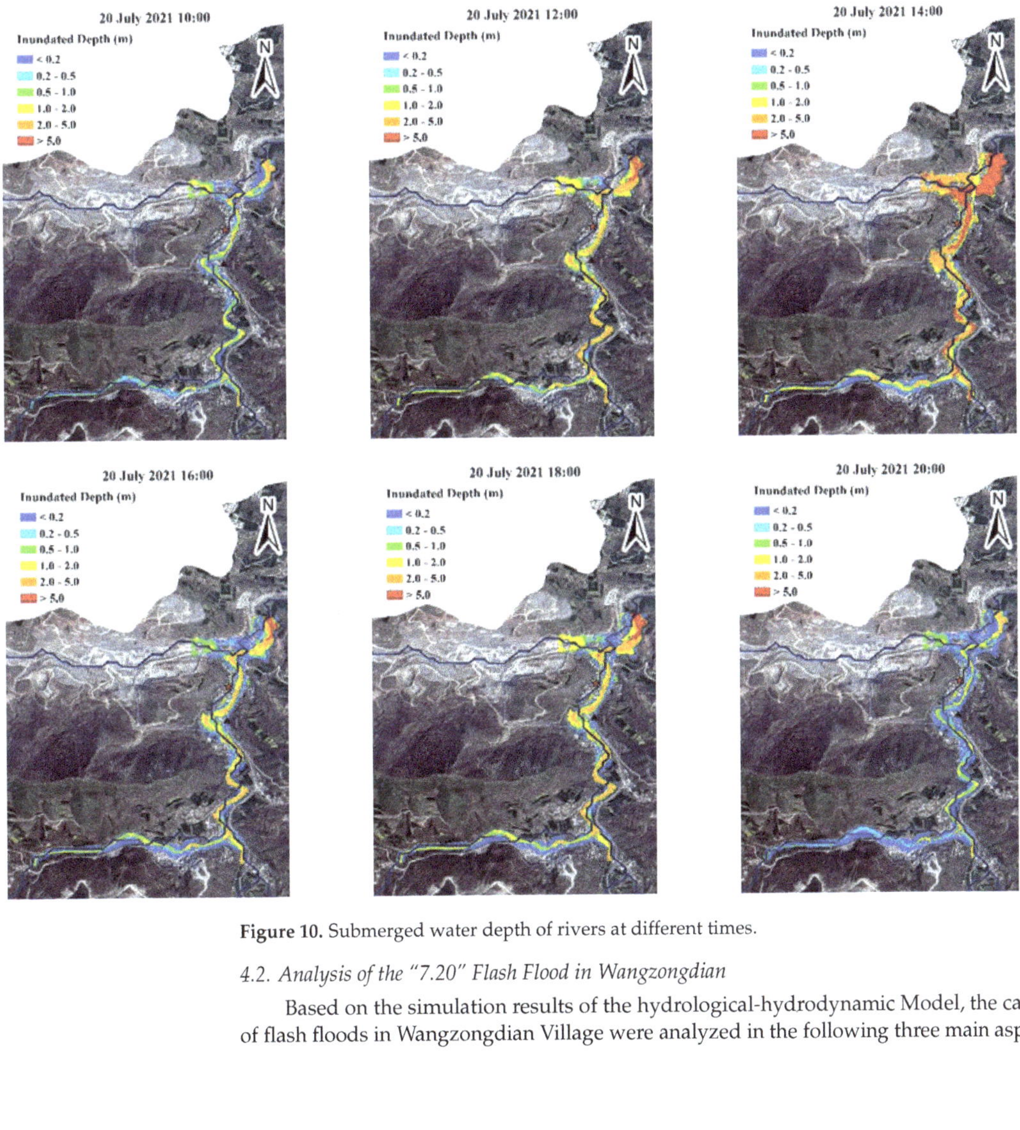


**Figure 10.** Submerged water depth of rivers at different times.

### 4.2. *Analysis of the "7.20" Flash Flood in Wangzongdian*

Based on the simulation results of the hydrological-hydrodynamic Model, the causes of flash floods in Wangzongdian Village were analyzed in the following three main aspects:

(1) On 20 July 2020, the total rainfall in the watershed of Wangzongdian Village reached the highest value of 353 mm, with a rainfall intensity of 58.5 mm/h. The flood peak discharge caused by short-term heavy rainfall and the rapid increase in water level were the direct causes of this disaster.
(2) Due to the special geographical location of Wangzongdian Village, surrounded by mountains on all sides, water flowed down its surrounding mountains and converged here during heavy rainfall. Flooding encountered in the upper four channels further increased the flood flow and exacerbated the flood disaster.
(3) There was some bare land in the upper reaches of the Wangzongdian watershed. Under the influence of high-intensity rainfall, flash floods carried a large amount of sediment down the river, causing serious damage to villages and infrastructure along the lower reaches of the river. Figure 11 shows the damage to the road and dams by the flash flood. It can be seen that many roads and dams in the upper reaches were damaged, and even continuous dam failures occurred. Due to the silting and shrinkage of river channels, multiple houses, bridges, roads, and other cross-channel buildings led to water blocking and water level rising, while the peak discharge caused by the water blocking and collapse of roads and bridges had devastating effects on the downstream.

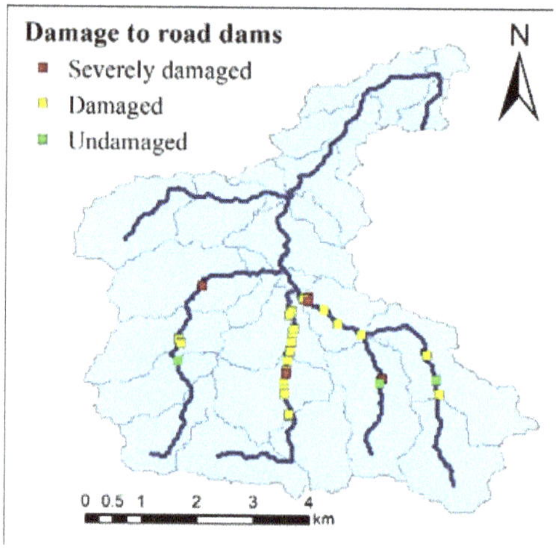


**Figure 11.** Distribution of damage to the upstream road dams in Wangzongdian Village.

### 4.3. Discussion

In this study, a distributed hydrological model and a two-dimensional hydrodynamic model were used to simulate the "7.20" flash flood event in Wangzongdian village based on the two-phase flash flood disaster review strategy. Post-flood estimates of maximum peak discharge, which is very valuable information available for small watersheds, are used to evaluate the accuracy of the simulation results. Up to now, a few studies, Sun et al. [35], used the KW–GIUH model to analyze the disaster mechanism of the "8·16" flash flood disaster that occurred in the Zhongdu River basin. Before the simulation calculation, the parameters of the KW–GIUH model need to be calibrated based on the historical flood data of the basin. In practice, the problems of imperfect monitoring facilities and insufficient measured data in small watersheds in mountainous areas often make it difficult to determine model parameters.

In order to simulate the evolution of natural disasters and make an assessment by setting the flood water sources in line with the flow discharge, a new distributed hydrological model named FFMS was developed. Hao et al. [27] reproduce the flooding process and the consistency of the flow discharge to explain the underlying reason for the disaster formation and evolution of the "8.12" Flash Flood in Liulin River Basin based on the FFMS model with an error of +13% compared with post-disaster investigation data. However, it can be seen from the simulation results of two different models in the "7.20" flash flood event that when a dam break caused by an extreme rainstorm destroys the natural process of basin flow generation and flood evolution, distributed hydrological simulation often fails to perform disaster recovery well, while hydrodynamic simulation results can better reflect the formation mechanism and evolution process of a flood, and the calculation results are more accurate.

Other scholars also do research on the "7.20" flash flood event. Liu et al. [36] analysis of the cause of the flash flood based on geological analysis. It is believed that the short-duration heavy rainfall and dam break are the main causes of the flash flood event, which is consistent with the result analysis based on model calculation. Therefore, the two-stage recovery strategy proposed in this paper can provide guidance for flood process reproduction and disaster mechanism analysis intuitively and can also apply to other regions in the future.

## 5. Conclusions

The reenactment simulation for extreme natural disasters such as flash floods is often plagued by data collection and modeling strategy, and there are few widely applicable reenactment approaches available for practical implementation. In this paper, the fast reenactment modeling strategy of flash flood disasters based on hydrological and hydrodynamic models was put forward, which can be used to carry out different degrees of flash flood disaster assessment in different situations. The selected validation case, the "7.20" flash flood disaster in Wangzongdian, is a representative extreme flood event in 2021. The hydrodynamic model simulation result is about 782 $m^3/s$ under the consideration of dam failure, which was basically consistent with the disaster investigation results, with a relative simulation flood error of +8%. The causes of the disaster are complex, both due to extreme rainfall and local blockage and dam failure. Through the modeling analysis, it was preliminarily determined that on 20 July 2020, in the Wangzongdian watershed, due to the pooling of flood water and the collapse of subgrade water damming caused by extremely heavy rainfall, the high level of flood water surged over a short distance to Wangzongdian Village, eventually causing this serious flash flood disaster. The analysis verified the applicability and operability of the reenactment modeling strategy proposed in this paper. Anyway, the selected model and method were able to represent the actual situation of flooding in hilly areas of China, with relatively strong parameter stability and high simulation accuracy. The simulation strategy of flash flood disaster reenactment based on the hydrological-hydrodynamic model proposed in this paper can provide technical support for future flash flood disaster analysis and post-disaster reconstruction in China.

**Author Contributions:** C.L. (Changzhi Li) and Q.M.: Conceptualization, Methodology. W.W.: Writing—review & editing. S.H.: Program implementation, data curation, Writing—original draft preparation. L.W.: Software. J.T.: Investigation. C.L. (Changjun Liu): Project administration. All authors have read and agreed to the published version of the manuscript.

**Funding:** This research was funded by National Key Research Program (2019YFC1510603), Major science and technology projects of the Ministry of Water Resources (SKR-2022034).

**Institutional Review Board Statement:** Not applicable.

**Informed Consent Statement:** Not applicable.

**Data Availability Statement:** All authors made sure that all data and materials support published claims and comply with field standards.

**Acknowledgments:** The research presented in this paper has been carried out as part of the National Key Research Program (No. 2019YFC1510603), Major science and technology projects of the Ministry of Water Resources (SKR-2022034).

**Conflicts of Interest:** The authors declare no conflict of interest.

## References


1. Ding, L.; Guo, L.; Liu, C.; He, B. Technical progress of flash flood disasters prevention and control systems in China. *China Flood Drought Manag.* **2020**, *30*, 11–17.
2. Tang, W. Research on Simulation and Analysis Technology of Flash Floods in Small Mountain Watersheds. Master's Thesis, Nanjing University of Information Science and Technology, Nanjing, China, 2022.
3. Li, H.; Tan, G.; Wang, X. Advances in Study on Flash Flood Forecast and Warning. *J. China Hydrol.* **2014**, *34*, 12–16.
4. Xing, Z.; Ma, M.; Wen, L. Application of HEC-HMS model in mountain flood forecasting in data deficient areas. *J. China Inst. Water Resour. Hydropower Res.* **2020**, *18*, 8.
5. Wu, X.; Liu, C. Progress in Watershed Hydrological Models. *Prog. Geogr.* **2002**, *21*, 8.
6. Liu, C.; Xia, J.; Guo, S. Advances in distributed hydrological modeling in the Yellow River basin. *Adv. Water Sci.* **2004**, *15*, 495–500.
7. Shu, L.; Chang, Y.; Wang, J. A brief review of numerical distributed hydrological model SHUD. *Adv. Earth Sci.* **2022**, *37*, 680–691.
8. Zhao, J.; Liang, Z.; Liu, J. Variable runoff generation layer distributed hydrological model for hilly regions. *Adv. Water Sci.* **2022**, *33*, 429–441.
9. Liu, C.; Zhou, J.; Wen, L. Research on spatio temporally-mixed runoff model and parameter regionalization for small and medium-sized catchments. *J. China Inst. Water Resour. Hydropower Res.* **2021**, *19*, 99–114.
10. Jiang, C.; Zhou, Q.; Shen, Y. Review on hydrological and hydrodynamic coupling models for flood forecasting in mountains watershed. *J. Hydraul. Eng.* **2021**, *52*, 14.
11. Godunov, S. A finite difference method for the numerical computation of discontinuous solutions of the equations of fluid dynamics. *Mat.-Eskij Sb.* **1959**, *47*, 271–306.
12. Berardi, L.; Laucelli, D.; Simeone, V.; Giustolisi, O. Simulating floods in ephemeral streams in Southern Italy by full-2D hydraulic models. *Int. J. River Basin Manag.* **2013**, *1*, 1–17. [CrossRef]
13. Zhang, D.; Quan, J.; Ma, J. Two-dimensional numerical simulation for surface runoff in catchments based on Godunov scheme. *J. Hydraul. Eng.* **2018**, *49*, 787–794+802.
14. Liu, C.; Wu, J.; Gao, J. Study on evaluation method of mountainous flood disaster. *Yangtze River* **2017**, *48*, 13–16.
15. Zhang, Y.; Song, Y.; Dong, W. Research and Application of Mountain Hydrological Model Based on the Mechanism of Subsurface Storm Flow. *Adv. Eng. Sci.* **2022**, 1–11. Available online: https://kns.cnki.net/kcms/detail/51.1773.TB.20220415.1133.002.html (accessed on 5 January 2023).
16. Anquetin, S.; Braud, I.; Vannier, O.; Viallet, P.; Boudevillain, B.; Creutin, J.; Manus, C. Sensitivity of the hydrological response to the variability of rainfall fields and soils for the Gard 2002 flash-flood event. *J. Hydrol.* **2010**, *394*, 134–147. [CrossRef]
17. Lin, Y.; Wang, X.; Bao, W. Flood Simulation and Submergence Analysis of Small Mountain Watershed Based on 1D and 2D Coupled Hydrodynamic Model. *Water Power* **2022**, *48*, 10–14.
18. Segura-Beltrán, F.; Sanchis-Ibor, C.; Morales-Hernández, M.; Bussi, G.; Ortiz, E. Using post-flood surveys and geomorphologic mapping to evaluate hydrological and hydraulic models: The flash flood of the Girona River (Spain) in 2007. *J. Hydrol.* **2016**, *1*, 310–329. [CrossRef]
19. El-Saoud, A.; Othman, A. An integrated hydrological and hydraulic modelling approach for flash flood hazard assessment in eastern makkah city, saudi arabia. *J. King Saud Univ.-Sci.* **2022**, *34*, 102045. [CrossRef]
20. Xingyang Municipal Water Resources Bureau. *Preliminary Verification Report on the "7.20" Rainstorm in Xingyang City*; Xingyang Municipal Water Resources Bureau: Zhengzhou, China, 2021.
21. Investigation Group on Disaster of State Council of China. *Investigation Report on the "7.20" Extraordinary Rainstorm Disaster in Zhengzhou City, Henan Province*; Investigation Group on Disaster of State Council of China: Beijing, China, 2022.
22. Liu, C.; Lv, J.; Zhai, X. Risk simulation and comparative analysis of "21·7" heavy rainfall and flood in Henan Province. *Express Water Resour. Hydropower Inf.* **2021**, *42*, 8–17.
23. Yellow River Engineering Consulting Co., Ltd. *Investigation Report on "7.20" Flood in Wangzongdian in Cuimiao Town, Xingyang City*; Yellow River Engineering Consulting Co., Ltd.: Zhengzhou, China, 2021.
24. Tian, J.; Fan, S.; Wang, W. Comparison between FFMS Model and HEC-HMS Model Applied in Rainstorm and Flood Simulation. *J. Yangtze River Sci. Res. Inst.* **2021**, *38*, 59–65.
25. Guo, L.; Ding, L.; Sun, D.; Liu, C.; He, B.; Liu, R. Key techniques of flash flood disaster prevention in China. *J. Hydraul. Eng.* **2018**, *49*, 1123–1136.
26. Liu, C. A new generation of hydrological model driven by artificial intelligence and big data and its application in flood forecasting and early warning. *China Flood Drought Manag.* **2019**, *29*, 11+22.
27. Hao, S.; Wang, W.; Ma, Q.; Li, C.; Wen, L.; Tian, J.; Liu, C. Analysis on the Disaster Mechanism of "8.12" Flash Flood in Liulin River Basin. *Water* **2022**, *14*, 2017. [CrossRef]
28. Talbot, C.; Ogden, F. A method for computing infiltration and redistribution in a discretized moisture content domain. *Water Resour. Res.* **2008**, *44*, W08453. [CrossRef]
29. Lai, W.; Ogden, F.; Steinke, R.; Talbot, C. An efficient and guaranteed stable numerical method for continuous modeling of infiltration and redistribution with a shallow dynamic water table. *Water Resour. Res.* **2015**, *51*, 1514–1528. [CrossRef]

30. Zhang, X.; Zhou, J.; Wen, L. Application of spatio-temporal variable source mixed runoff model to flood simulation of small watersheds: A Case study of four small watersheds in Sichuan and Gansu Province. *J. Water Resour. Water Eng.* **2021**, *32*, 80–90.
31. Liu, C.; Wen, L.; Zhou, J. Comparative analysis of hydrological and hydrodynamic calculation method for flash flood in small watershed. *J. China Inst. Water Resour. Hydropower Res.* **2019**, *17*, 262–270+278.
32. Ma, J.; Yu, H. Features and applications of the flood analysis software IFMS/Urban. *China Water Resour.* **2017**, *5*, 74–75.
33. Yu, H.; Ma, J.; Zhang, D. Application of IFMS Urban software in urban flood risk mapping. *China Flood Drought Manag.* **2018**, *28*, 13–17.
34. Wang, S.; Tao, F.; Li, G.; Xie, H.; Ten, Y.; Gu, S. Comparative analysis of the calculation of maximum flow of embankment dam-breaking with different empirical formula. *J. Qinghai Univ. (Nat. Sci. Ed.)* **2012**, *30*, 38–41.
35. Sun, T.; Yang, P.; Xu, Z.; Wang, Y.; Wang, X. Analysis on The Disaster Mechanism of "8.16" Flash Flood in Zhongdu River basin. *Adv. Eng. Sci.* **2021**, *53*, 132–138.
36. Liu, C.; Huang, S. Research on "7.20" mountain torrents and geological disasters in Zhengzhou city, Henan Province of China. *J. Eng. Geol.* **2022**, *30*, 931–943.

water

MDPI

Article

# Application of a Regionalization Method for Estimating Flash Floods: Cuautepec Basin, Mexico

Maritza Arganis [1,*], Margarita Preciado [2], Faustino De Luna [1], Liliana Cruz [3], Ramón Domínguez [1] and Olaf Santana [1]

1 Instituto de Ingeniería, Universidad Nacional Autónoma de México, Mexico City 04510, Mexico
2 Instituto Mexicano de Tecnología del Agua, Jiutepec 62550, Mexico
3 Comisión Nacional del Agua, Mexico City 04340, Mexico
* Correspondence: marganisj@iingen.unam.mx; Tel.: +52-5-556233600 (ext. 8644)

**Abstract:** A rainfall regionalization method based on variation coefficient was applied with a variant in the construction of flash flood hyetographs with several return periods using the flash flood shape of the historical event that occurred in September 2021 in the Tlalnepantla River basin, Mexico, that caused severe damage to population and its infrastructure in a few hours. The historical flash flood was simulated with a semi-distributed model in the free software HEC-HMS in order to obtain the outflow hydrograph, and the flood plains were obtained with Iber and Hec-Ras 2d software that simulate free surface flow with a two-dimensional analysis. With photographs of the site, it was possible to locate traces of water that were contrasted with they calculated depths; they were concordant. Synthetic design storms were then simulated to estimate their potential consequences on the site.

**Keywords:** Tlalnepantla River; flash floods; hyetograph shape; Hec Ras 2d; Dorrigo diagram; regionalization

**Citation:** Arganis, M.; Preciado, M.; Luna, F.D.; Cruz, L.; Domínguez, R.; Santana, O. Application of a Regionalization Method for Estimating Flash Floods: Cuautepec Basin, Mexico. *Water* **2023**, *15*, 303. https://doi.org/10.3390/w15020303

Academic Editors: Xiekang Wang, Philippe Gourbesville, Changjun Liu and Athanasios Loukas

Received: 14 November 2022
Revised: 2 January 2023
Accepted: 4 January 2023
Published: 11 January 2023



## 1. Introduction

According to the National Weather Service [1], a flash flood is defined as runoff that occurs after the first 3 to 6 h after a heavy rain or some other cause. On the other hand, catastrophic floods are associated both with the intensity of the event and with the magnitude of the material damage they cause [2].

Flash floods have been the subject of studies going back nearly 25 years or more. Year after year they rise again, for example, in the face of different world events that have taken place over time; additionally, they appear in the recurrence of catastrophic floods.

At the global level, flash floods have been a study object for several decades; for example, Garzón et al. [3] make a description of morphology that occurs in terrain and natural channels before torrential floods occur, reporting some case studies highlighting ravines and river fans of basins in Spain (highlighting the Abanico of Arás where a terrible catastrophe occurred in 1996). They emphasize that the sense of security some flood protection works give must be distrusted, since they can be susceptible to failure in a flash flood event.

Aroca [4] analyzed the importance of characterizing initial abstraction prior to the runoff process associated with flash floods. Using the Green Amp method [5,6], comparing it with the curve number, generating the hydrograph with the SCS method [7], and using a semi-distributed hydrological model to simulate using Hec-Hms 4.0 software, he concluded in his study that the Green Amp method is better than the curve number method because it uses basin parameters, while the curve number method is reported as a method that is still empirical and was originally developed for basins of a specific country.

Karbasi et al. [8] developed a regional model for estimating loss of life due to flash flood in residential areas in the Kan watershed, Iran. They applied a model based on hydraulic variables such as depth, velocity, and rise rate of the water, as well as evacuation parameters, including available time for evacuation and fraction of people evacuated. They concluded that the most significant factors affecting fatalities, number of, were evacuation parameters, including evacuation time and fraction of evacuated people. From a local sensitivity analysis, the lead time between observed flood at the early flood warning station and the arrival time at the area of interest were distinguished as the most important input for evaluating the number of fatalities.

Aristizábal et al. [9] present a quantification of torrential floods and the population affected by torrential floods in various places in Colombia; they carried out torrential flood classification in terms of flash floods, debris flow, and debris flooding, reporting the last item as that with the greatest capacity for destruction. Due to the fact that there is no unanimity on the definition of torrential floods, the authors concluded that it is difficult to propose a methodology for threat assessment.

Bilasco et al. [10] present a methodology for the strategic management of floods to mitigate their damage in areas that anthropogenic activity has been invading for food production purposes; the analysis was carried out in the context of governance in the digital age. Their approach consists of a system that includes analysis and flooding, modeling, risk assessment, and diffusion of flood effects to the population.

Shuvo et al. [11] analyzed two flash floods due to monsoon events in Bangladesh using a coupled atmospheric–hydrological numerical weather prediction (NWP) model, namely the weather research and forecasting (WRF) model. Evaluating the goodness of the results with the mean square error and the Nash–Sutcliffe criterion, they found that the NWP model is applicable for flash flood prediction in the basins analyzed, despite the short duration of such events.

When making reviews and effect reconstructions caused by flash floods or making new simulations of events associated with a return period, the hyetograph shaping problem design has its particularities [12,13], since the traditional method of alternating blocks does not lead to the typical shape for storms associated with flash floods.

In this paper, a rainfall regionalization method was successfully applied [14] to obtain a design of flash floods associated with different return periods, but little variation in the hydrograph shape was obtained based on an ordering that takes into account the case of a historical event that occurred in the Tlalnepantla River basin in Mexico State (that is, the traditional alternating block method was not applied). These designed flash floods were simulated in a semi-distributed model using the HEC-HMS software [15], obtaining depth and velocity maps with the Iber software and Hec Ras 2d; additionally, the simulation for the historical flash flood was made to compare the results, in addition to contrasting them with photographic evidence of the water mark level.

The work is organized in the following parts: the introduction presented here, the methodology in which the procedures used and data from the study site are described, and the results and discussion, in addition to the conclusions derived from the research.

## 2. Methodology

### 2.1. Rainfall Regionalization Method Based on Variation Coefficient

This regionalization method [14] groups climatological stations that have daily precipitation historical data with more than 20 measurement years, based on the stations–year technique and variation coefficient (Equation (1)) for maximum precipitation series grouped in intervals after annual data sorting.

$$CV = \frac{Sx}{\overline{x}} \tag{1}$$

where

$CV$ is the coefficient of variation of the maximum annual rainfall series;
$Sx$ is the standard deviation of the maximum annual rainfall series;
$\overline{x}$ is the mean of the maximum annual rainfall series.

The first step is to identify the set of stations after the quality and quantity of information is checked. Then, the stations–year technique is applied, that is, each station was modulated or normalized with respect to its historical average and a large record was built by placing one normalized station after another. Later, an analysis of frequencies to find the regional distribution function of best fit was performed, with the least standard error criterion of fit (EEA) being between the measured data $x$ and the calculated $\hat{x}$ with a distribution function with $p$ parameters, in an annual series of size $n$ [14] given by the Equation (2).

$$EEA = \left[ \frac{\sum_i^n (x_i - \hat{x}_i)}{n - p} \right]^{0.5} \tag{2}$$

With the regional distribution function obtained, regional factors associated with different return periods are obtained.

### 2.2. Estimation of the Design Storm from Regional Factors

To obtain a design storm associated with a return period, the following steps are used [14].

1. The historical average maximum annual rainfall of the basin is estimated from an isohyet map and the approximate location of the basin centroid.
2. The regional factor for the study site corresponding to the analyzed return period is selected.
3. The historical average maximum annual rainfall is multiplied by the regional factor, thus obtaining the total annual maximum accumulated rainfall of 24 h.
4. Rainfall previously obtained is affected by a reduction factor per area obtained with the basin area and the equation that corresponds to the study area.
5. One-hour total rainfall is estimated from twenty-four-hour total rainfall data, calculated in step 4 with the help of the convective factor corresponding to the analysis site.
6. Chen and Bell tables [16] are used to estimate precipitation values for durations between 1 and 24 h or for durations less than one hour or greater than 24 h.
7. The design mass curve for the selected $\Delta t$ is presented. The bars of the design hyetograph, not yet sorted, are defined for the duration of the selected storm.
8. To shape the design storm, a block ordering process is carried out (alternating blocks are traditionally used in statistical storms, but in the case of flash floods, a skewness in the hyetograph must be considered).

### 2.3. Construction of the Design Hyetograph Using Flash Flood Shape

The variant proposed in this research occurs in the way of placing the design bars of the hyetograph associated with a flash flood.

For this, the historical storm event of the study site that caused a flash flood was selected and its position occupied by bars of the hyetograph was identified, from the largest to the smallest. This order was used to order the statistically generated storm bars; with the above, it is possible to obtain non-centered hyetographs with a skewness that correspond to a behavior similar to the hyetographs of a flash flood.

### 2.4. Study Site and Data Set

Cuautepec, Edomex has a geography conducive to the occurrence of floods, many of them flash floods that, as they increase, generate flows that descend from the hill upper zones towards the urban area in the lower zones. Constructions located in particular on the left bank of the plain area in the central basin part are flooded when the Cuautepec River overflows, since there is a combined drainage system that carries rainwater and sewage, due mainly to the invasion of the floodplains and the reduction in the hydraulic area of the natural channels in such urban zones. Among dozens of school supplies, books, computers, benches, and children's games rescued from the mud and garbage that was dragged by the water current that caused the storm on 30 October 2009 in Cuautepec, teachers from the Gertrudis Armendariz and Josefa Ortiz de Domínguez schools exposed to officials, as well as federal and local authorities, that each year, during the rainy season, these centers are flooded by the overflow of the rivers that discharge sewage. In the Josefa Ortiz de Domínguez kindergarten, the teachers lamented the loss of the computer center, where five days later they could still see the 16 computers covered in mud and with traces of the mark of the water level reached by the water that flooded the school.

Seven houses for families that suffered irremediable structural damage were incorporated into the housing improvement program; these houses were located on the left bank of the Cuautepec River near the Precious Blood of Christ Church.

In this regard, José Pérez Castañeda, director of Clean and Urban Image of the Secretariat Services, said that the unit in his charge had so far disposed of 2751 waste tons, including solid waste and household items that were dragged by runoff that descended from the hills to the mild part where the community of Cuautepec is allocated.

Figure 1 shows on furniture the water depth mark reached by the flood inside a kindergarten and it is approximately 1.0 m (photograph from La Jornada newspaper of the event on 30 October 2009 [17]); unfortunately, there was no rainfall measurement, because a pluviograph was not installed in the area surrounding and the kindergarten is located next to the church aforementioned.

**Figure 1.** Water depth reached, observed on furniture at the affected site after flood event of 30 October 2009. Source: Google images.

At the end of August 2021 and during the first days of September, storm events occurred, some associated with Hurricane Grace [18,19], that caused severe flooding in several municipalities of Mexico State [20]. The contribution basin of the storm recorded between 3:40 p.m. on 6 September and 5:40 a.m. on 7 September 2021, with a total rainfall of 44.57 mm, is shown in Figure 2; the analyzed basin area is 15 km$^2$, with a main channel length equal to 8.24 km and a main channel slope (dimensionless) of 0.0428, which can lead to supercritical flow conditions. The OH Hydrological Observatory pluviograph of the Institute of Engineering (Figure 2) reported this event that caused a flash flood (Figure 3).

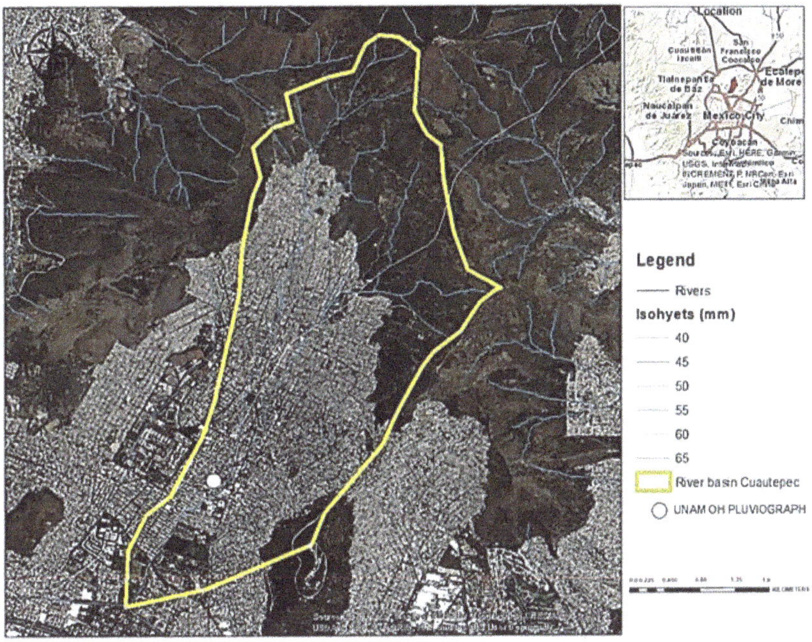


Figure 2. Location of the study site and the pluviograph of the OH, UNAM.

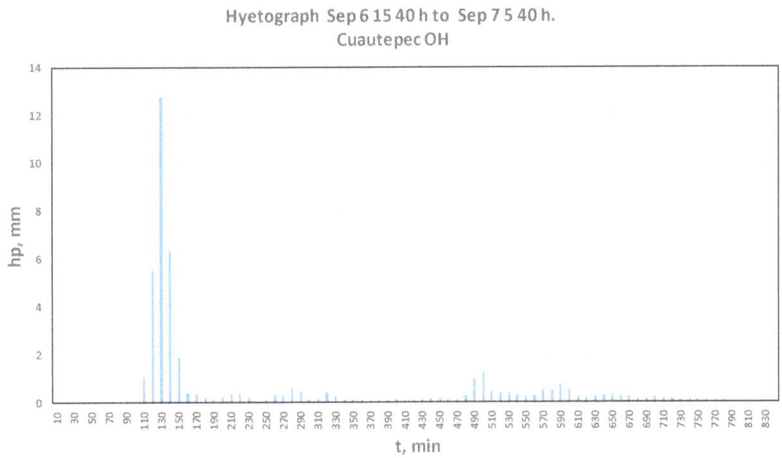


Figure 3. Flash flood hyetograph that occurred in September 2021 in Mexico State, Mexico.

Journalistic sources highlighted at least 2 deaths, nearly 100,000 affected inhabitants, and material damage in 19 different neighborhoods.

### 2.5. Hydrological Region for Mexico Valley River Basin

The entire Mexico Valley River Basin (MVRB) hydrology was considered. Since little variation was noted in the variation coefficients for the different stations analyzed in the study by Domínguez et al. [21] (Figure 4), in said study it was reported that the regional function of best data fit, using the procedure described in the methodology, was of the Double Gumbel type [22]; regional factors corresponding to different return periods are highlighted in Table 1.

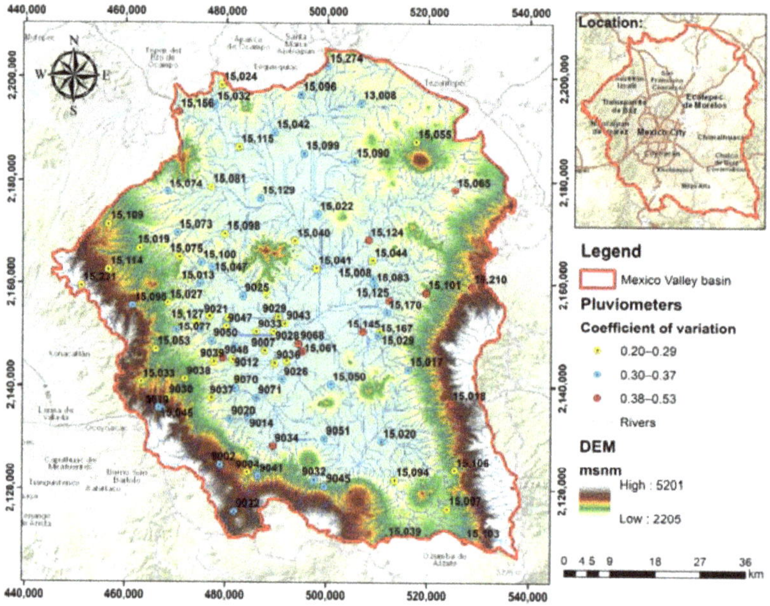


**Figure 4.** Coefficients map of variation for historical annual maximum daily rainfall in Mexico Valley. "Adapted from Ref. [21], 2022, SACMEX".

**Table 1.** Dimensionless regional factors for estimating annual maximum daily rainfall for different return periods Tr in years for Mexico Valley River Basin. "Information consulted from Ref. [21], 2022, SACMEX" Source: [21].

| Tr, Years | CVM (D-GUMBEL) |
|---|---|
| 2 | 0.94 |
| 5 | 1.2 |
| 10 | 1.39 |
| 20 | 1.59 |
| 50 | 1.88 |
| 100 | 2.09 |
| 200 | 2.29 |
| 500 | 2.56 |
| 1000 | 2.76 |
| 2000 | 2.97 |
| 5000 | 3.23 |
| 10,000 | 3.44 |

### 2.6. Area Reduction Factor Equation

The equation developed by Domínguez et al. [21] was used for the reduction factor per area as a function of the area $a$ of the basin (Equation (3)). The reduction factor per area takes into account the spatial distribution of rainfall.

$$FRA = -0.058\ln(a) + 1.10155 \tag{3}$$

$$\text{Para } 0 < a \leq 25\ \text{km}^2$$

### 2.7. Convective Factor

For this paper, a convective factor R equal to 0.61 was considered, based on an updated map from a study by Domínguez et al. [21] (Figure 5). The convective factor is the ratio of the rainfall that occurs in 1 h to the rainfall that occurs in 24 h.

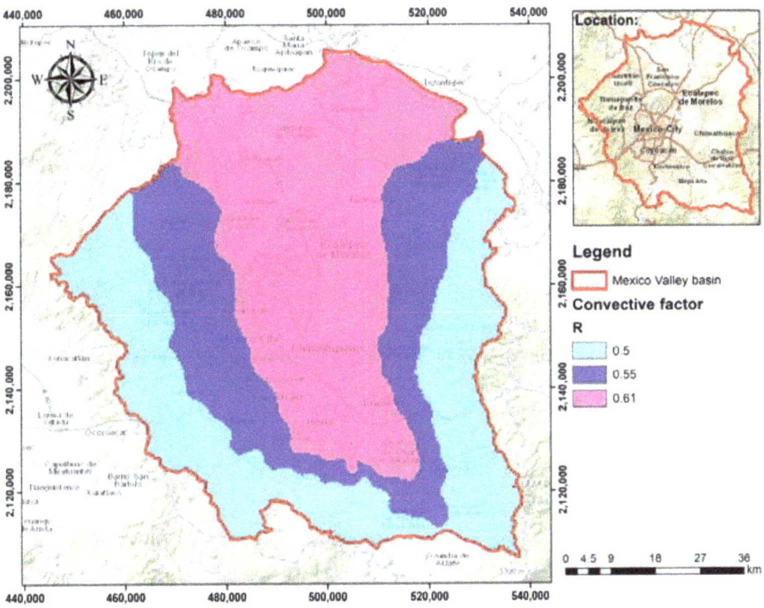


**Figure 5.** Convective factor R Map for Mexico Valley River Basin. "Adapted from Ref. [21], 2022, SACMEX".

### 2.8. K-Factors to Move to Durations Other than One Hour

K-factors were considered, corresponding to convective factor R, from the modified Chen and Bell [16] version for the study by Domínguez et al. [21]. The K factors allow for obtaining the rainfall that occurs in durations of less than one hour, with respect to the rainfall of one hour.

#### 2.8.1. Hec-HMS Software

The HEC-HMS (Hydrologic Engineering Center–Hydrologic Modeling System) model is a rainfall–runoff model developed by the Hydrologic Engineering Center HEC of the U.S. Army Corps of Engineers USACE that is designed to simulate the runoff hydrograph that occurs at a given point in the river network as a result of a rain event. The predecessor of this model, the HEC-1, was born as an event model and has been considered by many as the most versatile model [23] and probably the most widely used in this type of hydrological characterization of floods.

To calculate evapotranspiration losses, HEC-HMS has different methods. In this case, the method of the SCS Soil Conservation Service, also called the CN curve number, was chosen because it has quality digitized information on the use and type of soil. This method was developed by the SCS of the US Department of Agriculture, USDA, to estimate the losses (or abstractions) in a rain or downpour event [24], and today it is one of the most used in the professional field. In this method, the effective rain height is a function of the total precipitation volume and a loss parameter called the CN curve number. The curve number varies in the range from 0 to 100 and depends on factors that influence the generation of runoff in the basin: hydrological type of soil (hydrological group–drainage capacity); land use and management; soil surface condition; and antecedent moisture condition.

The starting hypotheses or conditioning factors from which the model is based are the following: the simulation is limited to rain events (event model) as a consequence of the application of the model itself to the simulation of floods; the modeling is based on simulating only the direct surface runoff, the base flow is estimated prior to the application of the model; and snow is not taken into account, as we started from events in which there was no snow.

#### 2.8.2. Iber

Iber is a two-dimensional mathematical model for river and estuary flow simulation developed from the collaboration of the Water and Environment Engineering Group, GEAMA (University of La Coruña); the Mathematical Engineering Group (University of Santiago de Compostela); and the Flumen Institute (Polytechnic University of Catalonia and International Center for Numerical Methods in Engineering) and promoted by the Center for Hydrographic Studies of CEDEX. Iber is a numerical model developed directly from the Spanish public administration in collaboration with the aforementioned universities and designed to be especially useful for the specific technical needs of hydrographic confederations in the application of current sectoral legislation on water. Iber's hydrodynamic module solves the two-dimensional St. Venant equations, incorporating the effects of turbulence and wind surface friction [25].

The shallow water equations and those of the k-ε model are solved using the finite volume method for two-dimensional unstructured grids. The numerical schemes used in Iber are especially suitable for modeling regime changes and dry–wet fronts (flood fronts). The discretization of the spatial domain is carried out with finite volumes in unstructured meshes, admitting these mixed ones formed by triangular and quadrangular elements. The convective flow is discretized using Godunov-type off-center schemes, specifically Roe's off-center scheme, as well as its depth to order 2 with a slope limiter to avoid oscillations in regions with local maxima or minima. The term that includes the bottom slope is discretized off-center in order to avoid spurious oscillations of the free shell when working with complex terrain. The rest of the source terms, including those of turbulent diffusion, are discretized with a centered scheme.

#### 2.8.3. Hec-Ras 2d

Hec-Ras 2d software, developed by the United States Corps of Engineers [26], has been evolving in its potentialities; Hec-Ras was initially conceptualized for flow in one direction, and already in its two-dimensional form it solves the equations of 2D flow at the free surface for shallow water. The latest versions have incorporated a module that includes the complete two-dimensional moment equations for complete shallow waters, occupies structured grids with digital elevation models that can have high resolution (Lidar type), and considers the Manning N coefficient taking into account different values depending on soil type [27].

## 3. Results and Discussion

### 3.1. Flash Flood Hyetographs Obtained for Different Return Periods

By applying the regionalization procedure before-described and considering an updated map of the historical annual maximum daily mean isohyets of Mexico Valley as well as the basin centroid (Figure 6), a rainfall of 50 mm was estimated for the study site; this rainfall times the regional factor for a given return period results in the design rainfall. Subsequently, said rainfall was reduced to take into account its spatial distribution using the area reduction factor, and later the total rain of one hour and shorter duration was estimated to obtain the mass curve of the design storm. With this information, design hyetographs were determined, considering a duration of 14 h and shaping hyetographs based on behavior of the historical flash flood that was taken as a base. These hydrographs can be seen in Figure 7.

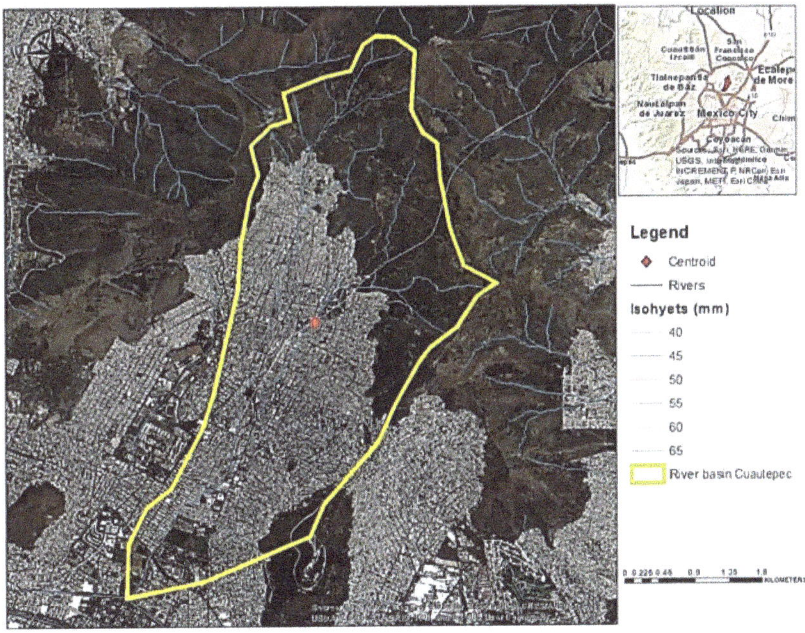


**Figure 6.** Estimation of the historical annual maximum daily average rainfall for the study site.

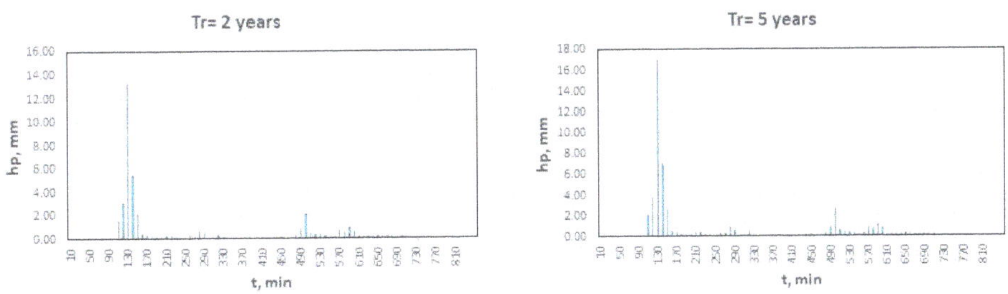


**Figure 7.** *Cont.*

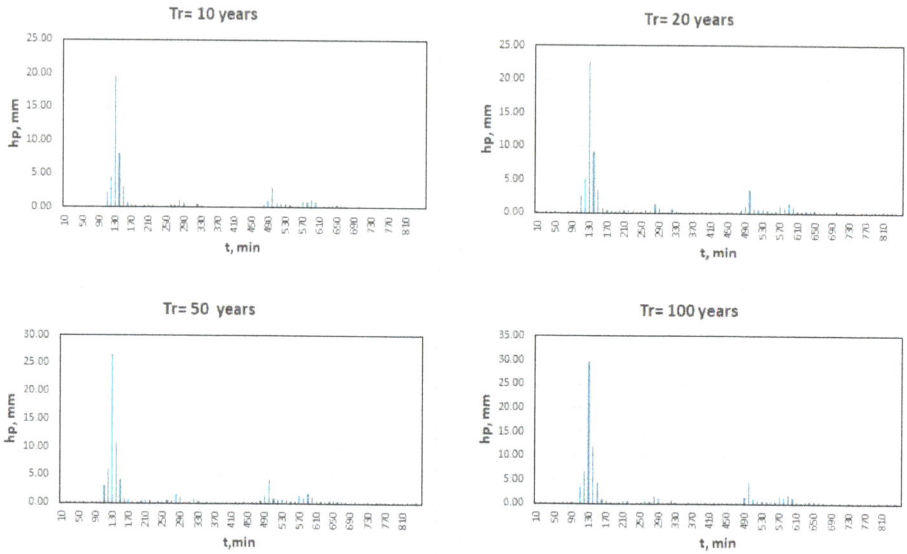


**Figure 7.** Hyetographs for flash floods. Cuautepec Basin Tr = 2, 5, 10, 20, 50 and 100 years.

## *3.2. Simulation of the Historic Flood Using Hec-HMS*

Flood simulation using the Hec-Hms semi-distributed model produced the hydrograph shown in Figure 8.

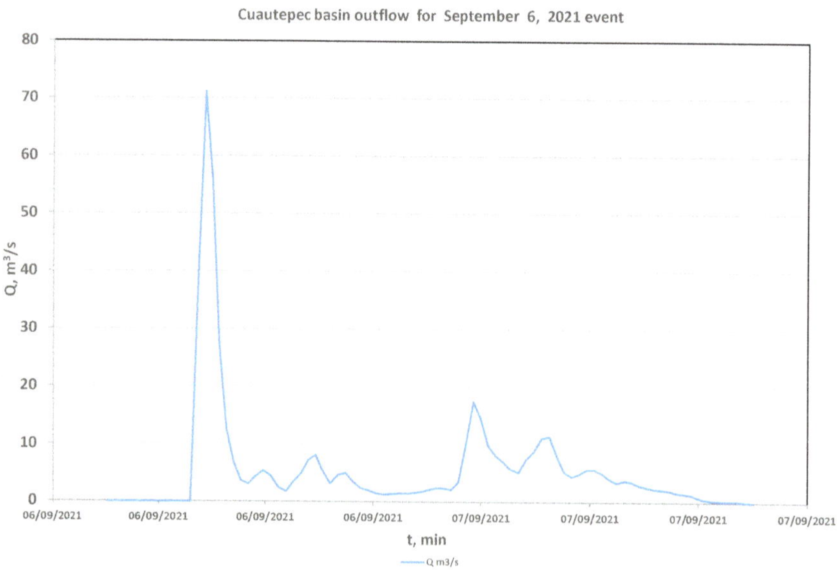


**Figure 8.** Flash flood hydrograph produced by the historic storm of September 2021. Estimated with Hec-HMS. Cuauhtepec Basin, Edomex.

### 3.3. Hydrographs Produced by Hyetographs for Different Tr Calculated with Hec-Hms

Hydrographs were obtained using Hec-Hms, and these were compared with the historical storm, as shown in Figure 9.

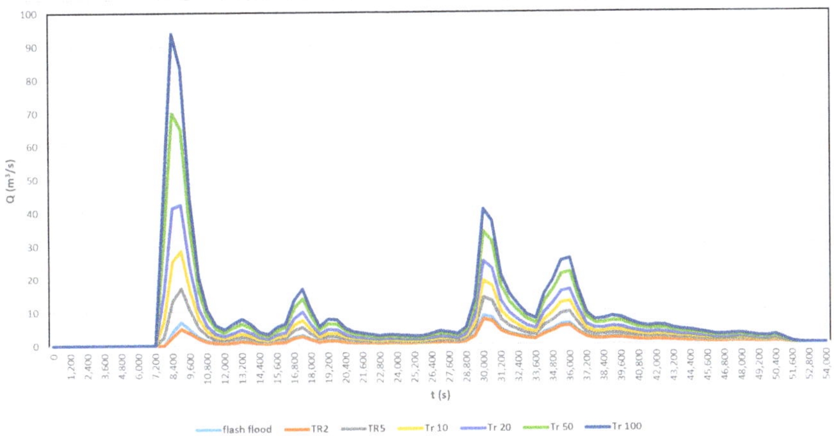


**Figure 9.** Estimated hydrographs for storms with different Tr and historical flash flood.

### 3.4. Depth and Velocity Maps of Historic Flash Flood

When using Iber software using the historical hydrograph produced by the flash flood obtained with Hec Hms and a calculation time of 24 h, depth and velocity maps created using a calculation time of 13,600 s (3.78 h) are presented in Figure 10; these result were partially due to the high calculation time required by the Iber software to perform processing.

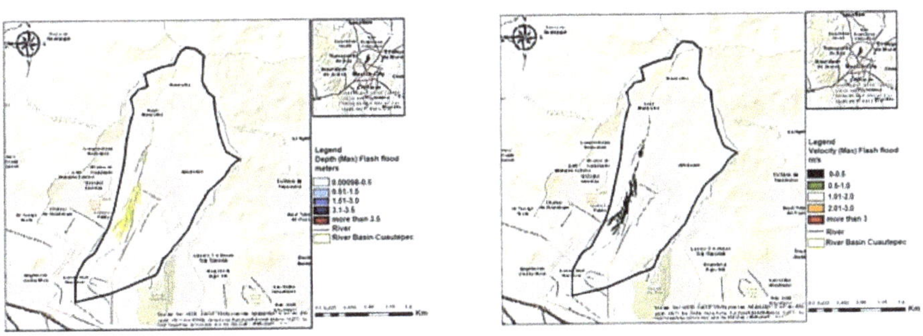


**Figure 10.** Depths and velocity maps, historical flash flood using Iber t = 13,600 s (3.78 h).

In contrast, when using the Hec-Ras 2d software, simulation was carried out for 7.5 h, and in a calculation time of approximately 23 min, in this case, the depth and velocity maps indicated in Figure 11 were obtained.

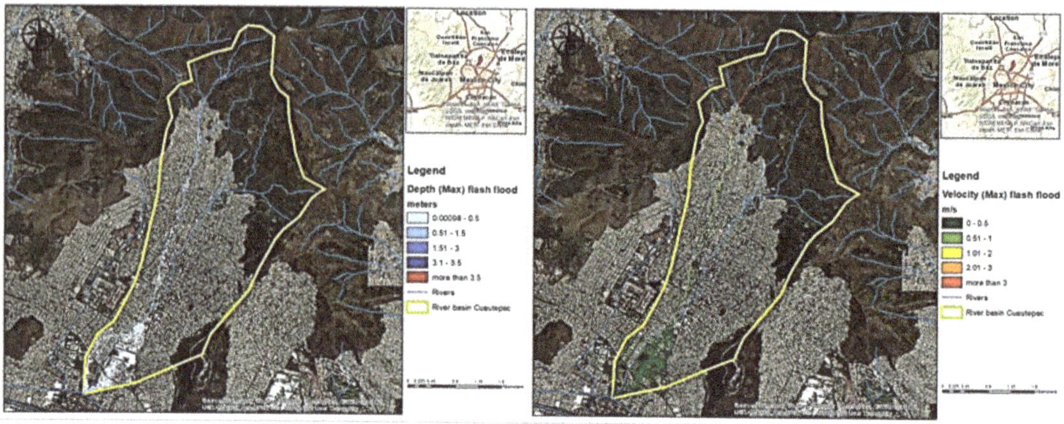


**Figure 11.** Depths and maximum velocity, maps for Historic flash flood using Hec-Ras2d.

The water depths reported using Iber in almost the first 4 h for the runoff process indicate highest values between 1.5 and 3 m and velocity between 3 and almost 9.5 m/s, which gives indications that possible supercritical flow occurred in the event, which could be intuited with main channel slope data (S = 0.0428).

The most extensive simulation (of 7.5 h) carried out by Hec-Ras 2d reported maximum depths of the order of 3.1 m and a maximum velocity that reached 3 m/s; after 6 h, water drained into the basin outlet. The Manning N coefficient was set with different values, according to soil type.

### 3.5. Flash Flood for Tr = 50 Years

The reported maps and velocities from simulation with Hec-Ras 2d for flash flood for Tr = 50 and Tr=100 years are shown in Figures 12 and 13.

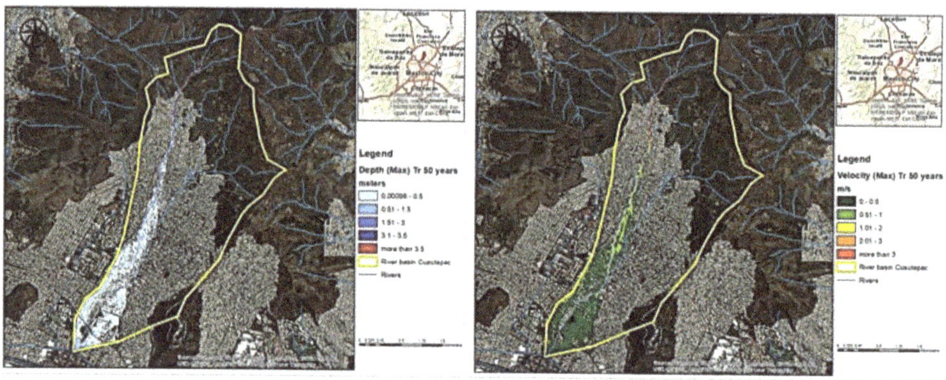


**Figure 12.** Depths and max velocity, maps for Tr 50 years using Hec-Ras2d.

In the case of floods with Tr = 50 and Tr = 100 years, maximum depths would reach 3.5 m or even greater, and velocities would also exceed 3 m/s, which does give an idea of a supercritical flow generated in the study site that increases risk in a flash flood event.

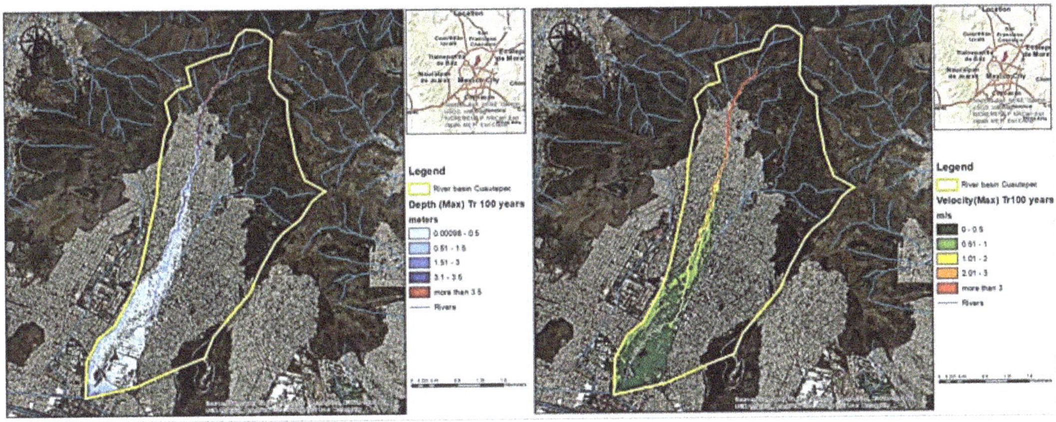


**Figure 13.** Maps of depths and maximum velocity for Tr 100 years using Hec-Ras2d.

When comparing historical results and floods with Tr = 50 and 100 years with the Dorrigo diagram [28–30], which is an indicator of the resistance to overturning (Figure 14), the depths and velocity that occurred in the historical event and for the Tr mentioned above lead to a high to very high probability of overturning or dragging in the water of the objects found in the passage of the current. The occurrence of human and material losses is inevitable in the face of said runoff event attributed to precipitation in the topography of the site that leads to the occurrence of dangerous supercritical flows.

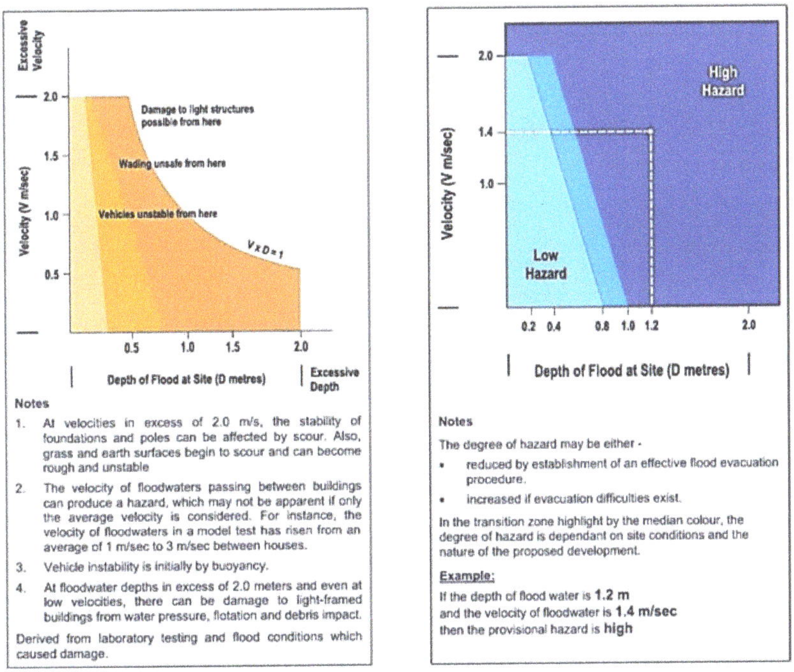


**Figure 14.** Hazard according to depth vs. velocity of water relationship and its categorization. Source: New South Wales, 2005.

For a finer estimate of the velocities that occurred in the analyzed event, it is advisable to carry out tests with different values of the calculation time interval, since instabilities can be generated in the results of the simulation algorithms. Hydrodynamic models that solve the Saint Venant equations with shock capturing [25,31] can also be used to consider supercritical flows and transitions to subcritical flows. On the other hand, in the case of constant high velocity conditions, even going down the water depth, the danger of tipping objects would still be high, as indicated by the Dorrigo diagram.

Regarding an event associated with a Tr of 100 years, it can be concluded that the braces reached in the area surrounding the kindergarten as well as the constructions on the left bank of the river according to the Hec-Ras 2d modeling are around 1 m, as shown in the evidence of Figure 1 and in the enlargement shown in Figure 15. For the velocity range from 1 to 3 m/s, if we apply Figure 13 in this specific case, it is concluded that for said tie rod there is a very high risk both of resistance to overturning and high danger, as defined in Figure 14.

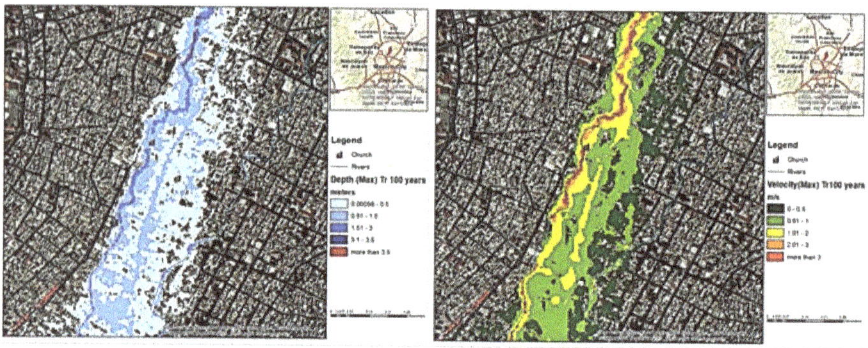


**Figure 15.** Floodplain expansion for flash flood with Tr = 100 years.

Additionally, the Tamez criterion [32] establishes that either a combination of depth of 1 m and a velocity of 1 m/s or the factor defined as the product of the depth by the velocity of 0.5 $m^2/s$ are sufficient to have a serious danger of loss of humans and damage to the household items of the population.

## 4. Conclusions

The regionalization application method based on variation coefficients with difference in hyetograph shape construction was useful to obtain direct runoff hydrographs proposed for different return periods. They were created from a historical flash flood from the study site, a kind of event that caused a lot of human and material damages. For both historical and statistical hydrograph simulation purposes, the free-use tool US Army Hec-Ras 2d was very useful since calculation times were relatively short, about 30 min, while the Iber software had problems in calculation time because so many hours were required to perform simulation that we finally decided to not continue to apply it in this investigation. Another inconvenience observed was that boundary conditions were fed with the hydrograph produced by a measured storm, that is, the Iber hydrological module could not be used and Hec-Ras 2d version did not request precipitation data; therefore, it is advisable to delve into improvements to hydrological modules to model rainfall–runoff scenarios. Depth maps and maximum depths obtained revealed that in the face of flash floods for return periods of up to 100 years, the probability of dragging obstacles in the water path is very high and the floodplain is also reflected as being wide. Due to rugged, steep topography in the upper basin, it is prone to flash floods, landslides, and debris, as well as supercritical flow development. Supercritical flows associated with high water velocities are really dangerous since they can cause great damage to infrastructure as well as people dragging, cars overturning, and human losses. Disorganized population growth and the territory, together with land use change, has encouraged construction in these highly dangerous areas. Therefore, the recommendation, in case there is already infrastructure that may be susceptible to being damaged by flash flood events, is the use of structural, such as velocity reducers that change roughness coefficient, temporary, or permanent measures to reduce future damage and continue promoting a culture of warning and prevention to the population.

**Author Contributions:** Conceptualization, M.A., M.P., F.D.L. and R.D.; methodology R.D., O.S., M.A. and M.P.; validation, M.P. and L.C.; data curation, F.D.L. and M.P.; writing—review and editing, M.A. and M.P.; visualization, F.D.L.; supervision, M.A., M.P. and F.D.L. All authors have read and agreed to the published version of the manuscript.

**Funding:** This research received no external funding.

**Institutional Review Board Statement:** Not applicable.

**Informed Consent Statement:** Not applicable.

**Data Availability Statement:** Not applicable.

**Conflicts of Interest:** The authors declare no conflict of interest.

## References

1. National Weather Service. 2022. Available online: https://www.weather.gov/phi/FlashFloodingDefinition (accessed on 5 November 2022).
2. Akstinas, V.; Meilutytė-Lukauskienė, D.; Kriaučiūnienė, J.; Šarauskienė, D. Features and causes of catastrophic floods in the Nemunas River basin. *Hydrol. Res.* **2020**, *51*, 308–321. [CrossRef]
3. Garzón, H.M.G.; Ortega, B.J.; Garrote, R.J. Las avenidas torrenciales en cauces efímeros: Ramblas y abanicos aluviales. Torrential floods in ephemeral streams: Wadies and alluvial fans. *Enseñanza Cienc. Tierra* **2009**, *17*, 264–276.
4. Aroca, J.E. Importancia de las Abstracciones Iniciales para la Génesis de Avenidas en Cuencas de Montaña. Master's Thesis, Universidad de Cantabria, Cantabria, Spain, 2014; 50p.
5. Green, W.H.; Ampt, G.A. Studies on soil physics: Part, I. The flow of air and water through soils. *J. Agric. Sci.* **1911**, *4*, 1–24.
6. Liu, J.; Zhang, J.; Feng, J. Green–Ampt model for layered soils with nonuniform initial water content under unsteady infiltration. *Soil Sci. Soc. Am. J.* **2008**, *72*, 1041–1047. [CrossRef]
7. Vannasy, M.; Nakagoshi, N. Estimating Direct Runoff from Storm Rainfall Using NRCS Runoff Method and GIS Mapping in Vientiane City, Laos. *Int. J. Grid Distrib. Comput.* **2016**, *9*, 253–266. [CrossRef]
8. Karbasi, M.; Shokoohi, A.; Saghafian, B. Loss of Life Estimation Due to Flash Floods in Residential Areas using a Regional Model. *Water Resour. Manag.* **2018**, *32*, 4575–4589. [CrossRef]
9. Aristizábal, E.; Arango, C.M.I.; García, L.I.K. Definición y clasificación de las avenidas torrenciales y su impacto en los Andes colombianos. *Cuad. Geogr. Rev. Colomb. Geogr.* **2020**, *1*, 242–258. [CrossRef]
10. Bilasco, S.; Hognogi, G.G.; Ros, C.S.; Pop, A.M.; Iuliu, V.; Fodorean, I.; Marian-Potra, A.C.; Sestras, P. Flash Flood Risk Assessment and Mitigation in Digital-Era Governance Using Unmanned Aerial Vehicle and GIS Spatial Analyses Case Study: Small River Basins. *Remote Sens.* **2022**, *14*, 2481. [CrossRef]
11. Shuvo, S.D.; Rashid, T.; Panda, S.K.; Das, S.; Abdul, Q.D. Forecasting of pre-monsoon flash flood events in the northeastern Bangladesh using coupled hydrometeorological NWP modelling system. *Meteorol. Atmos Phys.* **2021**, *133*, 1603–1625. [CrossRef]
12. Kim, E.; Choi, H. Assessment of Vulnerability to Extreme Flash Floods in Design Storms. *Int. J. Environ. Res. Public Health* **2011**, *8*, 2907–2922. [CrossRef] [PubMed]
13. Kong, F.; Huang, W.; Wang, Z.; Song, X. Effectof Unit Hydrographs and Rainfall Hyetographs on Critical Rainfall Estimates of Flash Flood. *Adv. Meteorol.* **2020**, *2020*, 2801963. [CrossRef]
14. Domínguez, R.R.; Carrizosa, E.E.; Fuentes-Mariles, G.E.; Arganis-Juárez, M.L.; Osnaya, J.O.; Galván, T.A. G Análisis regional para la estimación de precipitaciones de diseño en la República Mexicana Regional Analysis in approaching design rainfall in Mexican Republic. *Tecnol. Cienc. Agua.* **2018**, *9*, 5–29. [CrossRef]
15. Hec-HMS. 2022. Available online: https://www.hec.usace.army.mil/software/hec-hms/ (accessed on 3 November 2022).
16. Chen, C.L. Rainfall Intensity-Duration-Frequency Formulas. *J. Hydraul. Eng.* **1982**, *109*, 1603–1621. [CrossRef]
17. La Jornada. 2022. Available online: https://www.jornada.com.mx/2009/11/04/capital/032n1cap (accessed on 1 November 2022).
18. Milenio. 2021. Available online: https://www.milenio.com/politica/comunidad/declaran-desastre-natural-27-municipios-afectaciones-grace (accessed on 1 November 2022).
19. La Silla Rota. 2021. Available online: https://lasillarota.com/hidalgo/estado/2021/9/8/por-inundaciones-en-tula-el-mezquital-mil-casas-danadas-15-decesos-295643.html (accessed on 2 November 2022).

20. El País. 2021. Available online: https://elpais.com/mexico/2021-09-07/al-menos-dos-muertos-tras-una-fuerte-tormenta-en-ecatepec.html (accessed on 2 November 2022).
21. Domínguez, R.; Carrizosa, E.; Arganis, M.; Santana, A.; De Luna, F.; Mendoza, R.; Hernández, D.; González, S.; Vázquez, R.; Hernández, A. *Proyectos y Estudios para Mejorar la Infraestructura de Drenaje, Actualización del Manual de obras Hidráulicas de la Ciudad de México*; Technical Report; Informe Técnico para SACMEX: Mexico City, Mexico, 2022.
22. González. *Contribución al Análisis de Frecuencias de Valores Extremos de los Gastos Máximos en un Río*; Serie Azul; Instituto de Ingeniería, UNAM: Mexico City, Mexico, 1970.
23. Bedient, P.B.; Huber, W.C. *Hydrology and Floodplain Analysis*; Addison-Wesley: Boston, MA, USA, 1992.
24. Mockus, V. *National Engineering Handbook, Section 4: Hydrology*; United States Department of Agriculture (USDA), Soil Conservation Service (SCS): Washington, DC, USA, 1969.
25. Bladé, E.; Cea, L.; Corestein, G.; Escolano, E.; Puertas, J.; Vázquez-Cendón, E.; Dolz, J.; Coll, A. Iber: Herramienta de simulación numérica del flujo en ríos. *Rev. Int. Métodos Numéricos Cálculo Diseño Ing.* **2014**, *30*, 1–10. [CrossRef]
26. Available online: https://www.hec.usace.army.mil/software/hec-ras/ (accessed on 10 November 2022).
27. Costabile, P.; Costanzo, C.; Ferraro, D.; Macchione, F.; Petaccia, G. Performances of the New HEC-RAS Version 5 for 2-D Hydrodynamic-Based Rainfall-Runoff Simulations at Basin Scale: Comparison with a State-of-the Art Model. *Water* **2020**, *12*, 2326. [CrossRef]
28. NSW. *Floodplain Development Manual, the Management of Flood Liable Land*; News South Wales Government, Department of Infrastructure, Planning and Natural Resources: Sidney, NSW, Australia, 2005.
29. CONAGUA. *Lineamiento para la Elaboración de Mapas de Peligro por Inundación*; GASIR: Mexico City, México, 2014.
30. Velez, M.L.; Fuentes, M.; Rubio, G.H.; De Luna, C.F. Flood hazard maps as a tool for assessing the risk of damage to homes. In Proceedings of the XXIII National Hydraulic Congress, Puerto Vallarta, Jalisco, Mexico, 14 October 2014.
31. Musolino, G.; Ahmadian, R.; Xia, J.; Falconer, R.A. Mapping the danger to life in flash flood events adopting a mechanics based methodology and planning evacuation routes. *J. Flood Risk Manag.* **2019**, *1*, e12627. [CrossRef]
32. Nanía, L.S.; León, A.S.; García, M.H. Hydrologic-Hydraulic Model for Simulating Dual Drainage and Flooding in Urban Areas: Application to a Catchment in the Metropolitan Area of Chicago. *J. Hydrol. Eng.* **2015**, *20*, 04014071. [CrossRef]

water

MDPI

Article

# Experimental Study on Gully Erosion Characteristics of Mountain Torrent Debris Flow in a Strong Earthquake Area

Jiqin Zhang [1], Dengze Luo [2], Hongtao Li [1,3], Liang Pei [1,3] and Qiang Yao [1,3,*]

1 College of Water Resource and Hydropower, Sichuan University, Chengdu 610065, China
2 Yalong River Hydropower Development Company, Ltd., Chengdu 610000, China
3 State Key Laboratory of Hydraulics and Mountain River Engineering, Sichuan University, Chengdu 610065, China
* Correspondence: yaoqiang777@scu.edu.cn

**Abstract:** In recent years, as the frequency of debris flow outbreak in strong earthquake areas has increased and the scale has been expanding, in order to explore the erosion characteristics of debris flow, a lateral erosion flume model experimental device has been designed, and 18 groups of incomplete orthogonal experiments have been carried out, with a unit weight of debris flow of 1.6~2.0 g/cm$^3$, a content of fine particles in the accumulation of 0~28.82%, and a longitudinal slope gradient of the gully of 8°~20° as variables. The results show that the erosion width, erosion depth, and erosion volume decrease with the increase in fluid bulk density and increase with the increase in gully slope. When the longitudinal slope of the gully was 16°, the sediment with 11.40% fine particles had the strongest erosion effect, indicating that more or less fine particles are not conducive to the occurrence of lateral erosion of the gully. Finally, through multi-factor variance analysis, it was found that the order of the three factors on the gully lateral erosion degree from strong to weak is: debris flow unit weight, gully slope, and accumulation grading. The analysis results further showed that the unit weight of debris flow has the greatest impact on the erosion degree of the side slope, which is consistent with the experimental results. The research results have important reference significance for revealing the mechanism of lateral erosion and improving the level of debris flow disaster prevention in strong earthquake areas.

**Keywords:** debris flow; lateral erosion; strong earthquake area; model experiment; erosion pattern

**Citation:** Zhang, J.; Luo, D.; Li, H.; Pei, L.; Yao, Q. Experimental Study on Gully Erosion Characteristics of Mountain Torrent Debris Flow in a Strong Earthquake Area. *Water* **2023**, *15*, 283. https://doi.org/10.3390/w15020283

Academic Editor: Cristiana Di Cristo

Received: 20 November 2022
Revised: 30 December 2022
Accepted: 5 January 2023
Published: 9 January 2023



## 1. Introduction

After the "5.12 Wenchuan" earthquake in 2008, the mountains in the strong earthquake area were severely disturbed, adverse geological disasters occurred frequently, and the reserves of loose rock and soil mass generated by gully collapse and landslide increased sharply [1,2]. These deposits have poor particle sorting, a large pore ratio, and strong water permeability, which can easily initiate the formation of debris flows [3], resulting in an increase in the scale and frequency of debris flows [4]. This type of debris flow is mostly formed by the continuously increasing concentration of solids inside the fluid as the water in the channel erodes the channel bank. The experimental study of the lateral erosion of debris flow and the exploration of its disaster-causing mechanism can provide a better basis for the prediction and prevention of debris flow, so as to improve the level of debris flow disaster prevention and reduce the adverse impact of debris flow on gully erosion [5,6].

As hydraulic debris flow has a stronger erosive capacity to gullies, its occurrence ratio is increasing year by year [7]. The gullies are gradually deepened under the erosion of debris flow [8], and their boundary conditions change. The gully bank deposits are more likely to lose stability and fall into the gullies under the lateral erosion of water flow, which widens the gullies and sharply increases the scale of debris flow [9–11]. Through the analysis of debris flow data in Kansia Basin, Simoni et al. found that when the longitudinal slope of the river channel exceeds 16 degrees, the river channel will produce lateral erosion [12]. G. J. Hanson et al. studied the erosion resistance of different parts of the riverbed through an on-site spraying experiment [13]. Zhou [14] and others found that riverbed erosion, bank collapse, and river widening caused by erosion are the main reasons for the triggering and scaling up of debris flow in the lower reaches of Wenjia Valley. The 2017 Wenchuan Yangtang ditch debris flow showed a strong lateral erosion widening effect on the ditch bank during the flow through the circulation area, the widening width generally reached 8~10 m, and the amount of material source squared taken away by lateral erosion along the course was as high as $3.0 \times 10^5$ $m^3$ [15]. During the debris flow movement in the 2019 Xiazhuang Valley ditch, the foot of the terrace slope on both banks of the ditch channel gradually panned back under the lateral erosion effect of the mudslide, which made the soil deposited on the shore at an early stage collapse in block form to converge in the debris flow [16]. It can be seen that material generated by lateral erosion of gully banks plays an increasingly important role in the formation of post-earthquake debris flows [17–19]. Through the relevant literature, we know that there are many factors affecting gully erosion, such as geology, landform, elevation, slope, soil characteristics, rainfall, etc. [6,20,21], which will have a great impact on gully erosion. In addition, Paramita Roy et al. also assessed the importance of gully erosion according to the influencing factors of gully erosion and drew the gully erosion sensitivity map of the Hinglo River Basin of eastern India, which can help the land management department to control the potential erosion area in advance, and also make full use of land resources to promote the sustainable development of the basin [22]. Pan Huali et al. [15] analyzed the erosion pattern and factors influencing the movement of debris flow in the channel and found that lateral erosion is particularly strong at the concave bank of the channel, and the foot of the channel bank is continuously retreating under the continuous erosion of debris flow, which makes the channel bank form a suspended body and collapse under the action of gravity, thus widening the channel laterally. Zhu Xinghua et al. [23] gave calculation formulas for lateral erosion occurrence, occurrence type, and transport rate through theoretical analysis and field in situ experiments, and classified lateral bank erosion types into dumping, falling, and slip collapse, and performed mechanical equilibrium analysis for each mode by analyzing the form of lateral bank fractures. Qu YP et al. [24] investigated the channel initiation mechanism and movement characteristics of hydraulic-type debris flows under different combinations of flow velocity, flow capacity, and longitudinal slope gradient in a hydraulic-type debris flow wash-out scale model experiment, but the influence of debris flow density on the results was not considered during the experiment. Zhao Yanbo et al. [25] explored the rule of erosion depth of debris flow from the gully bed under the conditions of gully slope, debris flow unit weight, and gully bed deposit gradation through a flume test. Chen Jing et al. [9] considered the erosion rate and depth of debris flow with different unit weights under different accumulation water contents and different slopes in a flume experiment of influencing factors of debris flow bottom erosion; however, neither Zhao Yanbo et al. [25] nor Chen Jing et al. [9] conducted in-depth research on the erosion width and volume of the accumulation body.

To date, many researchers have studied the erosion mechanism of debris flow through experiments, theories, and numerical simulation. However, their research objects are mainly focused on undercutting erosion. Therefore, the lateral erosion of gullies needs to be studied further [20,26,27]. Aiming at the problems of insufficient research on the mechanism of lateral erosion of debris flow and incomplete consideration of relevant influencing factors, this paper, on the basis of previous experimental research, carries out 18 groups of incomplete orthogonal experiments to explore the influence of the volume weight of debris flow, the slope of channel, and the gradation of deposits on lateral erosion and carries out analysis of the data on the shape, width, and square of deposits after erosion. The influence of different factors on the lateral erosion degree of debris flow is studied. The results are of great significance for further theoretical research on the side erosion of debris flow, prediction of debris flow disasters, and improvement of the recognition and prevention of debris flow in strong earthquake areas.

## 2. Experiment Scheme Design

### 2.1. Experimental Apparatus and Equipment

The independently designed debris flow bed flume model experimental device [28] that can change the slope from 0 to 45° is used for the experiment. The experimental model is composed of four parts: a lifting device, mixing device, flume device, and stacking platform, as shown in Figure 1. The left side of the model device is a 6 m long and 0.3 m wide water tank, to stabilize the flow pattern; the first 4.2 m is set as the fixed bed section and the last 1.8 m is set as the dynamic bed section.

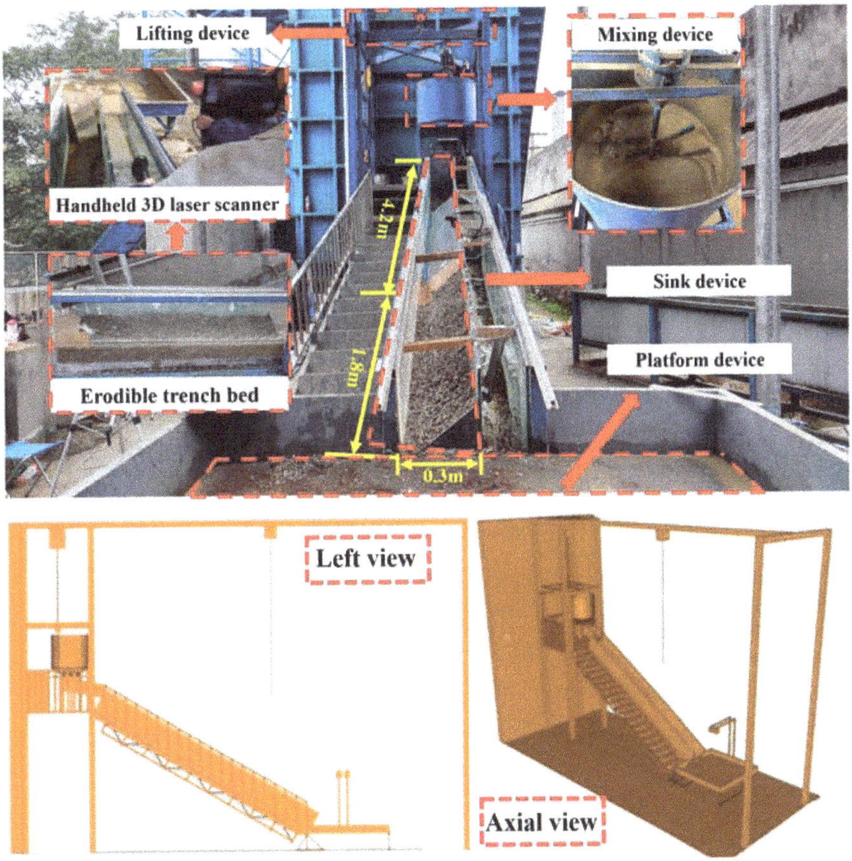


**Figure 1.** Experimental device.

### 2.2. Experimental Parameters

This experiment belongs to the generalized flume model experiment. The unit weight of debris flow in the strong earthquake area ranges from 1.6 g/cm$^3$ to 2.0 g/cm$^3$, the longitudinal slope of the gully varies from 7.6° to 42.8°, and the average longitudinal slope is 21.8° [24,29]. In order to reflect the flow characteristics of debris flow in the field, the volume weight range of debris flow configured in the experiment is consistent with field data. At the same time, considering the water infiltration under different longitudinal slopes of the gully in order to minimize the impact of water infiltration on the unit weight of debris flow before erosion of the gully; referring to the relevant literature, the variation range of the longitudinal slope of the experimental gully is 8° to 20° [30].

Three kinds of debris flow with different unit weights, namely, rarefaction (1.6 g/cm$^3$), transition (1.7 g/cm$^3$), and viscosity (1.8 g/cm$^3$), are set up in the experiment. The materials for preparing the debris flow are natural river sand and clean water, and bentonite is added to improve the viscosity and workability of the debris flow, to prevent the debris flow from stratification. The proportions of debris flow with different unit weights are shown in Table 1. To consider the erosion effect of debris flow on different graded accumulations, three types of accumulations with 25% water content and different contents of fine particles are selected in the experiment. The content of fine particles in the accumulations at different levels is shown in Table 2.

**Table 1.** Composition of debris flow.

| Density | Water | Soil | Sand |
|---|---|---|---|
| 1.6 g/cm$^3$ | 0.35 | 0.1 | 0.55 |
| 1.7 g/cm$^3$ | 0.29 | 0.08 | 0.63 |
| 1.8 g/cm$^3$ | 0.24 | 0.06 | 0.7 |

**Table 2.** Fine particle content of accumulation.

| Gradation Types | Grading I | Grading II | Grading III |
|---|---|---|---|
| Content of fine particles (<1 mm) | 11.40% | 0.00% | 28.82% |

### 2.3. Experimental Scheme

In the experiment, the unit weight of debris flow, the longitudinal slope of the gully, and the gradation of accumulation are used as variables to study the influence of each variable on the lateral erosion of debris flow. Due to a large number of experimental variables, the experimental scheme is set by the method of incomplete orthogonality. Given the widespread existence of narrow and steep gully-type debris flow in the strong earthquake area, this type of gully is severely eroded and has a strong disaster-causing capacity. Its outbreak frequency increases significantly after an earthquake, and the slope of the gully is concentrated at about 16°. Therefore, the experiment takes 16° as the main slope to study the erosion law of different accumulation gradations under this slope. In the pre-experiment process, it was found that the erosion capacity of the debris flow with a unit weight of 1.8 g/cm$^3$ was weak, which was mainly represented by siltation. Therefore, the erosion process was only studied under the condition of the 20° longitudinal slope. The test scheme is shown in Table 3.

**Table 3.** Experimental scheme.

| Number | Longitudinal Slope Gradient (°) | Fluid Density (g/cm$^3$) | Gradation of Accumulation |
|---|---|---|---|
| 1 | 20 | Water | Grading I |
| 2 | 16 | Water | Grading I |
| 3 | 12 | Water | Grading I |
| 4 | 8 | Water | Grading I |
| 5 | 20 | 1.7 | Grading I |
| 6 | 20 | 1.8 | Grading I |
| 7 | 20 | 1.6 | Grading I |
| 8 | 16 | 1.6 | Grading I |
| 9 | 12 | 1.6 | Grading I |
| 10 | 8 | 1.6 | Grading I |

**Table 3.** *Cont.*

| Number | Longitudinal Slope Gradient (°) | Fluid Density (g/cm$^3$) | Gradation of Accumulation |
|---|---|---|---|
| 11 | 16 | 1.7 | Grading I |
| 12 | 12 | 1.7 | Grading I |
| 13 | 16 | 1.7 | Grading II |
| 14 | 16 | 1.6 | Grading II |
| 15 | 16 | Water | Grading II |
| 16 | 16 | 1.6 | Grading III |
| 17 | 16 | 1.7 | Grading III |
| 18 | 16 | Water | Grading III |

## 3. Experimental Study on Lateral Erosion of Erodible Gully Bed

In natural debris flow gullies, due to the erodibility of the gully bed, the lateral erosion of debris flow will be affected by the undercutting of the gully. Exploring lateral erosion regularity of erodible gully beds is beneficial to further deepen our understanding of lateral erosion of debris flow in strong earthquake areas, eliminate potential hazards in advance, and reduce potential natural disasters in the future, which is of great significance to reduce the safety risks of gully basins [31,32].

### *3.1. Erosion Form Analysis*

After scouring, the hand-held 3D laser scanner is used to obtain the point cloud data of the final erosion form, and Geo magic software is used to process the measured 3D point cloud data, establish a scouring form model, and obtain the section curve information at the maximum erosion width and depth.

(1) Erosion degree analysis of debris flow unit weight change

The morphology and 3D comparison of the gully channel after erosion of grade I accumulation by the different unit weight of debris flow under 20° longitudinal slope conditions are shown in Figure 2, and the characteristic cross-section of the gully channel after erosion is shown in Figure 3.

The results show that the maximum erosion width of 1.6 g/cm$^3$ debris flow is 111.6 mm, the maximum erosion depth is 96.1 mm, and the erosion volume is 22,338.27 cm$^3$. The maximum erosion width of 1.7 g/cm$^3$ debris flow is 90.70 mm, the maximum erosion depth is 104.7 mm, and the erosion volume is 9263.56 cm$^3$. The 1.8 g/cm$^3$ debris flow has no lateral erosion, and the erosion depth of 10.4 mm is generated at the front end of the gully, and 39.8 mm thick siltation is generated at the rear end of the gully, with the siltation volume of 2057.83 cm$^3$. The maximum erosion width of clear water is 192.2 mm, the maximum erosion depth is 91.3 mm, and the erosion volume is 28,556.99 cm$^3$. As can be seen from the erosion pattern under the action of 1.6 g/cm$^3$ debris flow, the gully presents a cut and pull trough at the bottom of the whole section, and the bank slope at the pull trough is steepened by lateral erosion. The 1.7 g/cm$^3$ debris flow forms a punch hole at the front end of the gully, and the erosion at the rear end is weakened; 1.8 g/cm$^3$ debris flow shows siltation in the process. In this process, water eroded the entire side bank slope, and a large number of side bank material sources accumulated in the ditch, reducing the original side bank slope, widening the bottom of the ditch, and forming a maximum thickness of 88.9 mm at the position after x = 1300 mm. The erosion degree of different fluids under the condition of an erodible gully bed is in the order of strong to weak: water, 1.6 g/cm$^3$ debris flow, 1.7 g/cm$^3$ debris flow, 1.8 g/cm$^3$ debris flow.

(**a**) Water

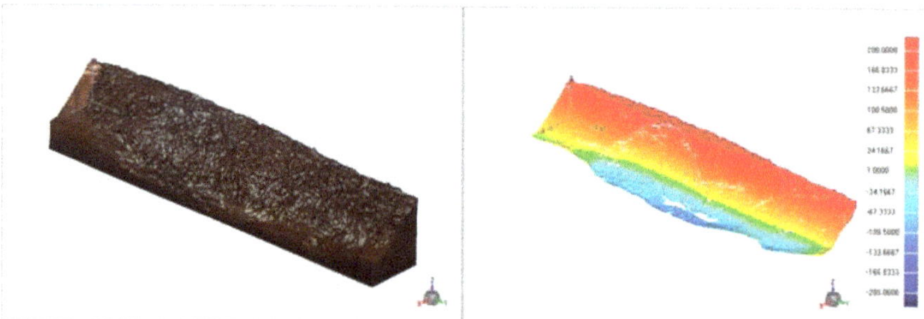

(**b**) 1.6 g/cm³ Debris flow

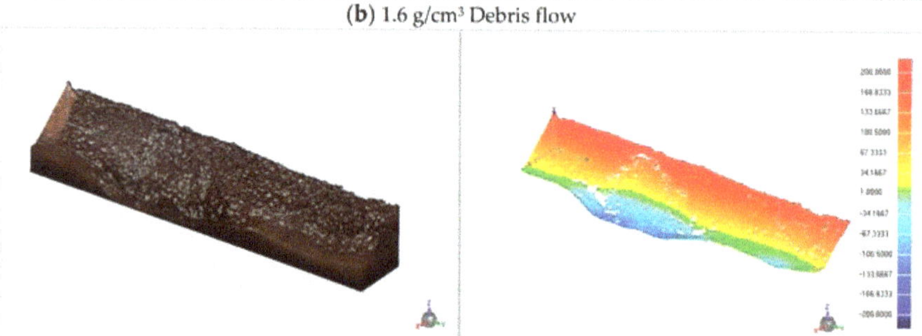

(**c**) 1.7 g/cm³ Debris flow

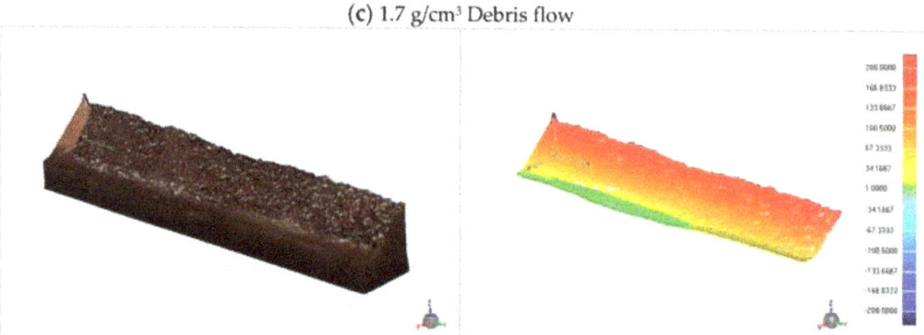

(**d**) 1.8 g/cm³ Debris flow

**Figure 2.** Comparison diagram of debris flow erosion with different unit weights.

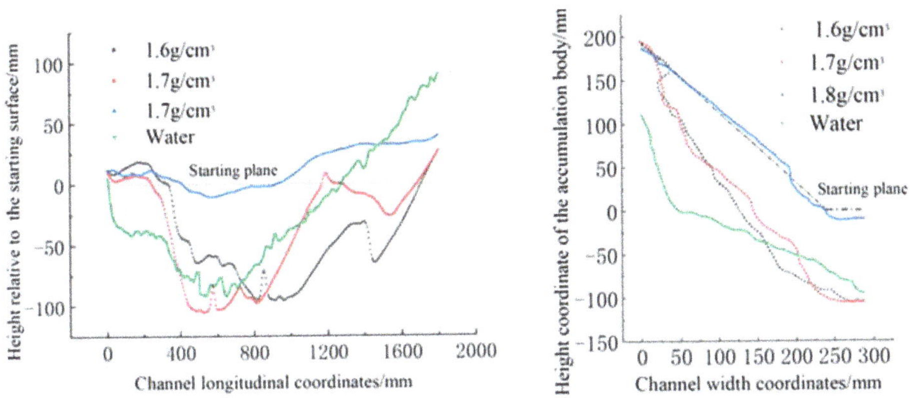


(a) (b)

**Figure 3.** Characteristic cross-section after erosion of debris flow with different capacity. (**a**) Cross-section at maximum erosion depth. (**b**) Cross-section at maximum erosion width.

(2) Analysis on erosion degree of gully longitudinal slope change

The morphology and 3D comparison of the gully channel after erosion of the graded I accumulation by different longitudinal slope conditions with a unit weight of 1.6 g/cm$^3$ debris flow are shown in Figure 4, and the characteristic cross-section of the gully channel after erosion is shown in Figure 5.

The analysis of the erosion of the lateral gully bank accumulation at different slopes showed that when the slope is 8°, the debris flow only forms scour holes at x = 300 mm, the maximum erosion width is 68.2 mm, the maximum erosion depth is 57.6 mm, and the erosion volume is 2244.01 cm$^3$; the rest of the locations are largely free of erosion. When the slope rises to 12°, the erosion degree gradually increases. At this time, the debris flow forms scour pits at x = 300 mm and x = 1200 mm, the maximum erosion width is 86.6 mm, the maximum erosion depth is 71.4 mm, and the erosion volume is 6347.80 cm$^3$. When the longitudinal slope of the gully is 16°, the debris flow forms a scour hole at the accumulation body x = 400 mm, and then a backward-pulling slot to form erosion on the entire section, which eventually leads to the steep side slope, with the maximum erosion width of 108.2 mm, the maximum erosion depth of 110.0 mm, and the erosion volume of 16,283.31 cm$^3$. When the gradient increases to 20°, under the action of debris flow, the gully is characterized by a full-section lower cut groove, the maximum erosion width is 111.6 mm, and the amount of erosion is 22,338.27 cm$^3$. Due to the instability of the bank slope at the position of the lower cut groove, particles flow into the eroded gully, resulting in the final erosion depth of 96.1 mm. In the actual erosion process, the erosion depth of debris flow under this unit weight reached 110.0 mm at the gully bed boundary. From the erosion situation, under the condition of the erodible gully bed, the erosion degree gradually increases with the increase in slope.

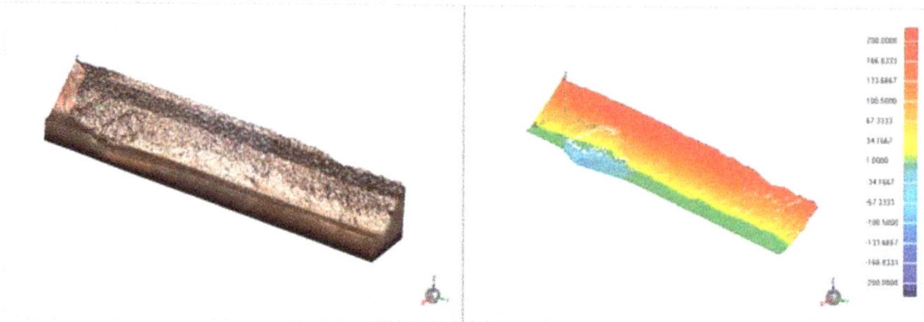

(**a**) 8° Longitudinal slope

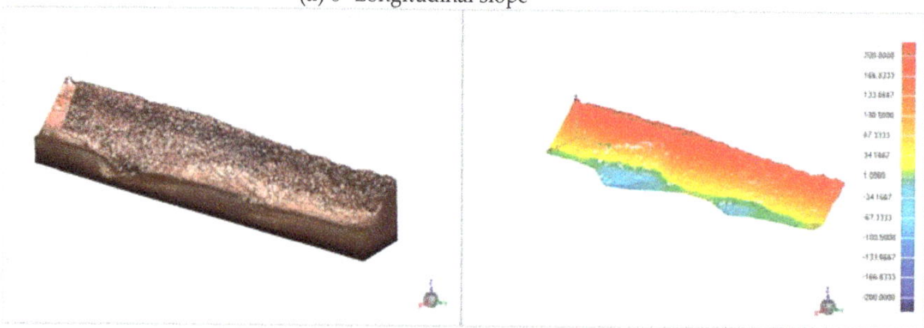

(**b**) 12° Longitudinal slope

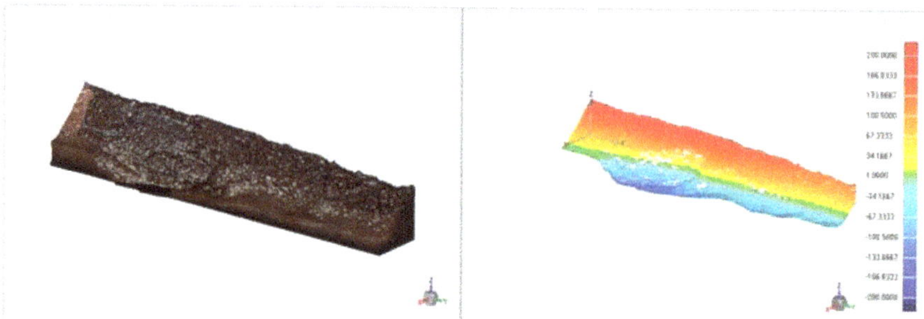

(**c**) 16° Longitudinal slope

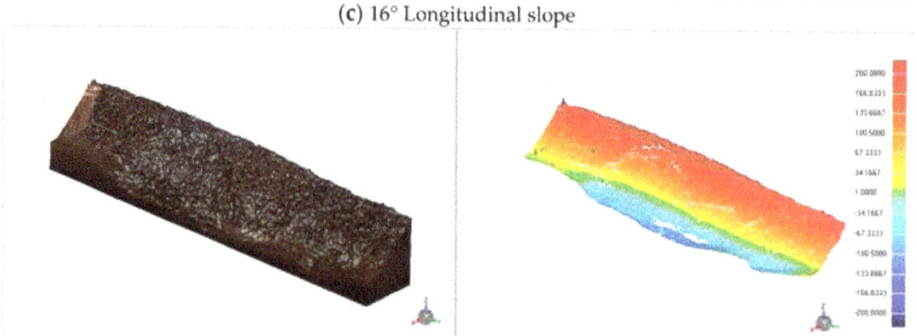

(**d**) 20° Longitudinal slope

**Figure 4.** Comparison diagram of debris flow erosion with different longitudinal slopes.

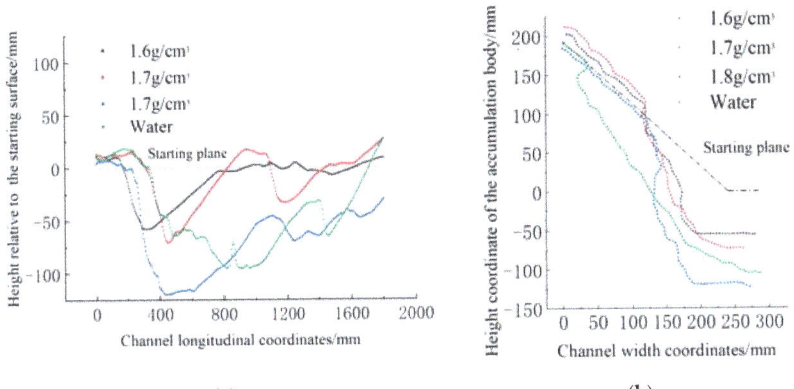


**Figure 5.** Characteristic cross-sections after debris flow erosion under different longitudinal slope conditions (**a**) Cross-section at maximum erosion depth. (**b**) Cross-section at maximum erosion width.

(3) Erosion degree analysis of different graded accumulations

The morphology and 3D comparison of the gully channel after erosion of different graded accumulation by the 1.6 g/cm$^3$ capacity debris flow under 16° longitudinal slope conditions are shown in Figure 6, and the characteristic cross-section of the gully channel after erosion is shown in Figure 7.

Under the condition of the erodible gully bed, the 1.6 g/cm$^3$ bulk weight debris flow produced a large degree of lateral erosion of the different graded accumulations, all accumulations reaching the boundary limit at the depth of erosion. In terms of erosion width of accumulations, the erosion extent of debris flow to graded I accumulations is the largest, with the erosion width reaching 108.2 mm, while the maximum erosion width of graded II and graded III accumulations are 83.5 mm and 48.3 mm, respectively. There is little difference in erosion volume among accumulations, with the erosion volume being 16,283.31 cm$^3$, 20,172.77 cm$^3$, and 18,932.91 cm$^3$, in turn. In terms of erosion pattern, the erosion pattern of debris flow to the three graded deposits is that scouring pits are formed near the channel longitudinal x = 450 mm, then scouring pits backward-pull grooves to form erosion across the whole section, which finally leads to the steepening of the entire side bank slope. The degree of erosion of the different graded accumulations shows that the fine grain content has a greater effect on the width of erosion and a smaller effect on the volume of erosion.

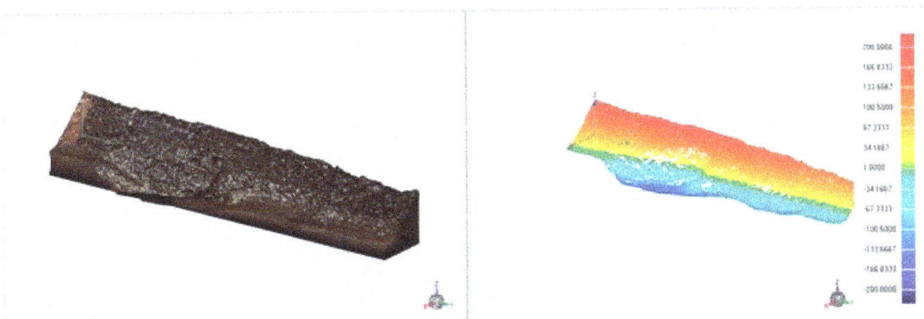

(**a**) Grading I

**Figure 6.** *Cont.*

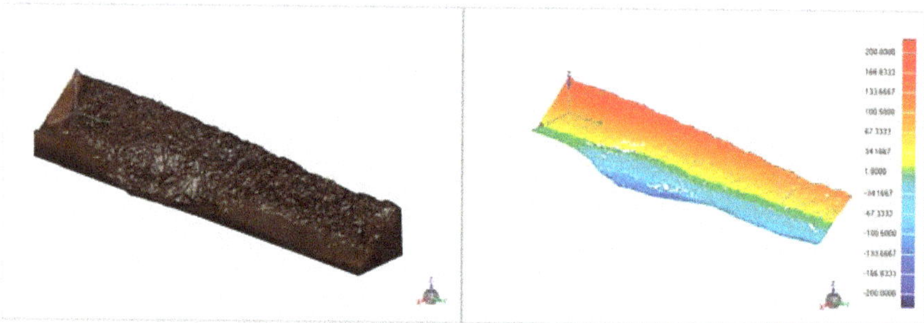

(b) Grading II

(c) Grading III

**Figure 6.** Comparison diagram of erosion of accumulation bodies with different gradations.

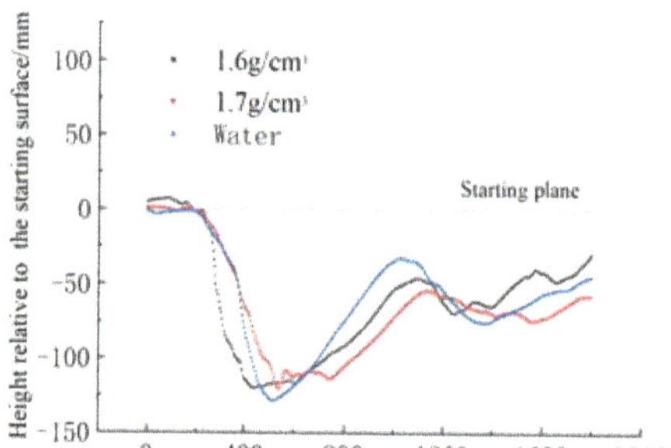


(a)

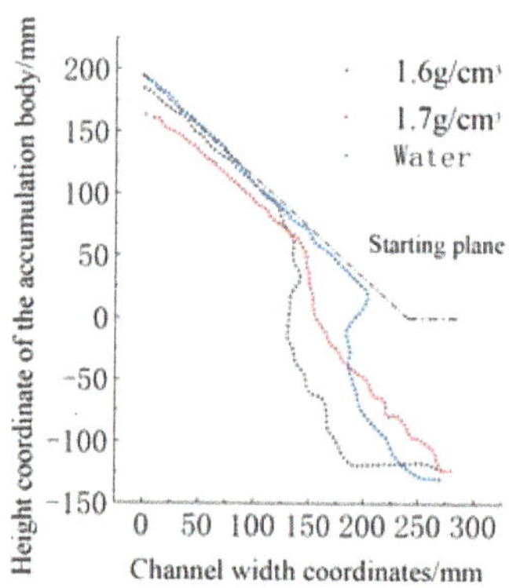


(b)

**Figure 7.** Characteristic cross-sections of different graded mounds after erosion by debris flow (**a**) Cross-section at maximum erosion depth.(**b**) Cross-section at maximum erosion width.

*3.2. Analysis of the Width, Depth, and the Volume of Erosion*

(1) Erosion width

Under the condition of the erodible gully bed, the erosion width under different slope gradients is shown in Figure 8a. The erosion width of water and the 1.6 g/cm$^3$ debris flow are positively correlated with the channel slope. The 1.7 g/cm$^3$ debris flow does not show lateral erosion from 12° to 16°, but lateral erosion occurs when the slope is increased to 20°, suggesting a critical lateral erosion slope in the range of 16° to 20° for this volume of debris flow. The erosion widths of fluids with different bulk densities are in the following order: water, 1.6 g/cm$^3$ debris flow, 1.7 g/cm$^3$ debris flow, indirectly indicating that fluid viscosity has a greater influence on the erosion width.

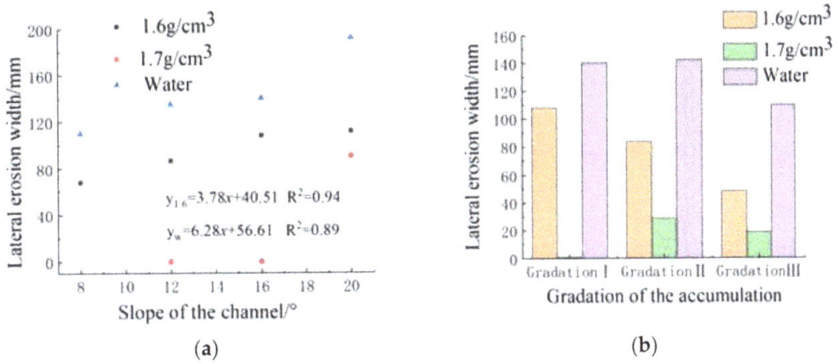


**Figure 8.** Lateral erosion width. (**a**) Erosion width under different slope gradients. (**b**) Erosion width under different graded accumulations.

The erosion widths of the different graded accumulations under 16° longitudinal slope conditions are shown in Figure 8b. Water and 1.6 g/cm$^3$ debris flow erode each channel greatly, while 1.7 g/cm$^3$ debris flow has a strong viscosity, low turbulence in the flow process, and low erosion width to the three graded deposits. The grade III accumulations have a higher content of fine particles than the other two grades, are more cohesive, and ultimately suffer the least erosion.

(2) Erosion depth

The depth of erosion on different slopes is shown in Figure 9a. As the maximum erosion depth of the experimental design is 110 mm, the erosion depth reached the limit due to the influence of boundary conditions during the experiment; however, from the overall change degree, the erosion depth of each experimental group still increases with the increase in slope. However, under the condition of water and a 16° longitudinal slope, the final erosion depth is less than 12° longitudinal slope. This is because the channel accumulation collapses and slides at the end of the experimental process, causing the accumulation to block the bottom of the channel. The subsequent inflow is not enough to move the blockage body, making the final measured erosion depth smaller, but the actual erosion depth reaches the bottom of the channel.

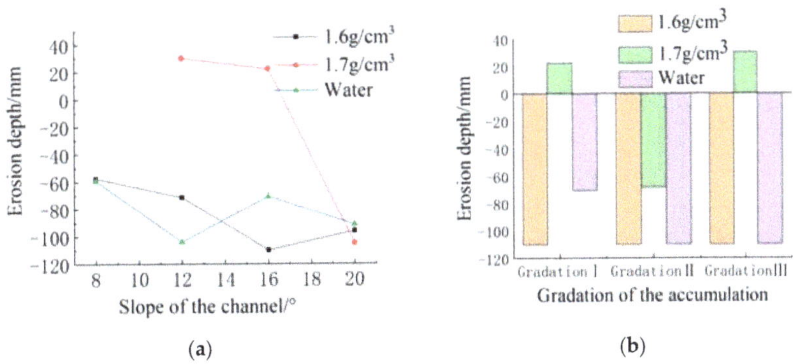


**Figure 9.** Lateral erosion depth. (**a**) Erosion depth under different slope gradients. (**b**) Erosion depth under different graded accumulations.

The depth of erosion of the different graded accumulations is shown in Figure 9b. Under the condition of the 16° longitudinal slope, the maximum erosion depth of water and the 1.6 g/cm$^3$ debris flow reaches the boundary value. The 1.7 g/cm$^3$ debris flow is silted under the conditions of grading I and grading III, and only under the conditions of grading II is incised, forming a 68.2 mm deep scour pit. The gradation II accumulation has a low content of fine particles, low cohesion, and large soil pores, which makes it easier for the debris flow to penetrate deeper into the mound, making the soil subject to undercutting erosion by floating forces. A correlation can be observed between the production of undercutting erosion and the fine particle content.

(3) Volume of erosion

The total amount of erosion for each experimental group at different slopes is shown in Figure 10a. The clear water and 1.6 g/cm$^3$ debris flow eroded under the condition of the 8° longitudinal slope, and the amount of erosion was positively correlated with the longitudinal slope gradient, with a high degree of correlation; for the 1.7 g/cm$^3$ debris flow, under the conditions of 12° and 16° longitudinal slopes, only slight siltation occurs; under the conditions of 20° longitudinal slopes, erosion occurs in the gully, and the amount of erosion and siltation meets the exponential function relationship.

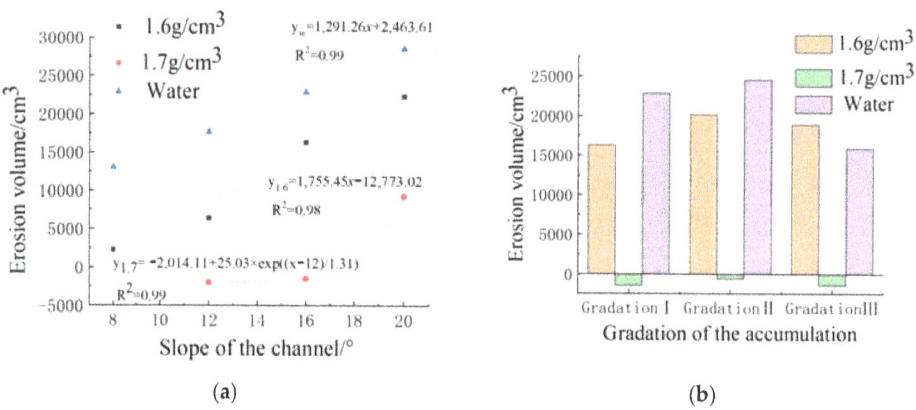


**Figure 10.** Lateral erosion volume. (**a**) Erosion volume under different slope gradients. (**b**) Erosion volume under different graded accumulations.

The total amount of erosion for different gradations is shown in Figure 10b. At the 16° longitudinal slope, 1.6 g/cm$^3$ debris flow and clear water washed through the graded accumulations, while 1.7 g/cm$^3$ debris flow silted up, and at this longitudinal slope, there was little difference in the total erosion of the different graded accumulations, and the gradation characteristics set in this experiment had little effect on the total erosion.

### 3.3. *Multi-Factor ANOVA on Factors Influencing Side Erosion*

When using SPSS software, the significance level is set as $p < 0.05$ and the larger the F value, the more significant the effect of the corresponding factor on the dependent variable; when the significance $p > 0.05$, the effect is not significant. With lateral erosion width and erosion volume as dependent variables and fluid density, slope, and accumulation grading as fixed factors, the Duncan model was selected to compare and analyze the main effects of each factor, and the results were as follows.

The impact analysis of each factor on the erosion volume and erosion width is shown in Table 4. Under the condition of the erodible gully bed, the volume of debris flow erosion and erosion width are affected by the same order of factors. The factor that had the greatest impact on the degree of erosion was fluid unit weight, followed by the gradient of the gully longitudinal slope. It was further proved that under the condition of the erodible gully bed, the content of fine particles has no significant impact on the degree of gully erosion.

**Table 4.** Test for effect of each influencing factor.

| Factor | Erosion Volume | | Erosion Width | |
|---|---|---|---|---|
| | F | Significance $p$ | F | Significance $p$ |
| Density | 53.2988 | 0.000 | 35.450 | 0.000 |
| Slope | 7.994 | 0.007 | 9.224 | 0.004 |
| Graded | 1.922 | 0.202 | 2.883 | 0.108 |

## 4. Discussion

During the course of lateral erosion of channel-type debris flow in the strong earthquake area, the volume weight of debris flow, the longitudinal slope of the channel, and the gradation of deposits are three important factors affecting its lateral erosion. Taking these three factors as variables to carry out a tank model experiment is particularly important for studying the width, depth, and square of debris flow, and subsequent debris flow control and disaster prediction.

The volume weight of debris flow, the longitudinal slope of the channel, and the gradation of deposits have significant effects on the lateral erosion of debris flow. According to the model test and the results of data analysis, the factors influencing the width and volume of debris flow erosion are unit weight of debris flow, channel slope, and gradation of deposits, in turn. In addition, the gully slope provides power for debris flow, and the increase in the slope will also reduce the stability of the side bank slope, which will eventually lead to an increase in erosion volume and erosion width with the increase in the slope. In addition, the gully side bank is more prone to erosion damage. At the same time, it is observed in the test that the flow pattern of 1.6 $g/cm^3$ debris flow is relatively disordered, and it has a strong scouring ability to the side bank slope, while the 1.7 $g/cm^3$ debris flow is mainly laminar flow in the fixed bed section. After entering the side bank accumulation experimental area, the flow pattern evolves into a disturbed turbulent flow, and with the increase in slope, the turbulence degree intensifies, and the erosion to the gully gradually increases. At the same time, it was observed that the flow pattern of 1.6 $g/cm^3$ debris flow is relatively disordered, which has a strong scouring ability on the side bank slope, while the 1.7 $g/cm^3$ debris flow is mainly manifested as laminar flow in the fixed-bed section; the flow pattern evolved into the disturbed turbulent flow after entering the experimental area of side bank accumulation, the degree of turbulence increased with the increase in slope, and the erosion of the channel was gradually enhanced.

The results of erosion under erodible gully bed conditions show that when the slope is gentle, the debris flow will preferentially produce undercutting erosion of the gully bed during the flow through the gully, making the bank slope steeper, and thus inducing lateral erosion in the way of bank slope collapse and instability. The further intensification of erosion as the slope increases is due to the strong undercutting erosion, the expansion of the scale of the washout pit triggering a larger scale of lateral bank destabilization, and the more powerful transport capacity of the debris flow itself, which induces a large amount of channel material transport and eventually leads to the expansion of the erosion degree. The erodible ditch bed has a low degree of lateral erosion on the foot of the bank slope and a small erosion width, which makes it easier to produce a plugging effect when the loss of the lateral bank slope occurs, amplifying the scale of debris flow. Therefore, in the debris flow control project for the erodible gully bed, the prevention of undercutting erosion should be the main method, and on this basis, the protection of the bank slope foot should be strengthened.

This study, similar to other studies, has some limitations that cannot be ignored in the future [33], such as the constraints of the experimental site. The sink model used in this experiment is small and it is difficult to satisfy all similarity laws, so geometric similarity and boundary similarity were mainly considered during the experiment. In the future, similar conditions such as debris flow movement and dynamics can be further considered by carrying out larger-scale model experiments. Due to the limitation of model size, the thickness of the experimentally set erodible layer is 110 mm, and the undercutting erosion reached the model boundary during the erosion process, but from the overall degree of change, the erosion depth of each experimental group still shows an increasing trend as the slope increases. In the future, we can deepen the thickness of the erodible layer in the gully bed and study the effect of undercutting erosion on the lateral erosion of the gully channel to further reveal the mechanism of debris flow erosion.

## 5. Conclusions

Based on the unit weight of debris flow, the longitudinal slope of the channel, and the gradation of deposits as variables, 18 groups of lateral erosion model experiments of debris flow were carried out. The effect of each variable on the lateral erosion of debris flow was studied by collecting data on the scouring process and the channel morphology of debris flow during the experiment. The main conclusions are as follows:

(1) Both debris flow erosion volume and width increased with increasing gully slope, with a good linear fit correlation. This indicates that the gully side bank slopes are more susceptible to erosion damage under the increasing slope. The size of erosion volume and erosion width of the gully by different fluids are increased by clear water, 1.6 g/cm$^3$ mudflow, and 1.7 g/cm$^3$ mudflow, in order. In terms of the erosion volume and erosion width of the accumulation, the most severe erosion was observed for grade I. This indicates that a greater or lower number of fine particles are not conducive to the occurrence of lateral erosion in the gully, and this also provides direction for lateral erosion prevention and cure in the gully.

(2) Compared with rigid riverbeds, lateral erosion of erodible riverbeds changes from slope foot scouring to undercutting erosion, which leads to instability of the riverbank. In addition, the lateral erosion pattern of the debris flow can be summarized as: the debris flow forms a wash pit through undercutting erosion, and mixes and shears in the wash pit to enable the side bank slope to form a critical surface, resulting in the side bank slope instability under the combined effect of debris flow infiltration, erosion, and gravity into the debris flow, and finally the scale of debris flow increases.

(3) Based on the multi-factor ANOVA analysis of experimental data, the total erosion and erosion width were influenced by each factor in the following order of magnitude under erodible trench bed conditions: fluid capacity, trench longitudinal slope, and fine particle content of the accumulation.

As a natural disaster, debris flow is widely distributed in some areas of the world with special topography and geomorphological conditions. By studying erosion characteristics and investigating the influence of different factors on erosion, we can provide a better basis for prediction and prevention of debris flow. In the future, it is also possible to combine the degree of significance of each factor on debris flow erosion to map the erosion sensitivity of the watershed, such that potential risk areas can be treated in advance to reduce the adverse effects of debris flow erosion and promote the sustainable development of the watershed. Therefore, it is essential to explore the erosion characteristics of debris flow.

**Author Contributions:** J.Z. conceived the manuscript; J.Z. and D.L. dealt with point cloud data and drafted the manuscript; H.L., L.P. and Q.Y. provided funding support and ideas; J.Z. and D.L. conducted a field investigation and provided field data; Q.Y., H.L. and L.P. helped to improve the manuscript. All authors have read and agreed to the published version of the manuscript.

**Funding:** This research was funded by the Sichuan Provincial International Science and Technology Col-laboration and Innovation Project, grant number 2022YFH0078; the National Key R&D Program of China, grant number 2018YFC1505402; the National Natural Science Foundation of China, grant number No. 51809188" and "The APC was funded by Qiang Yao".

**Data Availability Statement:** The data used to support the findings of this study are available from the corresponding author upon request.

**Acknowledgments:** The authors would like to thank Qiang Yao, Hongtao Li, Liang Pei, and Dengze Luo for their suggestions on the manuscript and data analysis.

**Conflicts of Interest:** The authors declare no conflict of interest.

## References

1. Cui, P.; Zhuang, J.Q.; Chen, X.C.; Zhang, J.Q.; Zhou, X.J. Characteristics and Countermeasures of Debris Flow in Wenchuan Area After the Earthquake. *J. Sichuan Univ.* **2010**, *42*, 10–19.
2. Xun, H.; Chuan, T.; Maohua, L.; Desheng, T.; Zhilin, J. Mechanism and Characteristics on Debris Flow Hazards in Yin Chang Gou area Triggered by Rainstorm on August 18, 2012. *J. Eng. Geol.* **2013**, *21*, 761–769.
3. Lyu, L. Research on the initiation and motion of gully debris flows in Tibetan Plateau. Ph.D. Thesis, Tsinghua University, Beijing, China, 2017.
4. Yang, H.; Yang, T.; Zhang, S.; Zhao, F.; Hu, K.; Jiang, Y. Rainfall-induced landslides and debris flows in Mengdong Town, Yunnan Province, China. *Landslides* **2020**, *17*, 931–941. [CrossRef]
5. Yan, H.; Guo, X.; Ge, Y.; Zhan, M.; Du, Y. The randomness of lateral erosion failure of debris flow gully deposits. *J. Nat. Disasters* **2020**, *29*, 85–97.
6. Chowdhuri, I.; Pal, S.; Arabameri, A.; Saha, A.; Chakrabortty, R.; Blaschke, T.; Pradhan, B.; Band, S. Implementation of Artificial Intelligence Based Ensemble Models for Gully Erosion Susceptibility Assessment. *Remote Sens.* **2020**, *12*, 3620. [CrossRef]
7. Zhu, X.; Cui, P.; Zhou, G.; Chen, H. Experimental study on the scouring pattern of dilute debris flow. *Shui Li Xue Bao* **2012**, *43*, 85–91.
8. Xiong, J.; Tang, C.; Gong, L.; Shi, Q.; Li, N. Evolution Indexes Selecting and Laws analyzing of Debris Flow Source in Strong Earthquake Area Source in Strong Earthquake Area. *Res. Soil Water Conserv.* **2020**, *27*, 360–365.
9. Chen, J. Experimental Study on the Influence Factors of Debris Flow. Master's Thesis, Chengdu University of Technology, Chengdu, China, 2017.
10. Guo, X.; Chen, X.; Song, G.; Zhuang, J.; Fan, J. Debris flows in the Lushan earthquake area: Formation characteristics, rainfall conditions, and evolutionary tendency. *Nat. Hazards* **2021**, *103*, 2663–2687. [CrossRef]
11. Wang, Z.; Hu, K.H.; Ma, C.; Li, Y.; Liu, S. Landscape change in response to multiperiod glacial debris flows in Peilong catchment, southeastern Tibet. *J. Mt. Sci.* **2021**, *18*, 567–582. [CrossRef]
12. Simoni, A.; Bernard, M.; Berti, M.; Boreggio, M.; Lanzoni, S.; Stancanelli, L.M.; Gregoretti, C. Runoff-generated debris flows: Observation of initiation conditions and erosion–deposition dynamics along the channel at Cancia (eastern Italian Alps). *Earth Surf. Process. Landf.* **2020**, *45*, 3556–3571. [CrossRef]
13. Hanson, G.J.; Simon, A. Erodibility of cohesive streambeds in the loess area of the midwestern USA. *Hydrol. Process.* **2001**, *15*, 23–38. [CrossRef]
14. Zhou, G.G.D.; Cui, R.; Tang, J.B.; Chen, H.Y.; Zou, Q.; Sun, Q.C. Experimental study on the triggering mechanisms and kinematic properties of large debris flows in Wenjia Gully. *Eng. Geol.* **2015**, *194*, 52–61. [CrossRef]
15. Feng, Q.; Xingmin, M.; Fuyun, G.; Yan, Z. Characteristics analysis of "8·7" debris flow disaster in Wen County disturbed by Wenchuan earthquake: A case study of Yangtang Catchment. *J. Lanzhou Univ.* **2021**, *57*, 376–381.
16. Luo, Y.; Tang, C.; Xiong, J.; Cheng, M.; Zhang, X. Cause Analysis of "8·20" Debris Flow and Forecast of River-blocking Range in Xiazhuang Gully of Wenchuan County, Sichuan Province. *Bull. Soil Water Conserv.* **2020**, *40*, 193–199.
17. Zhan, M.; Ge, Y.; Guo, X.; Yan, H.; Du, Y. Experimental study on failure process of lateral bank deposition debris flow gully with different conditions of compactness. *J. Nat. Disaster* **2019**, *28*, 143–151.
18. Guo, X.; Ge, Y.; Zhan, M.; Yan, H. Failure process of a lateral slope deposit and its effect on debris flood formation. *Bull. Eng. Geol. Environ.* **2022**, *81*, 324. [CrossRef]
19. Ge, Y.; Guo, X.; Zhan, M.; Yan, H.; Liao, Y.; Yu, B. Erosion Rate of Lateral Slope Deposit Under the Effects of Different Influencing Factors. *Front. Earth Sci.* **2021**, *9*, 670087. [CrossRef]
20. Chakrabortty, R.; Pal, S.C.; Santosh, M.; Roy, P.; Chowdhuri, I. Gully erosion and climate induced chemical weathering for vulnerability assessment in sub-tropical environment. *Geomorphology* **2022**, *398*, 108027. [CrossRef]
21. Chowdhuri, I.; Pal, S.C.; Saha, A.; Chakrabortty, R.; Roy, P. Evaluation of different DEMs for gully erosion susceptibility mapping using in-situ field measurement and validation. *Ecol. Inform.* **2021**, *65*, 101425. [CrossRef]
22. Roy, P.; Pal, S.C.; Arabameri, A.; Chakrabortty, R.; Pradhan, B.; Chowdhuri, I.; Lee, S.; Bui, D.T. Novel Ensemble of Multivariate Adaptive Regression Spline with Spatial Logistic Regression and Boosted Regression Tree for Gully Erosion Susceptibility. *Remote. Sens.* **2020**, *12*, 3284. [CrossRef]
23. Zhu, X. Study on erosion characteristics and routing of debris flow along the channel. Ph.D. Thesis, Graduate School of Chinese Academy of Sciences, Beijing, China, 2013.
24. Qu, Y. The research on dynamic characteristics of urgent steep-channel debris flows in meizoseismal area. Ph.D. Thesis, Chengdu University of Technology, Chengdu, China, 2016.
25. Zhao, Y.; You, Y.; Liu, J.; Lin, X. Experimental study on scouring depth of cohesive debris flow gully bed. *Shui Li Xue Bao* **2012**, *43*, 92–97.

26. Le, M. *Dynamic Mechanism of Gully-type Debris Flow and Its Numerical Simulation*; China Institute of Water Resources and Hydropower Research: Beijing, China, 2019.
27. Li, T.; Fuller, T.K.; Sklar, L.S.; Gran, K.B.; Venditti, J.G. A Mechanistic Model for Lateral Erosion of Bedrock Channel Banks by Bedload Particle Impacts. *J. Geophys. Res. Earth Surf.* **2020**, *125*, e2019JF005509. [CrossRef]
28. Ji, J. Study on the identification method and scour pattern of narrow and steep gully type debris flow. Master's Thesis, Sichuan University, Chengdu, China, 2019.
29. Zhao, B.; Yu, B.; Chang, M.; Yang, L. Characterization of narrow and steep debris flow gully. *J. Sediment Res.* **2021**, *46*, 61–67.
30. Ruidas, D.; Pal, S.C.; Islam, A.R.M.T.; Saha, A. Characterization of groundwater potential zones in water-scarce hardrock regions using data driven model. *Environ. Earth Sci.* **2021**, *80*, 24. [CrossRef]
31. Ruidas, D.; Pal, S.C.; Saha, A.; Chowdhuri, I.; Shit, M. Hydrogeochemical characterization based water resources vulnerability assessment in India's first Ramsar site of Chilka lake. *Mar. Pollut. Bull.* **2022**, *184*, 114107. [CrossRef]
32. Ruidas, D.; Pal, S.C.; Islam, A.R.M.T.; Saha, A. Hydrogeochemical Evaluation of Groundwater Aquifers and Associated Health Hazard Risk Mapping Using Ensemble Data Driven Model in a Water Scares Plateau Region of Eastern India. *Expo. Heal.* **2022**, 1–19. [CrossRef]
33. Pal, S.C.; Ruidas, D.; Saha, A.; Islam, A.R.M.T.; Chowdhuri, I. Application of novel data-mining technique based nitrate concentration susceptibility prediction approach for coastal aquifers in India. *J. Clean. Prod.* **2022**, *346*, 131205. [CrossRef]

*water* MDPI

*Article*

# Impacts of Grass Coverage and Arrangement Patterns on Runoff and Sediment Yield in Slope-Gully System of the Loess Plateau, China

**Wenfeng Ding [1,2,*], Xiekang Wang [3], Guanhua Zhang [1,2], Xi Meng [1,2] and Zhiwei Ye [1,2]**

1 Changjiang River Scientific Research Institute, Changjiang Water Resource Commission, Wuhan 430010, China
2 Research Center on Mountain Torrent Geological Disaster Prevention of Ministry of Water Resources, Wuhan 430010, China
3 State Key Laboratory of Hydraulics and Mountain River Engineering, Sichuan University, Chengdu 610065, China
* Correspondence: dingwf@mail.crsri.cn

**Abstract:** Both vegetation coverage rates and arrangement patterns have important influences on erosion. Very little previous research focuses on the impacts of spatial vegetation distribution patterns on erosion. The slope-gully system was taken as the research object, which is composed of a 5.0 m long hillslope with a slope gradient of 20° and a 3.0 m long gully slope with a gradient of 50°. A series of scouring experiments with two inflow discharges (3.2 L $min^{-1}$, 5.2 L $min^{-1}$) was carried out. The effects of the flow discharges, spatial grass arrangement patterns (US, MS, and DS represent the presence of grass covering on up-hillslope, middle-hillslope, and down-hillslope, respectively) and grass coverage rates (0%, 30%, 50%, 70%, and 90%) on runoff and sediment were studied in this paper. The results indicated that either runoff or sediment yielding was significantly decreased with the grass coverage rates increasing and with the variation of grass arrangement patterns on a hillslope. While grass coverage had more effectiveness in controlling erosion compared with runoff reduction, and DS can control erosion more effectively than US and MS erosion controlling. For the gully slope, erosion significantly increased with the grass coverage rates increasing no matter how the grass arrangement patterns on the hillslope. Therefore, both different grass coverage and different grass arrangement patterns have an influence on erosion processes; any research that only takes care of the single factor mentioned above is not enough to reveal the effects of grass on erosion. In the process of erosion control in the Loess Plateau, taking effective measures both on the hillslope and gully slope will be effective methods of reducing soil erosion.

**Keywords:** grass coverage rate; grass spatial arrangement patterns; slope-gully system; erosion

**Citation:** Ding, W.; Wang, X.; Zhang, G.; Meng, X.; Ye, Z. Impacts of Grass Coverage and Arrangement Patterns on Runoff and Sediment Yield in Slope-Gully System of the Loess Plateau, China. *Water* **2023**, *15*, 133. https://doi.org/10.3390/w15010133

Academic Editors: Achim A. Beylich and Bommanna Krishnappan

Received: 20 November 2022
Revised: 27 December 2022
Accepted: 28 December 2022
Published: 30 December 2022



## 1. Introduction

Soil erosion is a worldwide serious environmental problem for most farming lands in the world. It not only threatens the future development of agriculture and society but can also cause soil quality to decrease and land productivity to decline [1,2]. Particularly in the Loess Plateau of China, erosion amounts have increased with the loss of vegetation cover, where the area of soil loss reaches $43 \times 10^4$ $km^2$, with an average annual soil loss of 3720 t $km^{-2}$, at higher elevations. The amount of erosion has been reported to exceed even 10,000 t $km^{-2}$ [3,4]. How to recover vegetation to effectively reduce soil loss has become an essential environmental issue and has attracted increasing attention over the recent several decades.

Vegetation coverage has an important role in effectively controlling soil erosion by runoff and improving the ecological environment, as has been demonstrated by many researchers [5–10]. The importance of vegetative coverage in reducing soil erosion has also been reported in a variety of literature [11,12]. With the Universal Soil Loss Equation (USLE), soil loss rates are hypothesized to be the highest on bare soil and to decrease linearly with the increase in soil surface coverage [13]. However, Rogers and Schumm reported that the relationship between vegetation coverage and erosion varied considerably rather than with a simple linear relationship, particularly for low-cover conditions [14]. Cerda also indicated that exponential functions existed between erosion levels and that linear models are not generally applicable through either field or laboratory experimental studies [15].

Different parts of vegetation (canopy, understory, and plant roots) typically have differing roles in regulating the surface hydrological processes, a factor that results in the variation of the erosive forces of rainfall [16]. In order to reveal the mechanism of the vegetation reducing water erosion, much research works has been conducted to separate the effects of the aboveground biomass or plant canopy from the belowground biomass in runoff and sediment flux reduction [17–19]. In addition, the fact that the spatial distribution or patterns of vegetation have the differing erosion reduction roles has come to be realized in many parts of the world in recent years [15]. A few studies have revealed that different spatial vegetation arrangements on slope divide the slope into runon and runoff areas, and then into erosion and deposition areas acting as sediment sources and sinks, respectively [20]. Neibling and Alberts [21] reported that grassland buffer zone decreased more than 90% sediment yielding on 5.0 m long slope, and over 91% of the sediment was deposited at 0.6 m grass buffer area from top of the slope. Dillaha [22] compared the effects of different slope gradients on sediment reduction while keeping other factors constant, and found that the slope gradient was inversely proportional to the sediment capture. Runoff sediment concentration flowing out of vegetation buffer zone is a function of runoff sediment concentration flowing into vegetation buffer strip and buffer strip width. All these studies demonstrated that different vegetation coverage and different vegetation spatial arrangement are effective for sediment decreasing, also provided the basic theory to reveal the mechanism of vegetation reducing soil erosion.

Although there has been much research on vegetation reducing erosion, impacts of different vegetation coverage and different vegetation arrangement patterns on the slope-gully system has rarely been reported. Due to the particularity of the Loess Plateau, from the up-slope boundary to the gully edges, both erosion forms and intensity have shown significant vertical zonation. Significant impacts of runoff from upper slope on down-slope sediment transport process [23–25]. Therefore, some researchers have recognized that hillslope and gully slope are the basic topography units of Loess Plateau, any research including alone hillslope or gully slope cannot response the actual erosion process of the entire erosion system [26–28]. In addition, since the middle of the last century, two crucial problems have aroused controversy in many scholars. There are two main points of contention. One is which section in the slope-gully system is the key erosion controlling area, if only restore hillslope vegetation can achieve the goal of soil erosion controlling effective? The other one is which of the two parts, the hillsolpe or gully slope is the main source of sediment yield? Both of the two controversies indicate that the related research is still lacking. So, the objective of present study are: (i) to assess the impact of grass coverage degrees and spatial arrangements on runoff and sediment yield both on hillslope and gully slope; (ii) to analyze the relative contribution of hillslope and gully side account for the total erosion on entire hillslope-gully system; (iii) to understand the validity of the policy on soil and water conservations in Loess Plateau.

## 2. Materials and Methods

### 2.1. Soil Sample Collection

The soil used in this study was collected from Zhengzhou, Henan province, China. The content of clay, silt and sand is 10.8%, 28.16%, and 61.04%, respectively. The soil is classified as Alfisol according to the U.S. Soil Taxonomy. The pH value of soil is 7.2. The natural consolidated soil has a bulk density about 1.3 g $cm^{-3}$ and with an organic matter content of 1.93%. The soil texture information is listed in Table 1. The soil was taken from a cultivated land with a depth of 0–0.3 m, and the soil was evenly taken down along the top of the slope. Sufficient amount of soil was transported back to the laboratory for experiments. All the soil used in study was air-dried.

**Table 1.** Soil basic information in this study.

| Soil Type | Soil Texture | Soil Particle-Size Distribution (g kg$^{-1}$) | | | pH ($H_2O$) | OM (%) |
|---|---|---|---|---|---|---|
| | | Sand | Silt | Clay | | |
| Loessic soil | Sandy loam | 61.04 ± 7.14 | 28.16 ± 4.20 | 10.80 ± 3.90 | 7.2 ± 0.26 | 1.93 ± 1.23 |

Values represent means ± SD (standard deviation).

## 2.2. Experimental Setup

According to earlier investigation and statistics results gathered by Jiang et al. [29], from the top of the hillslope to the edge of the gully and the bottom of the gully, both the erosion forms and the intensity demonstrate significant vertical zonation. Most of the hillslope gradients are about 20°, while the slope gradients of the gully are dominantly between 40° and 60° [29]. The Chinese Soil and Water Conservation Law specified 25° as the maximum tillage gradient for prohibiting farming. The slope gradient of 20° is almost the maximum slope gradient for cultivation. Take the above factors into consideration, the hillslope gradient was set at 20°, which is also a general gradient for cultivated land on the study area. The gully slope gradient designed in this study was 50 °. According to the previous research, the length ratio of the hillslope to the gully slope was from 1.4 to 2.0, and the intermediate value of 1.67 was taken in this study. Therefore, the length of hillslope was 5.0 m, and the gully slope length was 3.0 m. The horizontal projection length of the hillslope and gully side are 4.7 m and 1.93 m, respectively (Figure 1).

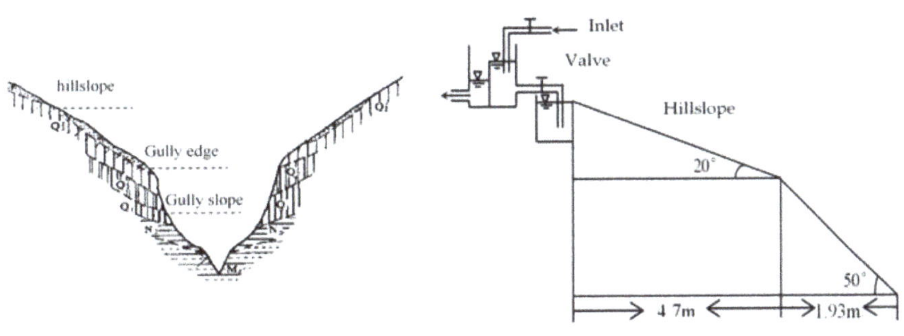


(A) (B)

**Figure 1.** Schematic diagram of the hillslope-gully system of the Loess Plateau (**A**) and the soil box designing diagram (**B**).

The runoff scouring experiment was applied in this study. The constant water head is used to control the inflow discharge. Before each test, the inflow discharges were calibrated at the outlet of the soil box by using a flow meter. The two inflow discharges of 3.2 and 5.2 L min$^{-1}$ were used in the present experiments.

## 2.3. Experiment Preparation and Procedure

The soil box was 2.0 m width and 0.5 m depth. In order to ensure the consistency of the experimental conditions, the 2.0 m wide soil box is divided into four plots with the same width of 0.5 m by PVC board. Therefore, a control plot with no grass cover (CK), and three additional plots were constructed in order to stand for differing grass arrangement patterns on slope, which were then considered as the: up-slope (US), middle-slope (MS) and down-slope (DS), respectively (Figure 2).

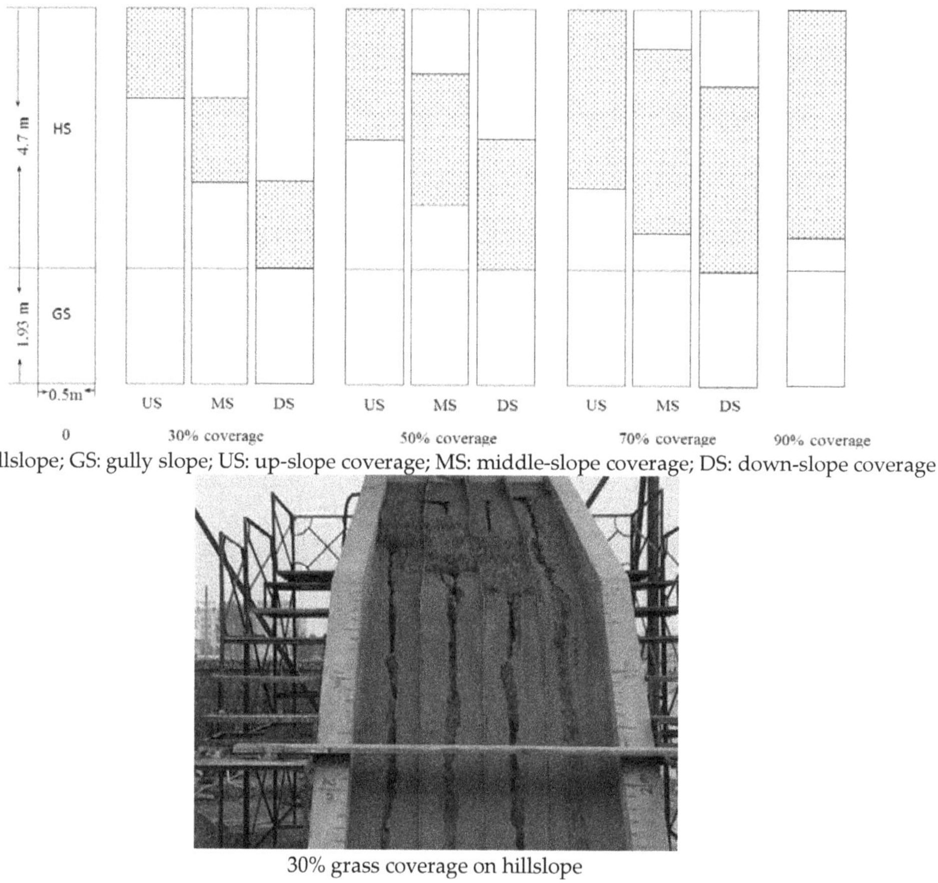


**Figure 2.** Schematic diagram of different grass coverage rates and different grass spatial arrangement patterns on soil box.

Before experiments, all the soil used in the study passed through a 10 mm sieve and removed gravel, animal, and plant residues in order to ensure homogeneity. A layer of fine sand 0.1 m thick was laid at the bottom of the plot to simulate drainage conditions. The soil was put in successive layers of 0.1 m thickness, a total thickness of 0.3 m, and a bulk density of 1.35 g cm$^{-3}$ was packed. Wild buffalograss (*Buchloe dactyloides*), which is a kind of native grass in the Loess Plateau, was selected as the target species. Sow the grass seeds into the plot to ensure that each plot's soil surface is covered with uniform grass. Each experimental plot used a similar planting density. The grass coverage rate is calculated by determining the amount of grass area, which accounts for the total hillslope surface area. It has been widely suggested that the critical coverage rate for vegetation affecting soil erosion is about 50% (Zhang et al., 2012). In order to minimize the number of experiments while ensuring that there are significant differences in erosion between different treatments, a 20% grass coverage rate interval was chosen. Consequently, experimental treatments that designed based on the above rules, while different grass arrangement patterns were designed in this study. In the present research, a total of 5 kinds of grass coverage rates were set, which were 0, 30%, 50%, 70%, and 90%, respectively. Except for 0% and 90% grass coverage rates, there were three spatial grass arrangement patterns, as mentioned above, on the hillslope of each plot, while there was no grass cover set up on the gully slope for each treatment. Two days before the experiment, each plot soil surface was pre-wetted uniformly by applying 20 mm/h rainfall intensity for 30 min to ensure the same soil water content. Therefore, we could keep the initial condition of every experiment as consistent as possible. The plot soil was repacked with new soil for the next experiment, and twice experiments were conducted for each inflow discharge that was applied for approximately half an hour.

For each scouring experiment, clear water was imported from the upper end of the plot. Considering that there have been many studies on light rain and moderate rain conditions in the past, this study focuses on erosion reduction under the slope grass arrangement for heavy rain conditions. Inflow discharge of 3.2 and 5.2 L·min$^{-1}$ was selected. The inflow discharges correspond to the farmland runoff generated under the typical local rainstorms with rainfall intensities of 100 mm h$^{-1}$ and 150 mm h$^{-1}$. The experiments continued for about 20 min. During the experiment, runoff and sediment samples were collected continuously every minute with a 10 litters bucket, and the flow velocities were measured by using the dye tracing method. The time for the tracer traveling via a fixed distance (2.0 m) was recorded according to the color-front propagation. The measured time values mentioned above, multiplied by the theoretical value of 0.67, were used to calculate the runoff mean velocities. After the experiment, the sediment in sampling buckets was transferred to iron boxes and oven-dried at 105 °C for 24 h until a constant mass was achieved and weighed. To acquire the gully slope erosion, the rill length, rill width, and rill depth of each rill in the gully slope were measured by steel rule at intervals of 10 cm along the latitude of the gully slope.

### 2.4. Data Calculation and Analysis

The sediment reduction due to grass coverage rate (%) could be calculated using the following equation [12]:

$$E_s = \frac{S_{ck} - S_g}{S_{ck}} \times 100\% \quad (1)$$

where $S_{ck}$ is the sediment yield in the bared plot (Kg), and $S_g$ is the sediment yield in the plot of different grass coverage rates and different grass arrangement patterns on the hillslope (Kg).

For runoff reduction due to grass coverage rate (%) calculations [12], the equation was

$$E_r = \frac{R_{ck} - R_g}{R_{ck}} \times 100\% \quad (2)$$

where $R_{ck}$ is the runoff generation in the bared plot (L), and $R_g$ is the runoff generation in the plot of different grass coverage rates and different grass arrangement patterns with the hillslope condition (L).

According to the measurement of rill length, width, and depth, and taking the rill section to be rectangular, the rill volume and rill erosion can be calculated with the following equation:

$$M_i = \sum_{i=1}^{n} \overline{W_i} \times \overline{H_i} \times l \times \rho \quad (3)$$

where $M_i$ is the erosion of any selected rill segment on the gully slope (Kg); $\overline{W_i}$ is the average width of any selected rill segment on the gully slope (m); $\overline{H_i}$ is the average depth of any selected rill segment on the gully slope (m); l is the actual selected rill length (m); $\rho$ is the soil bulk density of gully slope (Kg·m$^{-3}$).

Analysis of variance (ANOVA) was used to detect treatment effects on measured variables. Significant differences between treatments for runoff and soil loss rate were determined using the PLSD (Protected Least Significant Difference) procedure for a multiple range test at the 0.05 significance level. All tests were performed using the statistical program SPSS 17.0.

## 3. Results

### *3.1. Runoff and Sediment Reduction*

Both the runoff and sediment yield of the slope-gully system under the different grass coverage rates and different grass arrangement patterns were summarized in Table 2 and Figure 3.

**Table 2.** Runoff and sediment yielding on the slope-gully system under the different grass coverage and flow discharges.

| Flow Discharge ($L{\cdot}min^{-1}$) | Grass Coverage (%) | Runoff (L) | Erosion (kg) |
|---|---|---|---|
| | 0 | 68.17 a | 26.80 a |
| | 30 | 66.08 b | 23.76 b |
| 3.2 | 50 | 63.62 c | 23.10 c |
| | 70 | 58.89 d | 22.81 d |
| | 90 | 43.08 e | 16.46 e |
| | 0 | 115.24 a | 58.64 a |
| | 30 | 114.39 a | 52.01 b |
| 5.2 | 50 | 109.23 ab | 39.77 c |
| | 70 | 111.37 b | 35.83 d |
| | 90 | 103.30 c | 33.02 d |

Mean values of the same letter in the same columns are not significantly different at $p = 0.05$ level using the least significant difference method.

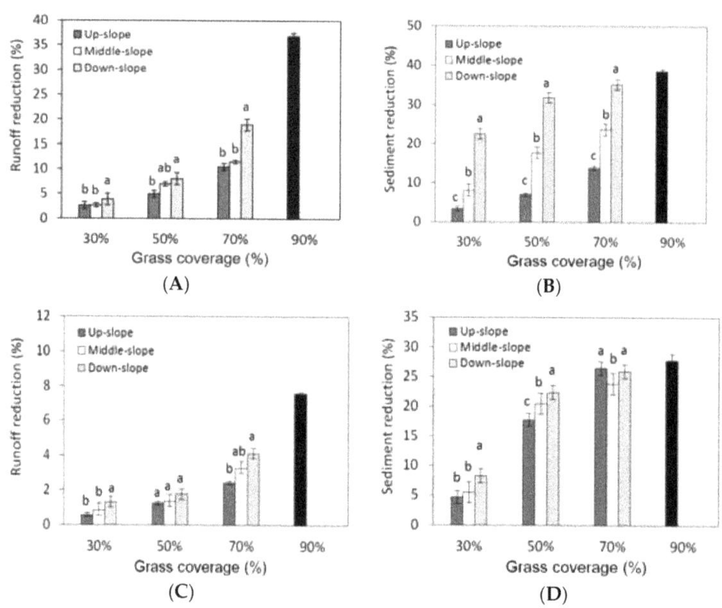


**Figure 3.** Runoff and sediment reduction of the slope-gully system under the different grass coverage, different grass spatial arrangement patterns, and different flow discharges (Mean values of the same letter in different columns are not significantly different at $p = 0.05$ level using the least significant difference method). (**A**) ($Q = 3.2\ L{\cdot}min^{-1}$). (**B**) ($Q = 3.2\ L{\cdot}min^{-1}$). (**C**) ($Q = 5.2\ L{\cdot}min^{-1}$). (**D**) ($Q = 5.2\ L{\cdot}min^{-1}$).

As can be observed in Table 1, either the runoff volume or the erosion mass was significant decreased with the grass coverage rates increasing and with the variation of grass distribution patterns on slope, indicating that both the grass coverage rates and the grass arrangement patterns have significant effect on runoff and sediment yield on slope-gully system. Runoff volume ranged from 43.08 to 68.17 L and erosion mass ranged from 16.46 to 26.80 Kg with the grass coverage rate decreasing from 90% to 0%, and runoff volume were reduced by approximately 3% to 36.8% and erosion mass were reduced by approximately 27.8% to 38.6% compared with the bare plot under the flow discharge 3.2 L·min$^{-1}$ condition, respectively (Table 1, Figure 1).

When the flow rate was 5.2 L min$^{-1}$, the changing trend of runoff and erosion with the vegetation coverage rate from 90% to 0% was the same, but the role of reducing runoff and sediment by grass coverage rate was smaller than that under the flow discharge of 3.2 L·min$^{-1}$. These indicated that grass was a very important factor in controlling soil erosion, for it could be observed that the runoff decreasing was lower than the rates of sediment reduction, indicating that the grass coverage had higher effectiveness in controlling soil erosion as compared to reducing runoff. As the grass coverage rate increased, sediment yields and runoff decreased dramatically. These results are consistent with the findings of other researchers who similarly found that vegetative cover significantly reduced runoff and sediment losses [11,30,31].

By comparing the rates of runoff and sediment reduction under the differing vegetation arrangement patterns on the slopes, it can be observed that the down-slope ($D_S$) had the lowest runoff and greatest runoff reduction in comparison with the up-slope ($U_S$) and middle-slope ($M_S$). However, no significant differences in runoff reduction were detected between the $U_S$ and $M_S$, although there exists a significant difference in the sediment reduction between the $U_S$ and $M_S$. In essence, these results indicated that the $D_S$ is more effective than either the $U_S$ or $M_S$ in both soil and water conversation under these established experimental conditions.

### *3.2. Contributions of Hillslope and Gully Slope to Slope-Gully System*

According to the results gathered by means of Equation (3), the erosion of the gully slope was calculated (Table 3). As the information in Table 3 indicates, there exists quite a difference between the flow discharge 3.2 L·min$^{-1}$ and 5.2 L·min$^{-1}$ for the gully slope erosion, as well as a large amount of additional erosion occurring on the gully slope with the flow discharge 3.2 L·min$^{-1}$, while only a small amount of additional erosion occurs on the gully slope with the flow discharge 5.2 L·min$^{-1}$. Under the flow discharge 3.2 L·min$^{-1}$ conditions, the erosion mass of the gully slope has a range from 1.89 Kg (with no grass coverage on the hillslope) to 10.53 Kg (90% grass coverage on the hillslope), demonstrating a significant increase with the grass coverage rate increasing. In opposition, for the flow discharge 5.2 L·min$^{-1}$, the erosion mass of the gully slope ranged from 16.26 Kg (with no grass coverage on the hillslope) to 23.93 Kg (30% grass coverage on the hillslope) and then fell to 17.95 Kg (90% grass coverage on the hillslope), demonstrating a significant decrease with the grass coverage rate increasing.

The grass arrangement patterns on hillslopes have a significant influence on gully slope erosion. By comparing the erosion of the gully slope under the different grass arrangement patterns on the hillslope, it is easy to see that the erosion from the gully slope for the $D_S$ condition had the lowest value in comparison with $U_S$ and $M_S$ condition under the flow discharge 3.2 L·min$^{-1}$, while the erosion from the gully slope for $D_S$, $M_S$ and $U_S$ condition has no obvious variation. The results above indicated that flow discharges, grass coverage rates, and grass arrangement patterns all affected the gully slope erosion. This supported Chen's results, which showed that the runoff discharge and sediment concentration from the upper hillslope are the important factors impacting the sediment yielding on the gully slope [24].

**Table 3.** Gully slope erosion under the different flow discharges, different grass coverage rates, and different grass arrangement patterns.

| VCR (%) | FD (L·min$^{-1}$) | VDP | Erosion (kg) | VCR (%) | FD (L·min$^{-1}$) | VDP | Erosion (kg) |
|---|---|---|---|---|---|---|---|
| 0 | 3.2 | | 1.89 | 0 | 5.2 | | 16.26 |
| | | Up-slope | 4.99 b | | | Up-slope | 23.93 b |
| 30 | 3.2 | Middle-slope | 6.98 a | 30 | 5.2 | Middle-slope | 23.79 b |
| | | Down-slope | 6.87 a | | | Down-slope | 26.7 a |
| | | Up-slope | 10.23 a | | | Up-slope | 21.25 a |
| 50 | 3.2 | Middle-slope | 11.93 a | 50 | 5.2 | Middle-slope | 18.6 b |
| | | Down-slope | 6.51 b | | | Down-slope | 20.4 a |
| | | Up-slope | 12.02 a | | | Up-slope | 22.99 a |
| 70 | 3.2 | Middle-slope | 12.5 a | 70 | 5.2 | Middle-slope | 19.36 b |
| | | Down-slope | 10.01 b | | | Down-slope | 19.08 b |
| 90 | 3.2 | | 10.53 | 90 | 5.2 | | 22.95 |

Mean values of the same letter in the same columns are not significantly different at $p = 0.05$. VCR: vegetation coverage rate; FD: flow discharge; VDP: vegetation distribution pattern.

The percentage of the gully slope erosion accounts for the total erosion of the slope-gully system ranged from 7% to 64% under the flow discharge 3.2 L·min$^{-1}$ and from 28% to 60% under the flow discharge 5.2 L·min$^{-1}$, respectively (Figure 4). This indicated that flow discharge has an important effect on the relationship between hillslope erosion and gully slope erosion. Tang et al., Gong and Jiang, and Chen also researched slope-gully system erosion processes and analyzed the percentage of gully slope erosion accounting for the total erosion of the slope-gully system by using field plot measuring data, which indicated that the percentage of the gully slope erosion ranged between 32% and 80% [24,32,33]. Jiang further pointed out that the gully side area accounts for 53% of the total area condition. In the case of a rainstorm, the sediment yield from the gully slope evenly accounts for 62% of the total erosion [29]. Especially for high flow years, the ratio of gully side erosion accounted for the erosion of slope-gully system was 63%. In the present research, the gully side area accounts for about 40% of the total area of the slope-gully erosion system, and the average erosion of the gully side account for 40% of the total erosion for the flow discharge of 3.2 L·min$^{-1}$ and 50% for the flow discharge of 5.2 L·min$^{-1}$, respectively. The reason for the difference between the present study and Jiang's research is that the present study was studied by scouring experiments. Under the natural rainfall condition, runoff actually increases with the increase in the down-hillslope position. The runoff detachment ability and soil loss on the gully slope will be greater than the above ratio. This indicates that further research is needed through rainfall experiments to accurately determine the location of severe erosion and the corresponding gull slope erosion percentage [26].

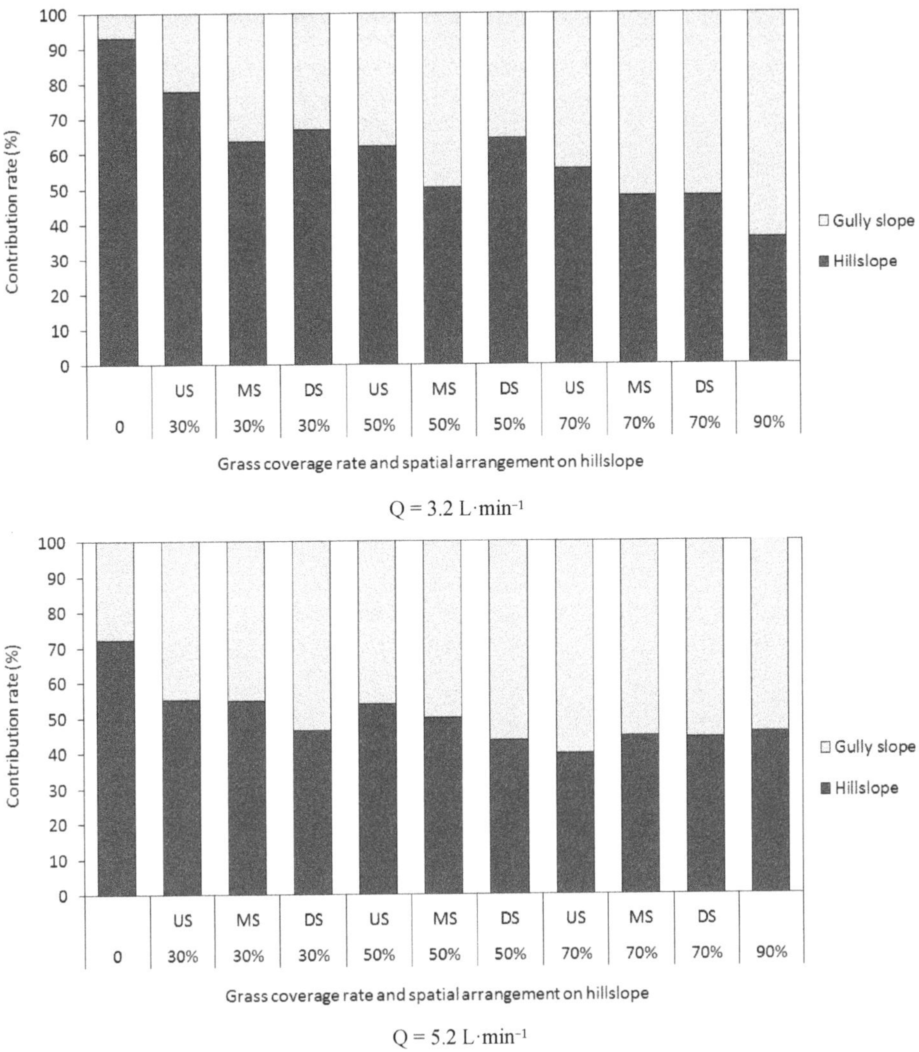


**Figure 4.** Contributions of the hillslope and the gully side to erosion reductions of the slope-gully erosion system.

## 4. Discussion

### *4.1. Effects of Grass Coverage and Arrangements on Runoff and Sediment Reduction on Hillslope*

The data of this study showed that the runoff and sediment yield on hillslopes are affected by slope grass coverage rates, grass arrangement patterns, and flow discharges. The flow discharges have a significant effect on the runoff and sediment yield, and the runoff and sediment yield increases with the increase in flow discharges. The effects of grass coverage rates on runoff were significantly different between 0% grass coverage rates and 70% and 90% grass coverage rates, while sediment yield was a significant difference under various grass coverage rates. The effects of grass arrangement patterns on runoff were not significant, but the effects on sediment yield were significant. The total runoff volumes have no obviously different among the three different grass arrangement patterns on the slope under the same grass coverage rates and the same flow discharges, but the sediment yield has distinct discrepancy among the three different grass arrangement patterns on the slope. The sediment yield on a hillslope with different grass arrangement patterns showed a trend of US > MS > DS. This is related to the fact that the greater the flow rate is, the smaller the change of the hydrodynamic parameters of the grass is, and the hydrodynamic parameters are an important factor in determining the erosion sediment yield. Therefore, the greater the flow rate is, the less the sediment reduction benefit is. Different grass coverage rates, the same grass arrangement patterns on hillslope sediment yield decrease with the increase in grass coverage rates.

In the last several decades, considerable studies have been conducted to evaluate the influence of vegetative coverage on soil erosion. The results suggested that soil erosion decreased with the vegetation coverage rates increasing [34,35]. In contrast, some other researchers indicated that soil erosion increased with the soil surface coverage rate increasing [36,37]. The above two contradictory results demonstrate that only researching the effect of vegetation coverage on erosion is not enough. The vegetation distribution patterns on the slope must be considered to reveal the relationship between the vegetation cover and erosion. In fact, a few studies have reported that the vegetation-patched patterns on the slope have a significant effect on erosion [15,20,38–40]. In our research, three factors (grass coverage, grass distribution patterns on slope, and flow discharges) were considered. The results showed that both grass coverage rates and grass arrangement patterns have significant influences on runoff and sediment yield on a hillslope. Compared with the effect of runoff reduction, grass coverage has better control affection on soil erosion. These results are consistent with previous researchers' results that vegetative cover significantly reduced runoff and sediment losses [11,15,30,31].

### 4.2. Effects of Grass Coverage and Arrangements on Runoff and Sediment Reduction on Gull Slope

The results indicate that the influence of the hillslope grass coverage rates and arrangement patterns on the gully slope soil loss is greater than on runoff (Table 4). For the same grass coverage rate and the same grass arrangement patterns, the sediment yield of the gully slope increases with the flow discharges increasing. Under the same flow discharges, the sediment yield of the gully slope did not decrease with the increase in the grass coverage rates but showed an increasing trend. The erosion of the gully slope under the condition of a 90% grass coverage rate is larger than that of the gully slope without grass coverage.

**Table 4.** Runoff and sediment yielding of hillslope under the different grass coverage and flow discharges.

| FD ($L\cdot min^{-1}$) | VCR (%) | VDP | Runoff (L) | Erosion (kg) | FD ($L\cdot min^{-1}$) | VCR (%) | VDP | Runoff (L) | Erosion (kg) |
|---|---|---|---|---|---|---|---|---|---|
| | 0 | - | 68.79 | 24.91 | | 0 | - | 115.05 | 42.38 |
| | | US | 67.83 a | 20.12 a | | | US | 114.41 a | 35.25 a |
| | 30 | MS | 67.56 a | 15.67 b | | 30 | MS | 113.50 a | 29.03 b |
| | | DS | 65.21 a | 13.87 c | | | DS | 111.13 a | 24.30 c |
| | | US | 61.64 a | 16.72 a | | | US | 106.87 a | 31.75 a |
| 3.2 | 50 | MS | 59.78 a | 12.12 b | 5.2 | 50 | MS | 103.76 a | 25.52 b |
| | | DS | 59.56 a | 11.77 b | | | DS | 103.71 a | 23.79 c |
| | | US | 56.85 a | 15.03 a | | | US | 111.58 a | 29.24 a |
| | 70 | MS | 58.82 a | 11.49 b | | 70 | MS | 108.19 a | 24.73 b |
| | | DS | 58.13 a | 9.37 c | | | DS | 107.94 a | 21.08 c |
| | 90 | - | 56.98 | 5.93 | | 90 | - | 90.81 a | 15.07 |

Mean values of the same letter in the same columns are not significantly different at $p = 0.05$. VCR: vegetation coverage rate; FD: flow discharge; VDP: vegetation distribution pattern.

Our results are contrary to those observed on vegetation plots [15,17]. These reported that sediment yield was negatively correlated with runoff rate in grassplots, and the runoff coefficient was negatively related to sediment concentrations on a Mediterranean hillslope with vegetation. The main reason for these differences is that the present study was conducted by using scouring experiments, while simulated rainfall experiments were used in both Pan's and Cerda's research. For the natural situation, when the rainfall experiment is used to study slope erosion, the runoff on the slope increases gradually along the downhill slope, and the runoff detachment capacity and erosion amount at the lower part of the slope also increases gradually. In the scouring experiment, the runoff on the slope decreased gradually along the slope due to infiltration, and the runoff detachment capacity and erosion amount in the lower part of the slope was less than those in the simulated rainfall experiment. Another reason for the difference between this study and Pan's study is that the grass distribution patterns used in this study are different from Pan's study. This study used different grass coverage rates and grass distribution patterns, while Pan's study used uniform grass coverage. This indicates that both vegetation coverage and distribution patterns have significant effects on erosion. However, further rainfall experiments are needed to determine the most accurate impact of grass coverage and distribution patterns on erosion.

While for the gully slope of the whole slope-gully system, erosion significantly increased with the grass coverage rate increasing on the slope no matter how the grass arrangement patterns on the hillslope. These results are consistent with some previous research results that showed that soil erosion rates increased with soil surface coverage increasing [37,38]. This can be explained as follow: due to the particularity of the Loess Plateau, from up-slope to hillslope boundary, then to gully edges, erosion forms and intensity showed significant vertical zonation. Many researchers reported that runoff from upper slopes has an important effect on erosion and sediment transport processes on the down-slope. An increase in runoff sediment concentration from uphill leads to a decrease in erosion in the down-slope [24], and an increase in runoff from the up-slope leads to an increase in erosion in the down-slope.

Table 5 shows the relationship between the gully side erosion and various influencing factors. From Table 5, we can see that three factors (flow discharge, runoff sediment concentration from upper slopes, and runoff unit stream power from upper slope) significantly correlate with the gully side erosion ($p < 0.05$), but the relationship between the grass coverage rate and gully side erosion appeared has no obvious correlation. Due to the sediment concentration of runoff on the hillslope being inversely proportional to the hillslope grass coverage, sediment concentration on the end of the hillslope decreased with the grass coverage rate increasing. That is to say, the sediment concentration of inlet flow decreases with the grass coverage rate increasing on the hillslope. Under this situation, the greater difference between the runoff sediment concentration and the sediment transport capacity, the more detachment of the runoff and the erosion on the gully side is greater. The above results showed that the spatial scale of the research object must also be considered in studying the effect of grass coverage rate and the spatial distribution pattern on erosion. The above-mentioned previous research was performed on a uniform slope, and the slope gradient is gentle. Our present experiments were conducted on very steep slope gradients, and vegetation coverage rate variation widely (from 0% to 90%), and the most important difference is that present experiments were conducted on a complicated hillslope-gully system. The gully side erosion has no apparent decrease with the increasing of the vegetation coverage rate on the hillslope. On the contrary, the gully side erosion increased with the increasing vegetation coverage rate on the hillslope. So, the different vegetation coverage, different vegetation distribution pattern, and different spatial scale of the research object all have an influence on erosion; any research that only takes care of the single factor mentioned above is not enough to reveal the effect of vegetation on erosion.

**Table 5.** The correlations between the gully slope erosion and effect factors.

| | Gully Erosion | Unit Stream Power | Flow Discharge | Sediment Concentration | Grass Coverage Rate on Slope |
|---|---|---|---|---|---|
| Gully erosion | 1 | 0.799 * | 0.91 * | −0.875 * | 0.447 |
| Unit stream power | | 1 | 0.841 * | −0.825 * | −0.647 |
| Flow discharge | | | 1 | −0.913 * | 0.314 |
| Sediment concentration | | | | 1 | 0.719 * |
| Grass coverage rate on slope | | | | | 1 |

* Significant at 0.05 level.

Further analysis can be concluded that just implementing soil and water conservation measures on the hillslope of the slope-gully system can reduce the total erosion to some extent, but, reducing erosion is mainly focused on hillslope sections, for gully side, its erosion not only does not reduced but also has a greater amount of increasing. Therefore, in the process of practical soil erosion controlling, taking effective measures both on the hillslope and gully side will be effective methods of reducing soil erosion.

## 5. Conclusions

The runoff and sediment yield of the hillslope-gully system with five grass coverage rates and three arrangement patterns were studied by scouring experiments with two runoff discharges. Results showed that both grass coverage rates and grass distribution patterns have significant effects on runoff and sediment yield on slope-gully erosion systems. Grass coverage had more effectiveness in controlling soil erosion compared with runoff-reducing effectiveness. However, for the gully side of the slope-gully erosion system, erosion significantly increased with the grass coverage rate increasing on the slope, no matter how the grass distribution pattern on the hillslope. Therefore, the different grass coverage, different grass distribution pattern, and different spatial scale of the research object all have an influence on erosion; any research that only takes care of the single factor mentioned above is not enough to reveal the effect of grass on erosion.

Gully side erosion was significantly correlated with the flow discharge, runoff sediment concentration from upper slopes, and runoff unit stream power from the upper slope ($p < 0.01$), but the relationship between the grass coverage rate and gully side erosion appeared to have no obvious correlation. Just implementing soil and water conservation measures on the hillslope of the slope-gully system can reduce the total erosion to some extent, but reducing erosion is mainly focused on hillslope sections. Gully side erosion not only does not reduce but also has a greater amount of increase. Therefore, in the process of practical soil erosion controlling, taking effective measures both on the hillslope and gully side will be effective methods of reducing soil erosion.

**Author Contributions:** Conceptualization, W.D.; methodology, W.D.; formal analysis, W.D., X.W., X.M., and Z.Y.; data curation, G.Z.; writing—original draft preparation, W.D., X.W., and G.Z.; writing—review and editing, W.D. and X.W. All authors have read and agreed to the published version of the manuscript.

**Funding:** This research was funded by the National Key R&D Program of China (No. 2019YFC1510705-05, and No. 2017YFC1502500), the National Natural Science Foundation of China (No.41271303, 52109002), and the basic scientific research business fee of scientific research institutes of Changjiang Academy of Sciences (CKSF2019185/TB).

**Acknowledgments:** We thank the editor and reviewers for their useful feedback that improved this paper.

**Conflicts of Interest:** The authors declare no conflict of interest.

## References


1. Tang, K.L. *Soil and Water Conservation in China*; Science Press: Beijing, China, 2004; p. 845. (In Chinese)
2. Zheng, F.L.; Stephen, D.M.; Huang, C.H.; Tanaka, D.L.; Darboux, F.; Liebig, M.A.; Halvorson, A.D. Runoff, soil erosion, and erodibility of conservation reserve program land under crop and hay production. *Soil Sci. Soc. Am. J.* **2004**, *68*, 1332–1341. [CrossRef]
3. The Loess Plateau in Central China. In *Ecological Restoration and Management*; UNESCO Office Beijing; Tsinghua University Press: Beijing, China, 2006.
4. Shi, H.; Shao, M.A. Soil and water loss from the Loess Plateau in China. *J. Arid. Environ.* **2000**, *45*, 9–20. [CrossRef]
5. Linda, C.; Ferdinand, B.; Alain, P. Vegetation indices derived from remote sensing for an estimation of soil protection against water erosion. *Ecol. Model* **1995**, *79*, 277–285.
6. Braud, I.; Vich, A.I.J.; Zuluaga, J.; Fornero, L.; Pedrani, A. Vegetation influence on runoff and sediment yield in the Andes region: Observation and modeling. *J. Hydrol.* **2001**, *254*, 124–144. [CrossRef]
7. Michaelides, K.; Lister, D.; Wainwright, J.; Parsons, A.J. Vegetation controls on small-scale runoff and erosion dynamics in a degrading dryland environment. *Hydrol. Process.* **2009**, *23*, 1617–1630. [CrossRef]
8. Zheng, F.L.; He, X.B. The effect of vegetation broken and recover on the soil erosion of the Loess Plateau. *J. Soil Water Conserv.* **2002**, *7*, 21–27. (In Chinese)
9. Joseph, P.H. Influence of vegetation cover and crust type on wind-blown sediment in a semi-arid climate. *J. Arid. Environ.* **2004**, *58*, 167–179.
10. Zhang, B.; Yang, Y.S.; Zepp, H. Effect of vegetation restoration on soil and water erosion and nutrient losses of a severely eroded clayey Plinthudult in southeastern China. *Catena* **2004**, *57*, 77–90. [CrossRef]
11. Gyssels, G.; Poesen, J.; Nachtergaele, J.; Govers, G. The impact of sowing density of small grains on rill and ephemeral gully erosion in concentrated flow zones. *Soil Tillage Res.* **2002**, *64*, 189–201. [CrossRef]
12. Zhou, Z.C.; Shangguan, Z.P. The effects of ryegrass roots and shoots on loess erosion under simulated rainfall. *Catena* **2007**, *70*, 350–355. [CrossRef]
13. Boer, M.; Puigdefábregas, J. Effects of spatially structured vegetation patterns on hill-slope erosion in a semiarid Mediterranean environment: A simulation study. *Earth Surf. Process. Landf.* **2005**, *30*, 149–167. [CrossRef]
14. Rogers, R.D.; Schumm, S.A. The effect of sparse vegetative cover on erosion and sediment yield. *J. Hydrol.* **1991**, *123*, 19–24. [CrossRef]
15. Cerda, A. The effect of patchy distribution of *Stipa tenacissima* L. on runoff and erosion. *J. Arid. Environ.* **1997**, *36*, 37–51. [CrossRef]
16. Vasquez, M.R.; Ventura, R.E.; Oleschko, K.; Hernández-Sandoval, L.; Parrot, J.F.; Nearing, M.A. Soil erosion and runoff in different vegetation patches from semiarid Central Mexico. *Catena* **2010**, *80*, 162–169. [CrossRef]
17. Pan, C.Z.; Shangguan, Z.P.; Lei, T.W. Influence of grass and moss on runoff and sediment yield on sloped loess surfaces under simulated rainfall. *Hydrol. Process.* **2006**, *20*, 3815–3824. [CrossRef]
18. Zhou, Z.C.; Shangguan, Z.P. Effect of ryegrasses on soil runoff and sediment control. *Pedosphere* **2008**, *18*, 131–136. [CrossRef]
19. Baets, S.D.; Poesen JGyssels, G. Effects of grass roots on the erodibility of topsoils during concentrated flow. *Geomorphology* **2006**, *76*, 54–67. [CrossRef]
20. Zhang, G.H.; Liu, G.B.; Zhang, P.C.; Yi, L. Influence of vegetation parameters on runoff and sediment characteristics in patterned Artemisia capillaris plots. *J. Arid. Land* **2014**, *6*, 352–360. [CrossRef]
21. Neibling, W.H.; Alberts, E.E. Composition and yield of soil particles transported through sod strips. *Am. Soc. Agric. Eng.* **1979**, *79*, 2065.
22. Dillaha, T.A.; Sherrard, J.H.; Lee, D.; Mostaghimi, S.; Shanholtz, V.O. Evaluation of vegetative filter strips as a best management practice for feedlots. *J. Water Pollut. Control. Fed.* **1988**, *60*, 1231–1238.
23. Chen, H. Effects of rainfall characteristics and runoff from up slope on erosion and yield sediment. *J. Soil Water Conserv.* **1992**, *6*, 17–23. (In Chinese)
24. Chen, H. *Study on the Slope and Gully Erosion in a Watershed*; China Meteorological Press: Beijing, China, 1993; pp. 166–179. (In Chinese)
25. Zheng, F.L.; Huang, C.H.; Norton, L.D. Vertical hydraulic gradient and run-on water and sediment effects on erosion processes and sediment regimes. *Soil Sci. Soc. Am. J.* **2000**, *64*, 4–11. [CrossRef]
26. Li, M.; Yao, W.Y.; Ding, W.F.; Yang, J.; Chen, J. Effect of grass coverage on sediment yield in the hillslope-gully side erosion system. *J. Geogr. Sci.* **2009**, *19*, 321–330. [CrossRef]
27. Lei, A.L.; Tang, K.L. Retrospect and prospect for soil erosion studies of ridge-hill-gully side system. *Bull. Soil Water Conserv.* **1997**, *17*, 37–43. (In Chinese)
28. Lei, A.L.; Tang, K.L.; Wang, W.L. Significant and character of conception of soil erosion chain. *J. Soil Water Conserv.* **2000**, *14*, 79–83. (In Chinese)
29. Jiang, D.L.; Zhao, C.X.; Chen, Z.L. Preliminary analysis of runoff and sediment source of small watershed in the middle reaches of the Yellow River. *Acta Geogr. Sin.* **1966**, *32*, 20–36. (In Chinese)
30. Castillo, V.M.; Martinez, M.M.; Albaladejo, J. Runoff and soil loss response to vegetation removal in a semiarid environment. *Soil Sci. Soc. Am. J.* **1997**, *61*, 1116–1121. [CrossRef]
31. Benito, E.; Santiago, J.L.; Blas, D.E.; Varela, M. Deforestation of water-repellent soils in Galicia (NW Spain): Effects on surface runoff and erosion under simulated rainfall. *Earth Surf. Process. Landf.* **2003**, *28*, 145–155. [CrossRef]

32. Tang, K.L.; Zheng, S.Q.; Xi, D.Q.; Sun, Q.F. Soil loss and treatment in sloping farmland in Xingzihe Watershed. *Bull. Soil Water Conserv.* **1983**, *3*, 43–48. (In Chinese)
33. Gong, S.Y.; Jiang, D.Q. The soil loss and treatment in a small watershed in loess hilly area in the middle reaches of Yellow River. *Sci. China* **1978**, *21*, 671–678. (In Chinese)
34. Wei, W.; Chen, L.D.; Fu, B.J.; Huang, Z.; Wu, D.; Gui, L. The effect of land uses and rainfall regimes on runoff and soil erosion in the semi-arid loess hilly area, China. *J. Hydrol.* **2007**, *335*, 247–258. [CrossRef]
35. Zhou, Z.C.; Shangguan, Z.P. Modeling vegetation coverage and soil erosion in the Loess Plateau Area of China. *Ecol. Model.* **2006**, *198*, 263–268. [CrossRef]
36. Ploey, D.J.; Savat, J.; Moeyersons, J. The differential impact of some soil factors on flow, runoff, creep, and rainwash. *Earth Surf. Process. Landf.* **1976**, *1*, 151–161. [CrossRef]
37. Morgan, R.P.C.; Finney, H.J.; Lavee, H.; Merritt, E.; Noble, C.A. Plant cover effects in hillslope runoff and erosion: Evidence from two laboratory experiments. In *Hillslope Processes*; Abrahams, A.D., Ed.; Allen and Unwin: London, UK, 1986; pp. 77–96.
38. Bochet, E.; Poesen, J.; Rubio, J.L. Mound development as an interaction of individual plants with soil, water erosion and sedimentation processes on slopes. *Earth Surf. Process. Landf.* **2000**, *25*, 847–867. [CrossRef]
39. Zhang, G.H.; Liu, G.B.; Wang, G.L. Effects of canopy and roots of patchy distributed artemisia capillaris on runoff, sediment, and the spatial variability of soil erosion at the plot scale. *Soil Sci.* **2012**, *177*, 409–415. [CrossRef]
40. Zhang, G.H.; Liu, G.B.; Wang, G.L. Effects of Caragana Korshinskii Kom. cover on runoff, sediment yield and nitrogen loss. *Int. J. Sediment Res.* **2010**, *25*, 245–257. [CrossRef]

water

MDPI

*Article*

# Flash Flood Susceptibility Mapping in Sinai, Egypt Using Hydromorphic Data, Principal Component Analysis and Logistic Regression

**Mustafa El-Rawy [1,2,*], Wael M. Elsadek [3] and Florimond De Smedt [4]**

1 Civil Engineering Department, Faculty of Engineering, Minia University, Minia 61111, Egypt
2 Civil Engineering Department, College of Engineering, Shaqra University, Dawadmi 11911, Ar Riyadh, Saudi Arabia
3 Civil Engineering Department, Faculty of Engineering, South Valley University, Qena 83523, Egypt
4 Department of Hydrology and Hydraulic Engineering, Vrije Universiteit Brussel, Pleinlaan 2, 1050 Brussels, Belgium
* Correspondence: mustafa.elrawy@mu.edu.eg

**Abstract:** Flash floods in the Sinai often cause significant damage to infrastructure and even loss of life. In this study, the susceptibility to flash flooding is determined using hydro-morphometric characteristics of the catchments. Basins and their hydro-morphometric features are derived from a digital elevation model from NASA Earthdata. Principal component analysis is used to identify principal components with a clear physical meaning that explains most of the variation in the data. The probability of flash flooding is estimated by logistic regression using the principal components as predictors and by fitting the model to flash flood observations. The model prediction results are cross validated. The logistic model is used to classify Sinai basins into four classes: low, moderate, high and very high susceptibility to flash flooding. The map indicating the susceptibility to flash flooding in Sinai shows that the large basins in the mountain ranges of the southern Sinai have a very high susceptibility for flash flooding, several basins in the southwest Sinai have a high or moderate susceptibility to flash flooding, some sub-basins of wadi El-Arish in the center have a high susceptibility to flash flooding, while smaller to medium-sized basins in flatter areas in the center and north usually have a moderate or low susceptibility to flash flooding. These results are consistent with observations of flash floods that occurred in different regions of the Sinai and with the findings or predictions of other studies.

**Keywords:** flash flood; flood hazard; morphometry; PCA; logistic regression; Sinai; Egypt

**Citation:** El-Rawy, M.; Elsadek, W.M.; De Smedt, F. Flash Flood Susceptibility Mapping in Sinai, Egypt Using Hydromorphic Data, Principal Component Analysis and Logistic Regression. *Water* **2022**, *14*, 2434. https://doi.org/10.3390/w14152434

Academic Editor: Luis Garrote

Received: 11 June 2022
Accepted: 1 August 2022
Published: 6 August 2022



## 1. Introduction

Flash floods are among the deadliest natural disasters in the world, responsible for 85% of inundations and a high death rate of more than 5000 people lost each year [1]. Egypt has experienced many flash floods with loss of life and serious damage to vital infrastructure and buildings, especially in the Sinai, the north coast and the Red Sea coast. Well-known examples are the 1979 flash flood in El-Quseir and Marsa Alam, which killed 19 and destroyed a coastal road along the Red Sea, the flash flood in Marsa Alam in 1991, the flash flood in Alexandria in 1993, which killed 21, and the flash flooding in Assiut in November 1994 resulting in loss of life and infrastructure [2]. Recently, in 26–27 October 2016, heavy rainfall in Ras Gharib on the Red Sea coast resulted in flash flooding that killed dozens and caused damage to infrastructure and property [3]. On 14 November 2019, heavy rains led to flooding in wadi El-Sukkari and further to the Idfu-Marsa Alam Road, fortunately without losses [4].

Sinai in particular is a flood-prone area where flash floods cause significant damage to infrastructure, displacement of populations and sometimes loss of life. An overview of major floods in the Sinai in the past is given by Abdel-Fattah et al. [5] and Omran [6]. Most devastating was the severe flooding in Wadi El-Arish on 17–18 January 2010, which resulted in six deaths and dozens of injuries and vital infrastructure and hundreds of houses destroyed [7]. Cools et al. [8] reported that out of 20 significant rainfall events over a 30-year period (1979–2010) in wadi Watir, located in Southeast Sinai, nine resulted in flash flooding, some of which caused severe damage to the coastal road between Nuweiba and Taba, which was completely washed away in some parts. More recently, a catastrophic flash flood in October 2016 caused many deaths and injuries and damage to roads and buildings in wadi Dahab in South Sinai [9]. El-Fakharany and Mansour [10] discussed dangerous flash flood locations in southern Sinai, including the wadis Werdan, Sedri, El-Aawag and Feiran.

Reliable and accurate data on flash flooding in arid environments are often lacking due to the remoteness and sparse habitation of such areas. Therefore, morphometric analyzes are often used instead to analyze the drainage behavior of such areas. Quantitative measurements of the geometrical features of drainage basins, such as basin size and shape, drainage network and relief form the basis of hydro-morphometric analysis as set forth in the classical works of Horton [11,12], Smith [13], Strahler [14,15] and Schumm [16]. Since the development of remote sensing (RS) and geographic information systems (GIS), hydro-morphological parameters can be easily obtained.

A comprehensive overview of past developments in the analysis of flood risk assessment has been provided by Diaconu et al. [17]. An overview of flash flood risk assessment based on morphometric analyses has been provided by Ali et al. [18], discussing flash flood hazard vulnerability and risk assessment approaches, uncertainties and challenges. Notable recent studies are as follows. Farhan et al. [19] presented a morphometric analysis and flash floods assessment for drainage basins of the Ras En Naqb Area, South Jordan. Mahmood and Rahman [20] studied the flash flood susceptibility modeling using geomorphometric and hydrological approaches in Panjkora Basin, Eastern Hindu Kush, Pakistan. Mahmood and Rahman [21] modeled the flash flood susceptibility using a geomorphometric ranking approach in the Ushairy Basin, in eastern Hindu Kush, Pakistan. Bhat et al. [22] presented a flood hazard assessment of the upper Jhelum basin using morphometric parameters. Obeidat et al. [23] performed a morphometric analysis and prioritized watersheds for flood risk management in the Wadi Easal Basin, Jordan. Pangali Sharma et al. [24] identified potential flash flood areas using a geomorphic approach in the East Rapti River Basin of Nepal.

Hydro-morphometric analyzes have been applied in studies for flood assessment in Egypt. Arnous et al. [25] reported that morphometric analysis strongly supported a high probability of flash flooding in the western part of the Gulf of Suez. Youssef et al. [26] used morphometric parameters to estimate the risk of flash flooding along the St. Katherine Road in southern Sinai. Abdel-Lattif and Sherief [27] used three morphometric parameters (bifurcation ratio, drainage density and stream frequency) to estimate flood hazard in wadi Sudr and wadi Wardan, Gulf of Suez. Elewa et al. [28] presented a hydro-morphometric analysis for the El-Arish Basin in northern and central Sinai to identify suitable sites for collecting runoff. Abdalla et al. [29] presented a geomorphometric classification of wadis along the southeastern Red Sea coast in Egypt, showing that most basins are highly susceptible to flash flooding. Abdel Ghaffar et al. [30] evaluated the flash flood hazard of wadi El-Arish by combining nine morphometric parameters using priority values and classification into high, medium and low hazard. Abuzied et al. [31] used morphometric parameters to evaluate flood sensitive basins and to map the flash flood susceptibility in the Nuweiba area, Sinai. Abdelkareem [32] derived a flash flood hazard map for the wadi Asyuti basin in the eastern desert of Egypt by ranking and combining morphometric parameters that favor higher flood peaks and runoff. Elsadek et al. [33] presented a morphometric analysis to estimate flood hazard in the wadi Qena basin. Abuzied and Mansour [34] combined normalized values of morphometric parameters to obtain hazard indices for sub-basins of wadi Dahab, Sinai. Elsadek et al. [35] investigated flood hazard in wadi Qena based on a combination of morphometric parameter ranking. Kamel and

Arfa [4] used a ranking method of 13 morphometric parameters to derive the degree of flash flood hazard of basins between Marsa Alam and Abu Ghuson along the Red Sea coast. Prama et al. [9] derived a flood hazard map for the Dahab region in southern Sinai, by an unweighted combination of normalized morphometric parameters. El-Fakharany and Mansour [10] evaluated morphometric parameters and the occurrence of flash floods in wadi El-Aawag in southwestern Sinai.

The main aim of this research is to quantify the risk of flash flooding in the Sinai Peninsula, Egypt. Various methods are used for this: (1) hydro-morphometric features are derived that are relevant for estimating the sensitivity to flash flooding using satellite images and spatial analysis tools; (2) principal component analysis is applied to reveal relationships between the different catchment properties and to identify the most significant hydro-morphometric parameters; (3) the probability of flash flooding is estimated by logistic regression using observed events of flash floods as dependent variable and the significant principal components of the hydro-morphometric parameters as explanatory variables; (4) prediction results are cross-validated to assess and prove the robustness of the modeling approach; (5) the logistic model is used to generate a map of flash flood probability for the Sinai Peninsula. The novelty of the approach consists primary in generating a flood susceptibility map for the entire Sinai Peninsula, which has not been presented before; previous studies [9,10,26–28,30,31,34] only considered only some local sub-basins. A complete flash flood susceptibility map for the Sinai Peninsula can be useful to authorities and decision makers for an overall impact assessment and can contribute to flash flood management and to the planning and implementation of mitigation measures. This study also aims to demonstrate how hydro-morphometric parameters can be used for flood risk assessment in Egypt through a combination of principle component analysis and logistic regression that is robust, reliable and validated. To date, most flood susceptibility studies conducted in Egypt have used a ranking method proposed by Davis [36] to derive flash flood hazard from hydro-morphometric data, by standardizing morphometric parameters, usually in the range of 1 to 5, and then combining and classifying them into groups ranging from lowest to highest risk level [25–35]. However, the number of parameters can vary, all parameters are treated with the same weight as if they have an equal impact on flooding, and the classification into hazard classes is done without rules or standards. In addition, the results are usually not validated.

## 2. Materials and Methods

### 2.1. Study Area

The Sinai Peninsula is located in the northeastern part of Egypt between 32.5–34.8° E and 27.8–31.3° N (Figure 1). It is about 61,000 km$^2$ in size and its largest dimensions are about 385 km from north to south and 210 km from west to east. Geographically, Sinai can be divided into three parts. The northern part consists of broad coastal plains with fossil beaches and extensive sand dunes, some of which are more than 100 m high. The main part is the wadi El-Arish basin which descends from an altitude of more than 900 m to the Mediterranean Sea and forms the largest valley of the Sinai Peninsula. The center is highland mainly composed of two limestone plateaus, El-Tih in the south and El-Egma in the north, where the sources of the wadi Al-Arish arise. The southern part consists of high and rugged mountain ranges of igneous rock, reaching more than 2400 m, with Mount Catherine at 2642 m above sea level being the highest point in Egypt.

Sinai is characterized by a Mediterranean climate in the north and an arid to semi-arid climate in the center and south. In general, summer is very hot and dry, while most rain falls in winter with occasional heavy rainfall combined with thunderstorms. The amount of precipitation decreases from north to south. Most rain falls in a narrow strip along the Mediterranean Sea with values of more than 200 mm/year; further inland the rainfall varies from 100 to 200 mm per year in the north, while in the center and south this usually less than 100 mm per year. Since the potential evaporation demand far exceeds rainfall, there are no real streams or rivers but only ephemeral riverbeds, referred to as wadis, which are generally dry but discharge drainage water after heavy rainfall usually in winter. Wadi El-Arish is the main drainage system to the Mediterranean in the north and center of Sinai. There are several smaller wadis in the south, some of which are known for flash floods, such as Watir, Dahab and Kid in the east which flow into the Gulf of Aqaba and Ras Sudr, Werdan, Feiran, Sedr, Gharandal and Meiar in the west which flow into the Gulf of Suez [6].

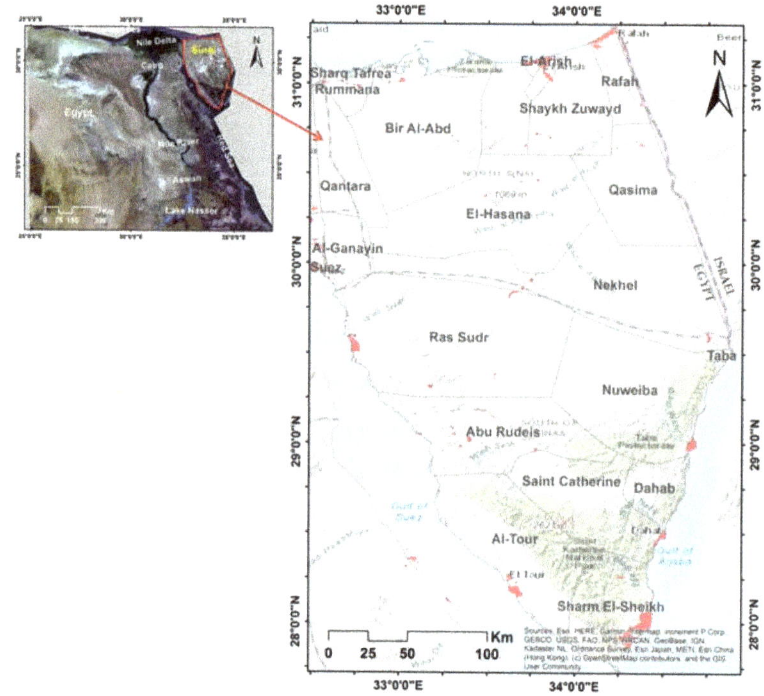


**Figure 1.** Location of the study area.

## 2.2. Hydro-Morphometric Parameters

An ASTER Global Digital Elevation Model (DEM) version 3 was downloaded from the NASA Earthdata website (Available online: https://search.earthdata.nasa.gov; accessed on 15 September 2021). The DEM has a latitude and longitude resolution of 1 arc-second (~30 m), and the elevation data have a resolution of 1 m and an accuracy of approximately 10 m [37]. Topographic elevations in the Sinai range from zero to 2612 m above mean sea level as shown in Figure 2. ArcGIS spatial analysis tools are utilized to delineate watersheds and determine their hydro-morphometric parameters. The drainage network is extracted from the DEM using the stream order method of Strahler [31] with a threshold of 4.5 km$^2$ for the upslope drainage area as the starting point of first order streams, which corresponds 5000 grid cells, a standard recommended by ArcGIS spatial analysis tools (Available online: https://pro.arcgis.com; accessed on 22 January 2022). Sub-basins are delineated based on stream orders and hydro-morphometric parameters are derived for each sub-basin using standard spatial analyses methods and equations as listed in Table 1.

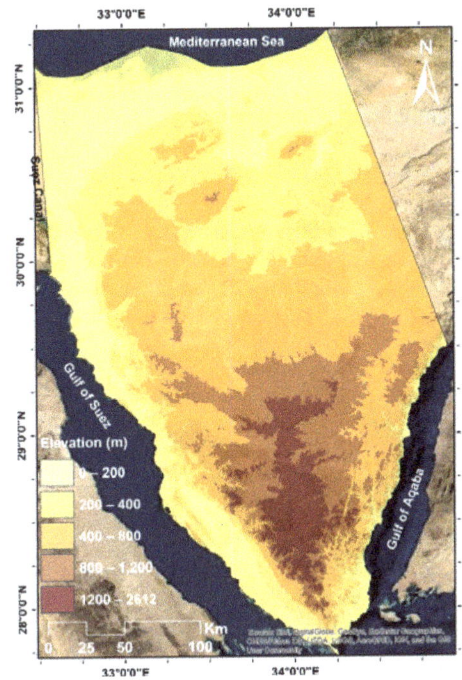


**Figure 2.** Topography of the Sinai derived from a digital elevation model from NASA Earthdata.

**Table 1.** Hydro-morphometric parameters used in this study.

| Parameter | Description | Determination | Reference |
|---|---|---|---|
| | Basin geometry | | |
| $A$ [$L^2$] | Area | Spatial analysis | - |
| $P$ [L] | Perimeter | Spatial analysis | - |
| $L_b$ [L] | Basin length | Spatial analysis | [11] |
| $F_f$ [-] | Form factor | $F_f = A/L_b^2$ | [11] |
| $C_c$ [-] | Compactness coefficient | $C_c = P/2\sqrt{\pi A}$ | [11,12] |
| $R_e$ [-] | Elongation ratio | $R_e = 2\sqrt{A/\pi}/L_b$ | [16] |
| | Drainage network | | |
| $S_u$ [-] | Stream order | Spatial analysis | [12] |
| $N_u$ [-] | Stream number | $N_u = \sum_1^n N_i$ | [15] |
| $R_b$ [-] | Bifurcation ratio | $R_b = (\sum_1^n N_i/N_{i+1})/(n-1)$ | [14] |
| $L_u$ [L] | Stream length | $L_u = \sum_1^n L_i$ | [14] |
| $D_d$ [$L^{-1}$] | Drainage density | $D_d = L_u/A$ | [11] |
| $F_s$ [$L^{-2}$] | Stream frequency | $F_s = N_u/A$ | [11,12] |
| $L_o$ [L] | Length of overland flow | $L_o = 1/2D_d$ | [11] |
| $R_t$ [$L^{-1}$] | Texture ratio | $R_t = N_1/P$ | [13] |
| | Relief | | |
| $R_f$ [L] | Basin relief | Spatial analysis | [16] |
| $R_r$ [-] | Relief ratio | $R_r = R_f/L_b$ | [16] |
| $R_n$ [-] | Ruggedness number | $R_n = R_f D_d$ | [38] |
| $S$ [°] | Mean slope | Spatial analysis | - |

Where $n$ is the number of stream orders in a basin, $N_i$ is the number of stream segments of order $i$, and $L_i$ is the length of stream segments of order $i$.

Three groups of parameters are considered. The basin geometry group includes six parameters:

- Area ($A$): surface of a drainage basin, which is a prime determinant of the total discharge [31]; large catchments receive more precipitation and have a higher peak discharge compared to smaller catchments.
- Perimeter ($P$): circumference of a drainage basin; there are no clear indications of direct significance for the hydrological regime, but it is used in the determination of other parameters.
- Basin length ($L_b$): maximum distance from the catchment boundary to the outlet; a good indicator of the concentration time of a flood wave [11].
- Form factor ($F_f$): ratio of the width to the length of a catchment and indicative of the flood regime [11]; large form factors lead to shorter lag times and a higher peak discharge.
- Compactness coefficient ($C_c$): ratio of the perimeter of the drainage basin to that of a circle of equal area; low values imply a shorter concentration time and a higher peak discharge [11,12].
- Elongation ratio ($R_e$): ratio of the diameter of a circle with the same area as the catchment area to the maximum catchment length [16]; low values mean less circular shape and longer flood concentration time.

The basin drainage network group includes eight parameters:

- Stream order ($S_u$): highest stream order in a basin according to the method designed by [14]; it is an indicative parameter of the basin dimensions, channel size and stream discharge.
- Stream number ($N_u$): total number of stream segments of all orders [15]; a high stream number is expected to imply faster peak flow.
- Bifurcation ratio ($R_b$): average ratio between the number of streams of one order and those of the next higher order [12]; indicative of the complexity of a catchment, but according to [14] less so for the flow regime, although [12] considers flooding more likely in catchments with a higher bifurcation ratio.
- Stream length ($L_u$): total length of all streams in a basin [11]; longer streams indicate a higher discharge producing capacity of a catchment area [39].
- Drainage density ($D_d$): length of streams per unit area; an indicator of infiltration and permeability of a drainage [11].
- Stream frequency ($F_s$): number of streams per unit area; although similar to drainage density, it has less hydrologic significance [11,12].
- Length of overland flow ($L_o$): the average length of overland flow is equal to the reciprocal of twice the drainage density [16]; low values indicate shorter flow paths, making the basin more prone to flash flooding.
- Texture ratio ($R_t$): total number of first order streams per basin circumference; indicates coarse, medium, or fine textured topography [40].

The basin relief group includes four parameters:

- Basin relief ($R_f$): height difference of the lowest and highest points of a basin and an essential indicator of surface runoff [16].
- Relief ratio ($R_r$): ratio of the basin relief to the basin length and a key element for understanding erosion and drainage [16].
- Ruggedness number ($R_n$): product of drainage density and basin relief; regions prone to flash flooding have higher ruggedness numbers, indicating high drainage density combined with steep slopes [38].
- Mean basin slope ($S$): major factor controlling infiltration and surface runoff and the resulting runoff rate and concentration time.

### 2.3. Principal Component Analysis

Before creating a predictive model, principal component analysis (PCA) [41] is used as an exploratory data analysis to identify the relevant information in the hydro-morphometric data set. PCA linearly transforms the data into orthogonal uncorrelated variables known as principal components (PCs), which preserve the total variance in the original data

$$PC_j = \sum_1^n a_{ij} x_i \quad j = 1, n \tag{1}$$

where $PC_j$ are the principal components, $a_{ij}$ are the scores of the linear transformation, $x_i$ are the standardized hydro-morphometric parameters, and $n$ is the number of parameters. Note that the hydro-morphometric parameters are standardized to remove any effect of scale and units of the observations by subtracting the sample mean and dividing by the sample standard deviation. It can be shown that the principal components are the eigenvectors of the correlation matrix of the data and that the associated eigenvalues give the variance explained by each eigenvector [41]. The eigenvectors are ordered in descending order of the eigenvalues and PCs with eigenvalues smaller than one are ignored because they contain less information than the original variables, which reduces the dimensionality of the data. The scores that relate the remaining PCs to the original parameters provide information about the impact and relevance of the original parameters on the overall information in the data set. In practice, correlation coefficients between the PCs and the original parameters are used to explain and interpret the strength of the relationships. Large (either positive or negative) correlation coefficients indicate that a parameter has a strong effect on that principal component. The interpretation is enhanced by Varimax rotation, which aligns the PCs in directions highlighting the relationships between the PCs and the observed data [41].

### 2.4. Logistic Regression

A logistic model is used to predict the probability of a flash flood

$$\text{logit}(p) = \ln\left(\frac{p}{1-p}\right) = c_0 + \sum\nolimits_1^m c_i y_i \tag{2}$$

where logit is the logistic function (natural logarithm of the odds), $p$ is the probability, $c_0$ is the model intercept, $c_i$ are the model coefficients, $y_i$ are the predictors (explanatory parameters), and $m$ is the number of predictors. The logistic model is used to predict the occurrence of a flash flood in a basin using observed characteristics of the basin as predictors. Since there are too many hydro-morphometric parameters to use as predictors, we will instead consider the significant PCs of the basin hydro-morphometric data as predictors. The logistic model predicts the flood probability which can be used to assess the susceptibility of a basin to flash flooding.

The model coefficients are estimated by logistic regression. For this, we use observations of flash floods reported in the literature [5,6,8–10]; basins where flooding has been observed are shown in Table 2. The observed probability of flooding is set to one for these basins, while for the other basins the probability is zero. Flash floods in Wadi El-Arish were excluded in the analysis because Wadi El-Arish consists of many sub-basins, while the exact location of the floods was usually not clearly observed or reported because Wadi El-Arish is so vast and sparsely populated.

The model coefficients are estimated by fitting the model to these observations using maximum likelihood, for which we use the *glm* generalized maximum likelihood fitting procedure of the R *Stats* package for statistical computing [42]. The goodness of fit is assessed by the deviance, a measure of the likelihood, and the quality of the model by the Akaike information criterion (AIC) [43], a trade-off between the goodness of fit and the complexity of the model. The significance of each predictor is verified by removing each predictor one at the time, re-estimating the model coefficients with the remaining predictors and comparing the resulting deviance and AIC with the original model. The reliability of the model is also verified by cross-validation where observed flood events are removed one by one, model coefficients are re-estimated by logistic regression with the remaining data and the flood probability is predicted for the removed event and compared to what was obtained with the original model.

**Table 2.** Basins where flash floods have been reported in the literature, used for calibration of the logistic model.

| Wadi | Basin No. |
|---|---|
| Sedri | 3 |
| Werdan | 9 |
| Ras Sudr | 12 |
| Watir | 54 |
| Dahab | 56 |
| Feiran | 58 |
| El-Aawag | 84 |
| Gharandal | 89 |
| Kid | 98 |

## 3. Results

### 3.1. *Drainage Catchments and Hydro-Morphometric Data*

The drainage network obtained from the DEM is shown in Figure 2 and consists of 112 sub-basins of different sizes and shapes, as shown in Figure 3. Large basins such as wadis El-Arish, Feiran, Dahab and Watir are subdivided to obtain a more or less uniform spatial distribution of the sub-basins over the area. Values of the hydro-morphometric parameters for each basin are given in Table S1 in the Supplementary Materials. An overview of the range of the hydro-morphometric parameters is given in Table 3, with minimum and maximum values and the mean and standard deviation necessary for standardization of the parameters in the PCA.

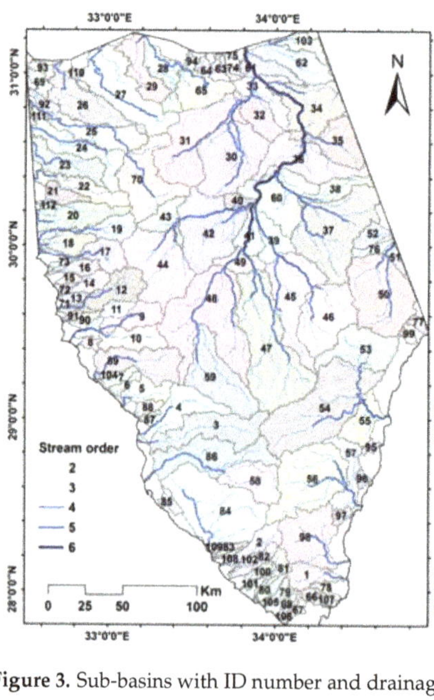


**Figure 3.** Sub-basins with ID number and drainage channels with stream order, excluding first order, derived from the digital elevation model.

**Table 3.** Range of the hydro-morphometric parameter, showing minimum, maximum, mean and standard deviation (SD).

| Parameter | Minimum | Maximum | Mean | SD |
|---|---|---|---|---|
| $A$ (km$^2$) | 19 | 2254 | 492 | 542 |
| $P$ (km) | 31 | 386 | 139 | 82 |
| $L_b$ (km) | 10.9 | 94.5 | 38.0 | 20.2 |
| $F_f$ | 0.07 | 0.52 | 0.26 | 0.10 |
| $C_c$ | 1.50 | 2.88 | 2.07 | 0.34 |
| $R_e$ | 0.30 | 0.81 | 0.56 | 0.11 |
| $S_u$ | 2 | 6 | 3.7 | 1.2 |
| $N_u$ | 3 | 194 | 42.1 | 45.1 |
| $R_b$ | 1.75 | 6 | 3.4 | 0.9 |
| $L_u$ (km) | 11 | 1122 | 229 | 250 |
| $D_d$ (km$^{-1}$) | 0.31 | 0.86 | 0.49 | 0.10 |
| $F_s$ (km$^{-1}$) | 0.04 | 0.16 | 0.09 | 0.02 |
| $L_o$ (km) | 0.58 | 1.60 | 1.06 | 0.19 |
| $R_t$ (km$^{-1}$) | 0.02 | 0.65 | 0.18 | 0.12 |
| $R_f$ (m) | 65 | 2595 | 806 | 585 |
| $R_r$ | 0.004 | 0.082 | 0.025 | 0.020 |
| $R_n$ | 0.03 | 1.31 | 0.39 | 0.29 |
| $S$ (°) | 1.50 | 21.40 | 6.20 | 4.69 |

### 3.2. Principal Component Analysis

The results of the PCA are presented in Tables 4 and 5. Table 4 lists the first eight eigenvalues, and the variance explained by each component, and cumulative variance, both expressed as a percentage of the total variance contained in the data. Significant values are indicated in bold. Only the first four PCs have eigenvalues greater than one, while the fifth eigenvalue is lower but very close to one, and together these account for 90% of the variation in the data. Therefore, the other PCs can be ignored. Table 5 lists the correlation coefficients between the first five PCs and the hydro-morphometric parameters after Varimax rotation; the corresponding scores are given in Table S2 in the Supplementary Materials. Significant values in Table 5 are shown in bold. The first component is highly correlated with several hydro-morphometric parameters: area ($A$), perimeter ($P$) and basin length ($L_b$) which are directly related to the size of a watershed, and with stream order ($S_u$), stream number ($N_u$), stream length ($L_u$) and texture ratio ($R_t$) which are also indirectly related to size. Thus, the first and most important principal component represents the effect of basin size and accounts for 37% of the variation in the data. The second component is strongly correlated with all relief parameters: basin relief ($R_f$), relief ratio ($R_r$), ruggedness number ($R_n$) and mean basin slope ($S$); this component accounts for 19% of the variation in the data. The third component, which accounts for 15% of the total variance, is strongly correlated with drainage density ($D_d$) and length of overland flow ($L_o$). This component thus represents the drainage capacity of a river basin. The fourth component accounts for 14% of the total variance and is strongly correlated with the form factor ($F_f$), compactness coefficient ($C_c$) and elongation ratio ($R_e$), so this component expresses the influence of the basin shape. The fifth component accounts for only 5% of the variance but is rather special in that it is only significantly correlated with the bifurcation ratio ($R_b$). Thus, this component represents the effect of stream bifurcation, which is apparently a unique basin property unrelated to any other hydro-morphometric parameter. Note that the stream frequency ($F_s$) is not significantly correlated with any of the PCs and thus contributes little to the information contained in the data. Principal component values for all basins are given Table S3 in the Supplementary Materials.

**Table 4.** Eigenvalue, variance (%) accounted for and cumulative variance (%) from a principal component analysis (significant values are indicated in bold).

| PC Number | 1 | 2 | 3 | 4 | 5 | 6 | 7 | 8 |
|---|---|---|---|---|---|---|---|---|
| Eigenvalue | **7.13** | **3.36** | **3.04** | **1.76** | **0.96** | 0.66 | 0.39 | 0.30 |
| Variance (%) | 0.40 | 0.19 | 0.17 | 0.10 | 0.05 | 0.04 | 0.02 | 0.02 |
| Cumulative (%) | 0.40 | 0.58 | 0.75 | 0.85 | 0.90 | 0.94 | 0.96 | 0.98 |

**Table 5.** Correlation coefficients between the first five principal components and the hydro-morphometric parameters after Varimax rotation (significant values are indicated in bold).

| Parameter | Principal Component | | | | |
|---|---|---|---|---|---|
| | $PC_1$ | $PC_2$ | $PC_3$ | $PC_4$ | $PC_5$ |
| $A$ | **0.94** | 0.04 | 0.21 | −0.05 | 0.15 |
| $P$ | **0.95** | −0.01 | −0.12 | −0.16 | 0.11 |
| $L_b$ | **0.95** | 0.06 | −0.18 | −0.11 | 0.12 |
| $F_f$ | 0.29 | −0.04 | **0.87** | −0.26 | 0.00 |
| $C_c$ | 0.04 | −0.16 | **−0.88** | 0.02 | 0.07 |
| $R_e$ | 0.29 | −0.07 | **0.86** | −0.30 | −0.03 |
| $S_u$ | **0.80** | −0.11 | 0.13 | −0.12 | −0.31 |
| $N_u$ | **0.93** | 0.05 | 0.26 | −0.01 | 0.15 |
| $R_b$ | 0.22 | 0.08 | −0.04 | −0.02 | 0.93 |
| $L_u$ | **0.94** | 0.01 | 0.21 | 0.03 | 0.15 |
| $D_d$ | −0.11 | 0.03 | −0.23 | **0.94** | −0.01 |
| $F_s$ | −0.39 | −0.03 | 0.28 | 0.41 | −0.33 |
| $L_o$ | 0.03 | 0.07 | 0.24 | **−0.94** | −0.01 |
| $R_t$ | **0.83** | 0.08 | 0.47 | 0.00 | 0.17 |
| $R_f$ | 0.26 | **0.94** | −0.02 | −0.05 | 0.10 |
| $R_r$ | −0.37 | **0.87** | 0.13 | 0.05 | −0.03 |
| $R_n$ | 0.18 | **0.91** | −0.10 | 0.27 | 0.12 |
| $S$ | −0.06 | **0.88** | 0.11 | −0.35 | −0.05 |

### *3.3. Logistic Regression*

Results of the logistic regression with the PCs as predictors are given in Table 6. The table shows the estimated model coefficients, the standard error, z-value (coefficient divided by standard error) and the probability that the predictor is statistically significant, for which it is common practice to prescribe a value less than 0.05 for a normal distribution. Note that this is only the case for $PC_1$ and $PC_4$ and not for $PC_3$ and $PC_5$, while $PC_2$ is very close to the threshold. However, the assumption of a normal distribution is not reliable if the sample size is small, as in this case. The next two columns in the table provide the deviation and AIC values, which indicate how well the model with the selected predictors fits the observations. The values corresponding to the intercept are for the null model, which is a logistic model with only an intercept and no predictors that is used as a reference to compare with other models. The values corresponding to the predictors are for excluding that predictor from the full model and the values on the last line are for the total full model. Both the deviance and AIC should be as small as possible. Comparison of the deviance and AIC obtained for the total model and for the null model shows a large difference, indicating that the predictors allow significant improvement in goodness of fit. Comparison of the deviance and AIC when one of the predictors is removed from the full model shows that all predictors are relevant and should not be removed from the model. The increase in deviance when one of the predictors is removed compared to the full model also indicates the importance of that predictor in the model. It follows that the order of importance of the predictors is: $PC_1$, $PC_4$, $PC_2$, $PC_5$ and $PC_3$, as given in the last column of Table 6.

**Table 6.** Regression results of the logistic model: estimated model coefficients, standard error, z-value and probability, deviance (Dev.), Akaike information criterion (AIC) and rank.

| Predictor | Estimate | Std. Err. | z-Value | Pr (>\|z\|) | Dev. | AIC | Rank |
|---|---|---|---|---|---|---|---|
| Intercept | −4.74 | 1.09 | −4.36 | $1.33 \times 10^{-5}$ | 62.6 | 64.6 | |
| $PC_1$ | 1.56 | 0.58 | 2.68 | 0.01 | 43.1 | 53.1 | 1 |
| $PC_2$ | 0.94 | 0.54 | 1.75 | 0.08 | 34.8 | 44.8 | 3 |
| $PC_3$ | 0.71 | 0.50 | 1.43 | 0.15 | 33.8 | 43.8 | 5 |
| $PC_4$ | −1.99 | 0.90 | −2.22 | 0.03 | 40.6 | 50.6 | 2 |
| $PC_5$ | 0.90 | 0.60 | 1.51 | 0.13 | 33.9 | 43.9 | 4 |
| Total | | | | | 31.5 | 43.5 | |

Results of the cross-validation where observed flood events are removed one by one, and the recalibrated model is used to predict the flood probability for the removed event are given in Table 7. This shows that after removing one of the observed flood basins, the model proves to be robust because the estimates of flood probability remain in the same range as predicted with the original model.

**Table 7.** Probability for flooding estimated with the logistic model and after cross-validation of the model.

| Basin No. | Probability Model | Cross Val. |
|---|---|---|
| 3 | 0.422 | 0.248 |
| 9 | 0.030 | 0.014 |
| 12 | 0.019 | 0.005 |
| 54 | 0.759 | 0.673 |
| 56 | 0.761 | 0.695 |
| 58 | 0.835 | 0.786 |
| 84 | 0.866 | 0.729 |
| 89 | 0.238 | 0.196 |
| 98 | 0.872 | 0.848 |

The flood probability predicted by the logistic model for all basins is shown in Table S4 in the Supplementary Materials. A comparison between the flood probability predicted by the logistic model and the observations is shown in Figure 4.

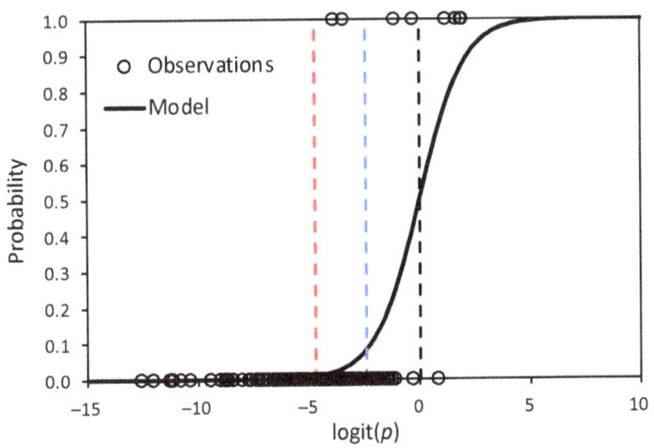


**Figure 4.** Predicted and observed flash flood probability against logit($p$): dots are observed flash flood probabilities, the solid black line represents the probability predicted by the logistic model, the red dotted line corresponds to the mean of the model predictions, logit($p$) = −4.74, the blue dotted line to the mean of the observations, logit($p$) = −2.44, and the black dotted line with logit ($p$ = 0.5) = 0.

The nine basins where flash floods were observed are shown in the upper part of the graph ($p = 1$) and the 103 basins where no floods were observed in the lower part ($p = 0$). The solid line represents the logistic model that fits the observations as closely as possible but has to compromise in the middle part of the graph where basins with observed flooding overlap with basins where no flooding has been observed. Since there are many more basins where no flooding has been observed, the logistic model is strongly conditioned by the non-flooding events, as can be clearly seen in the graph. This should be taken into account when evaluating the model results and selecting threshold values to identify the flood prone basins. The red dotted line in the graph represents the mean outcome predicted by the model given by the intercept, $\text{logit}(p) = c_0 = -4.74$ (Table 6). Note that all basin where floods have been observed are on the right side of this line, while all basins predicted by the model on the left of this line have a near zero predicted flooding probability. Basins predicted to the right of the red line are thus prone to flooding with a probability that increases the further they are from this line. The blue dotted line in the graph shows the average outcome of the observations, $\text{logit}(p) = \ln(9/103) = -2.44$. All basins plotted to the right of this line have a higher probability of flooding than is observed on average, and vice versa. Seven of the basins where flash floods have been observed are to the right of this line. Additionally, to the right of this line are 15 basins where no flooding has been observed, so these basins have features that indicate a higher probability of flash flooding than observed. The black dotted line represents $\text{logit}(p)$ equal to zero ($p = 0.5$). There are only six basins with a predicted $\text{logit}(p)$ greater than zero ($p > 0.5$) and thus very sensitive to flash flooding; five of these, wadis Kid, El-Aawag, Feiran upstream, Dahab and Watir, are basins where flooding has been observed and one, wadi Feiran downstream, where no flooding has been assumed but has similar characteristics to the other five.

The above considerations are used to classify the flash flood susceptibility of all basins. Flood sensitivity classes are defined as follows: $\text{logit}(p) < -4.74$ is low sensitivity to flooding, $-4.74 < \text{logit}(p) < -2.44$ is moderate sensitivity, $-2.44 < \text{logit}(p) < 0$ is high sensitivity and $\text{logit}(p) > 0$ is very high sensitivity. The resulting map with the sensitivity of all basins is shown in Figure 5.

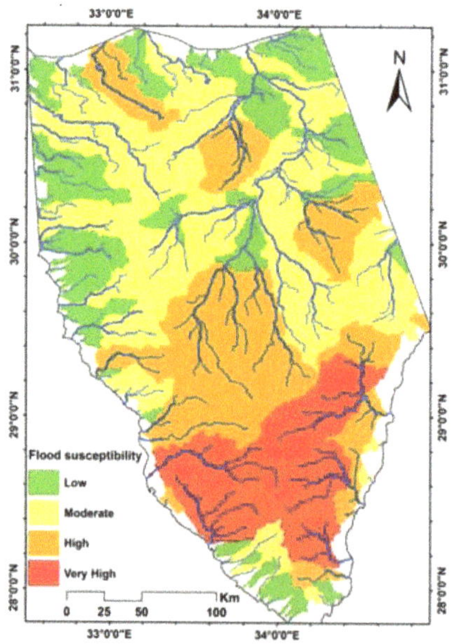


**Figure 5.** Susceptibility to flash flooding in the Sinai predicted with the probabilistic model (Equation (2)) using the principal components of the hydro-morphometric parameters as predictors.

## 4. Discussion

Results of the PCA (Table 5) indicate that the hydro-morphometric characteristics of the Sinai drainage basins can be combined into five groups that account for 90% of the variation of the data and have clear physical meaning. Ordered in decreasing importance in explaining the total variance, the first group includes all hydro-morphometric parameters related to size, the second group consists of parameters related to relief, the third group of parameters relates to basin shape, the fourth group of parameters relates to drainage capacity and the fifth group consists only of the bifurcation ratio. However, the importance of the basin characteristics on flash flood occurrence is not in order of explained variance in the data, but in a different order based on flash flood prediction, as shown by the logistic regression (Table 6). Most important in predicting flash flooding is the basin size, followed by drainage density, relief, bifurcation ratio and basin shape. The latter two are not statistically significant in the model regression, but nevertheless appear to be relevant in terms of likelihood combined with model complexity (AIC, Table 5), and therefore should not be removed from the model. The importance of a predictor is given by the regression coefficients (Table 5). It follows that basin size and drainage density are about two times more important than relief, stream bifurcation or basin shape.

Not surprisingly, the size of a basin is the most important factor in predicting flash flooding. In the case where local thunderstorms are more or less spatially random, the probability of an extreme thunderstorm in a large basin will be greater than in a small basin, increasing the risk of flash flooding in large basins. Equally important is the converse that less or no flash flooding is observed in small basins. The PC representing the drainage density is the second most important predictor of flash flooding, but surprisingly the regression coefficient is negative, so that the flood probability decreases with increasing drainage density, which is contradictory to what is commonly believed. The reason is that flash flooding and drainage density are examined here on a regional scale. Locally, higher drainage density may result in faster drainage, but this is not necessarily the case when comparing drainage densities of basins of different sizes and characteristics. The drainage density of basins expresses how channels and surrounding floodplains are spatially arranged, which is strongly determined by the shape, size and relief of the basin. A low drainage density can indicate large, compact basins with a strong relief, and on the other hand, a high drainage density can indicate small basins with flat elongated or dispersed floodplains. In the present case, the values of the drainage densities for all basins where flash flooding have been observed vary between 0.38 and 0.47, which is below average (Table 3); these basins are also large with a strong relief and compact shapes.

Relief is only the third most important factor after basin shape and drainage density, which is somewhat unexpected, as relief is usually considered one of the most important factors for flooding [18]. The fourth predictor relates to the bifurcation ratio, which appears to be a unique but minor factor in flash flood prediction that, however, is often ignored in other studies presented in the literature. The last and least important predictor relates to the shape of basins, where, as expected, compact basins are more prone to flooding than elongated bases.

Comparison of these results with results of similar studies in other countries or in Egypt [19–33] is fruitless, as all these studies were performed on a much smaller scale than the current study, usually only one basin, and no analysis of variance such as principal components was applied to establish relationships between hydro-morphometric parameters and to identify key parameters related to the total variance, and importantly, no observations of occurring flash floods were used to reveal the predictive power of the parameters to flooding. Therefore, this study shows that a large set of hydro-morphometric parameters can be reduced to a much smaller set without loss of information, indicating redundancy in the data. So, hydro-morphometric parameters are not independent and do not add more information by their number. Therefore, in flood sensitivity analyses, it makes no sense to combine correlated factors that express similar characteristics, as is done in traditional ranking methods where all parameters are combined with the same weight.

The flash flood susceptibility map (Figure 5) shows the spatial distribution into four categories: very high, high, moderate and low susceptibility. The very high sensitivity zone is located in the mountain ranges of the southern Sinai and consists of six major basins known for their flash flooding, such as wadis El-Aawag, Feiran and Kid and the upstream sub-basins of wadis Dahab and Watir, which in total represent an area of approximately 8000 km$^2$, or 15% of the total area of Sinai. The high sensitivity zone is mainly in the center of the Sinai Peninsula and some scattered areas further north. It encompasses 16 river basins, covering a total area of approximately 15,000 km$^2$ or 28% of the Sinai area, including some upstream sub-basins of wadi El-Arish, sub-basins of wadi Dahab and Watir, and the basins of wadis Sedri, Garf and Werdan which drain into the Gulf of Suez. The basins in the north are wadi El-Beada in Bir Al-Abd, which drains to the Mediterranean, and two sub-basins of wadi El-Harish in El-Hasana and Quasisma, respectively, which may be the source of the flash floods that have been observed in this wadi. The moderate sensitivity zone is mainly located in the north of the Sinai Peninsula and includes 36 watersheds with a total area of about 21,000 km$^2$ or 39% of the Sinai area. These are usually smaller to medium sized basins located in flatter areas. The low-sensitive zone comprises 45 catchments with a total area of approximately 10,000 km$^2$, or 18% of the Sinai area. These are usually very small basins on flat terrains with short drainage paths and few branches. Most are located in the north and center along the periphery of the Sinai Peninsula; some are also found in the southern part of Sinai. The resulting flash flood sensitivity map is largely consistent with flash flood observations that occurred in different regions of the Sinai and with the findings or predictions of other studies [6,9,10,31,34].

In particular, the map indicates the high probability of flash flooding in the mountainous basins of southern Sinai, as observed and reported in several publications, such as wadi Feiran [26], wadi Watir [8,31], wadi Dahab [9] and wadi El-Aawag [10]. The map also indicates the high or moderate sensitivity of some basins in southwestern Sinai draining to the Gulf of Suez as described in the literature, such as wadi Sedri [10], wadi Sudr and wadi Wardan [27]. By indicating the high sensitivity to flooding of some sub-basins in El-Hasana and Quasisma of wadi El-Arish, the map also sheds light on the possible origin of flash floods observed in El-Aris, as discussed by Moawad [7], Elewa et al. [28] and Abdel Ghaffar et al. [30]. Nevertheless, it is clear that these results can be improved as more accurate and detailed information on the characteristics of river basins and the occurrence of flash floods in the Sinai becomes available.

Flood susceptibility mapping presented by other studies conducted in Egypt or other countries [19–33] mainly used a classification method consisting of standardization of the morphometric parameters, usually in the range of 1 to 5, and combining the resulting scores without any ranking or weighting as if all parameters have an equal effect on flooding. In addition, there is no validation of the results with field data regarding the occurrence of flash floods, which makes it impossible to verify such simplifying assumptions and approach. In contrast, the method presented in the present study, combining principal component analysis and logistic regression, proves to be robust, reliable and validated and therefore superior to what has been presented before.

The drainage network extracted from satellite data and the sub-division of large basins introduces some bias regarding the range and magnitude of the derived hydro-morphometric parameters. Subdivision is, of course, necessary to get a map of the spatial distribution of flash flood susceptibility in the Sinai. A more detailed subdivision could improve the resolution and accuracy of such a map, but the uncertainty about the exact location of flash flood observations needed to optimize the logistic regression model could potentially cause more bias and uncertainty. In this regard, the current study is only a first attempt, and more research is needed to assess accuracy and improve results.

## 5. Conclusions

Thunderstorms in the Sinai Peninsula often lead to flash floods that can cause significant damage to infrastructure and sometimes even loss of life. Therefore, this study shows how to derive the susceptibility to flash flooding using hydro-morphometric characteristics of the watersheds. Subbasins of various sizes and shapes and their hydro-morphometric features are derived from a digital elevation model from NASA Earthdata. Principal component analysis reveals the relationships between the most important hydro-morphometric parameters and allows us to derive five significant principal components that explain 90% of the variation in the data and have clear physical significance: basin size, drainage density, relief, stream bifurcation and basin shape. This shows that hydro-morphometric parameters are not independent which makes traditional ranking methods to estimate flood sensitivity questionable.

The flash flood probability can be estimated by logistic regression using the significant principal components as predictors and the model coefficients estimated by fitting the model to flash flood observations using maximum likelihood. Cross-validation proves that the model is robust because the estimates of flood probability remain similar to those predicted with the original model. The model shows that the size of a basin is the most important factor in predicting flash flooding, followed by drainage density, relief, bifurcation ratio and basin shape. The logistic model can be used to classify all basins in Sinai into four classes: low, moderate, high and very high susceptibility to flash flooding. The resulting map indicates that the large basins in the mountain ranges of the southern Sinai have a very high susceptibility to flash flooding, several basins in the southwestern Sinai have a high or moderate susceptibility to flash flooding, some sub-basins of wadi El-Arish in the center have a high susceptibility to flash flooding, while smaller to medium-sized basins in flatter areas in the center and north usually have a moderate or low susceptibility to flash flooding. These results are consistent with observations of flash floods that occurred in different regions of the Sinai and with the findings or predictions of other studies.

**Supplementary Materials:** The following supporting information can be downloaded at: https://www.mdpi.com/article/10.3390/w14152434/s1, Table S1: Values of the hydro-morphometric parameters for each basin; Table S2: PCA scores after Varimax rotation; Table S3: Values of the principal components; Table S4: Predicted probability for flash flooding.

**Author Contributions:** Conceptualization, M.E.-R. and W.M.E.; methodology, M.E.-R. and F.D.S.; validation, M.E.-R. and F.D.S.; data curation, M.E.-R. and W.M.E.; writing—review and editing, M.E.-R. and F.D.S.; visualization, M.E.-R.; supervision, M.E.-R. All authors have read and agreed to the published version of the manuscript.

**Funding:** This research received no external funding.

**Institutional Review Board Statement:** Not applicable.

**Informed Consent Statement:** Not applicable.

**Data Availability Statement:** Not applicable.

**Conflicts of Interest:** The authors declare no conflict of interest.

## References

1. WMO. World Meteorological Organization: Flash Flood Guidance System (FFGS) with Global Coverage. 2016. Available online: https://community.wmo.int/hydrology-and-water-resources/flash-flood-guidance-system-ffgs-global-coverage (accessed on 1 May 2022).
2. El Gohary, R. Environmental Flash Flood Management in Egypt. In *Flash Floods in Egypt*; Negm, E.E., Ed.; Advances in Science, Technology & Innovation; Springer: New York, NY, USA, 2020; pp. 85–105.
3. Elnazer, A.A.; Salman, S.A.; Asmoay, A.S. Flash flood hazard affected Ras Gharib city, Red Sea, Egypt: A proposed flash flood channel. *Nat. Haz.* **2017**, *89*, 1389–1400. [CrossRef]
4. Kamel, M.; Arfa, M. Integration of remotely sensed and seismicity data for geo-natural hazard assessment along the Red Sea Coast, Egypt. Arab. *J. Geosci.* **2020**, *13*, 1195. [CrossRef]
5. Abdel-Fattah, M.; Kantoush, S.; Sumi, T. Integrated Management of Flash Flood in Wadi System of Egypt: Disaster Preventing and Water Harvesting. *Annu. Disas. Prev. Res. Inst. Kyoto Univ.* **2015**, *58*, 485–496.

6. Omran, E.-S.E. Egypt's Sinai Desert Cries: Flash Flood Hazard, Vulnerability, and Mitigation. In *Flash Floods in Egypt*; Negm, E.E., Ed.; Advances in Science, Technology & Innovation; Springer: New York, NY, USA, 2020; pp. 215–236.
7. Moawad, M.B. Analysis of the flash flood occurred on 18 January 2010 in wadi El Arish, Egypt (a case study). *Geomat. Nat. Haz. Risk.* **2013**, *4*, 254–274. [CrossRef]
8. Cools, J.; Vanderkimpen, P.; El-Afandi, G.; Abdelkhalek, A.; Fockedey, S.; El-Sammany, M.; Abdallah, G.; El-Bihery, M.; Bauwens, W.; Huygens, M. An early warning system for flash floods in hyper-arid Egypt. *Nat. Haz. Earth Syst. Sci.* **2012**, *12*, 443–457. [CrossRef]
9. Prama, M.; Omran, A.; Schröder, D.; Abouelmagd, A. Vulnerability assessment of flash floods in Wadi Dahab Basin, Egypt. *Environ. Earth Sci.* **2020**, *79*, 114. [CrossRef]
10. El-Fakharany, M.A.; Mansour, N.M. Morphometric analysis and flash floods hazards assessment for Wadi Al Aawag drainage Basins, southwest Sinai, Egypt. *Environ. Earth Sci.* **2021**, *80*, 168. [CrossRef]
11. Horton, R.E. Drainage-basin characteristics. *Eos Trans. Am. Geophys. Union* **1932**, *13*, 350–361. [CrossRef]
12. Horton, R.E. Erosional development of streams and their drainage basins; hydrophysical approach to quantitative morphology. *Geol. Soc. Am. Bull.* **1945**, *56*, 275–370. [CrossRef]
13. Smith, K.G. Standards for grading texture of erosional topography. *Am. J. Sci.* **1950**, *248*, 655–668. [CrossRef]
14. Strahler, A.N. Quantitative analysis of watershed geomorphology. *Trans. Am. Geoph. Union* **1957**, *38*, 913–920. [CrossRef]
15. Strahler, A.N. Quantitative Geomorphology of Drainage Basins and Channel Networks. In *Handbook of Applied Hydrology*; McGraw Hill Book Company: New York, NY, USA, 1964; p. 411.
16. Schumm, S.A. Evolution of drainage systems and slopes in badlands at Perth Amboy, New Jersey. *Bull. Geol. Soc. Am.* **1956**, *67*, 597–646. [CrossRef]
17. Diaconu, D.C.; Costache, R.; Popa, M.C. An overview of flood risk analysis methods. *Water* **2021**, *13*, 474. [CrossRef]
18. Ali, K.; Bajracharyar, R.M.; Raut, N. Advances and challenges in flash flood risk assessment: A review. *J. Geogr. Nat. Disast.* **2017**, *7*, 195. [CrossRef]
19. Farhan, Y.; Anaba, O.; Salim, A. Morphometric analysis and flash floods assessment for drainage basins of the Ras En Naqb area, south Jordan using GIS. *J. Geosci. Environ. Protect.* **2016**, *4*, 9–33. [CrossRef]
20. Mahmood, S.; Rahman, A. Flash flood susceptibility modeling using geo-morphometric and hydrological approaches in Panjkora Basin, Eastern Hindu Kush, Pakistan. *Environ. Earth Sci.* **2019**, *78*, 43. [CrossRef]
21. Mahmood, S.; Rahman, A. Flash flood susceptibility modelling using geomorphometric approach in the Ushairy Basin, eastern Hindu Kush. *J. Earth Syst. Sci.* **2019**, *128*, 97. [CrossRef]
22. Bhat, M.S.; Alam, A.; Ahmad, S.; Farooq, H.; Ahmad, B. Flood hazard assessment of upper Jhelum basin using morphometric parameters. *Environ. Earth Sci.* **2019**, *78*, 54. [CrossRef]
23. Obeidat, M.; Awawdeh, M.; Al-Hatouli, F. Morphometric analysis and prioritisation of watersheds for flood risk management in Wadi Easal Basin (WEB), Jordan, using geospatial technologies. *Flood Risk Manag.* **2021**, *14*, e12711. [CrossRef]
24. Pangali Sharma, T.P.; Zhang, J.; Khanal, N.R.; Prodhan, F.A.; Nanzad, L.; Zhang, D.; Nepal, P. A geomorphic approach for identifying flash flood potential areas in the East Rapti River Basin of Nepal. *ISPRS Int. J. Geo-Inf.* **2021**, *10*, 247. [CrossRef]
25. Arnous, M.O.; Aboulela, H.; Green, D. Geo-environmental hazards assessment of the north western Gulf of Suez, Egypt. *J. Coast. Conserv.* **2011**, *15*, 37–50. [CrossRef]
26. Youssef, A.M.; Pradhan, B.; Hassan, A.M. Flash flood risk estimation along the St. Katherine road, southern Sinai, Egypt using GIS based morphometry and satellite imagery. *Environ. Earth Sci.* **2011**, *62*, 611–623. [CrossRef]
27. Abdel-Lattif, A.; Sherief, Y. Morphometric analysis and flash floods of Wadi Sudr and Wadi Wardan, Gulf of Suez, Egypt: Using digital elevation model. *Arab. J. Geosci.* **2012**, *5*, 181–195. [CrossRef]
28. Elewa, H.H.; Ramadan, E.M.; El-Feel, A.A.; Abu El-Ella, E.A.; Nosair, A.M. Runoff water harvesting optimization by using RS, GIS and watershed modelling in Wadi El-Arish, Sinai. *Int. J. Eng. Res. Technol.* **2013**, *2*, 1635–1648.
29. Abdalla, F.; El-Shamy, I.; Bamousa, A.; Mansour, A.; Mohamed, A.; Tahoon, M. Flash floods and groundwater recharge potentials in arid land alluvial basins, southern Red Sea coast, Egypt. Inter. *J. Geosci.* **2014**, *5*, 971–982. [CrossRef]
30. Abdel Ghaffar, M.K.; Abdellatif, A.D.; Azzam, M.A.; Riad, M.H. Watershed characteristic and potentiality of wadi El-Arish, Sinai, Egypt. *Int. J. Adv. Remote Sens. GIS* **2015**, *4*, 1070–1091.
31. Abuzied, S.; Yuan, M.; Ibrahim, S.; Kaiser, M.; Saleem, T. Geospatial risk assessment of flash floods in Nuweiba area, Egypt. *J. Arid Environ.* **2016**, *133*, 54–72. [CrossRef]
32. Abdelkareem, M. Targeting flash flood potential areas using remotely sensed data and GIS techniques. *Nat. Haz.* **2017**, *85*, 19–37. [CrossRef]
33. Elsadek, W.M.; Ibrahim, M.G.; Mahmod, W.E. Flash flood risk estimation of Wadi Qena Watershed, Egypt using GIS based morphometric analysis. *Appl. Environ. Res.* **2018**, *40*, 36–45. [CrossRef]
34. Abuzied, S.M.; Mansour, B.M.H. Geospatial hazard modeling for the delineation of flash flood-prone zones in Wadi Dahab basin, Egypt. *J. Hydroinform.* **2019**, *21*, 180–206. [CrossRef]
35. Elsadek, W.M.; Ibrahim, M.G.; Mahmod, W.E.; Kanae, S. Developing an overall assessment map for flood hazard on large area watershed using multi-method approach: Case study of Wadi Qena watershed, Egypt. *Nat. Haz.* **2019**, *95*, 739–767. [CrossRef]
36. Davis, J. *Statics and Data Analysis in Geology*; Wiley: New York. NY, USA, 1975.

37. Abrams, M.; Crippen, R.; Fujisada, H. ASTER Global Digital Elevation Model (GDEM) and ASTER Global Water Body Dataset (ASTWBD). *Remote Sens.* **2020**, *12*, 1156. [CrossRef]
38. Melton, M. *An Analysis of the Relations among Elements of Climate, Surface Properties and Geomorphology*; Department of Geology, Columbia University, Technical Report, 11, Project NR 389-042; Office of Navy Research: New York, NY, USA, 1957. [CrossRef]
39. Strahler, A.N. Hypsometric (area-altitude) analysis of erosional topography. *Geol. Soc. Am. Bull.* **1952**, *63*, 1117–1142. [CrossRef]
40. Smith, K.G. Erosional processes and landforms in badlands national monument, South Dakota. *Bull. Geol. Soc. Am.* **1958**, *69*, 975–1008. [CrossRef]
41. Jolliffe, I.T. *Principal Component Analysis*, 2nd ed.; Springer: New York, NY, USA, 2002.
42. R Core Team. *R: A Language and Environment for Statistical Computing*; R Foundation for Statistical Computing: Vienna, Austria, 2017. Available online: http://www.R-project.org/ (accessed on 15 November 2021).
43. Akaike, H. A new look at the statistical model identification. *IEEE Trans. Autom. Control.* **1974**, *19*, 716–723. [CrossRef]

water

Article

# Driving Effects and Spatial-Temporal Variations in Economic Losses Due to Flood Disasters in China

Zhixiong Zhang [1,2], Qing Li [1,2], Changjun Liu [1,2,*], Liuqian Ding [1,2], Qiang Ma [1,2] and Yao Chen [1,2]

1 China Institute of Water Resources and Hydropower Research, Beijing 100038, China; zhixiongzhang1991@foxmail.com (Z.Z.); liqing@iwhr.com (Q.L.); dinglq@iwhr.com (L.D.); maqiang@iwhr.com (Q.M.); chenyao88520@126.com (Y.C.)
2 Research Center on Flood & Drought Disaster Reduction of the Ministry of Water Resources, Beijing 100038, China
* Correspondence: lcj2005@iwhr.com; Tel.: +86-10-6878-1214

**Abstract:** The economic loss caused by frequent flood disasters poses a great threat to China's economic prosperity. This study analyzes the driving factors of flood-related economic losses in China. We used the extended Kaya identity to establish a factor decomposition model and the logarithmic mean Divisia index decomposition method to identify five flood-related driving effects for economic loss: demographic effect, economic effect, flash flood disaster control effect, capital efficiency effect, and loss-rainfall effect. Among these factors, the flash flood disaster control effect most obviously reduced flood-related economic losses. Considering the weak foundation of flash flood disaster prevention and control in China, non-engineering measures for flash flood prevention and control have been implemented since 2010, achieving remarkable results. Influenced by these measures, the loss-rainfall effect also showed reduction output characteristics. The demographic, economic, and capital efficiency effects showed incremental effect characteristics. China's current economic growth leads to an increase in flood control pressure, thus explaining the incremental effect of the economic effect. This study discusses the relationship between flood-related economic loss and flash flood disaster prevention and control in China, adding value for the adjustment and formulation of future flood disaster prevention policies.

**Keywords:** economic losses from flood disasters; flash flood disaster control; Kaya identity; LMDI technique decomposition method

**Citation:** Zhang, Z.; Li, Q.; Liu, C.; Ding, L.; Ma, Q.; Chen, Y. Driving Effects and Spatial-Temporal Variations in Economic Losses Due to Flood Disasters in China. *Water* **2022**, *14*, 2266. https://doi.org/10.3390/w14142266

Academic Editor: Gwo-Fong Lin

Received: 20 June 2022
Accepted: 18 July 2022
Published: 20 July 2022



## 1. Introduction

Flooding has the highest frequency of all natural disasters worldwide. From 2000 to 2019, flooding accounted for 44% of the total number of natural disasters [1]. About two thirds of China's land area regularly face the threat of floods of different types and degrees of danger [2]. Moreover, the economic losses caused by these floods have seriously hindered the sustainable development of China's economic society [3]. Therefore, investigating the main factors affecting the economic losses related to flood disasters has important reference significance for flood disaster prevention and control policymaking in China.

Considering the economic losses caused by flood disasters, many researchers have performed in-depth analyses and research, obtaining corresponding research results. Among them, Jiang performed a comprehensive analysis of the characteristics of flood disaster losses in China from 1950 to 2016 and proposed that flood-related economic losses in China showed a downward trend [4]. Jiang mainly defines the concept and limitations of the indirect economic losses related to urban flood disasters, highlighting that the correlation between industrial loss and resources is a substantial part of the indirect economic loss caused by urban flood disasters [5]. Some researchers introduced the dynamic computable general equilibrium model for comprehensive disaster-related economic loss assessment and constructed an equilibrium model of rainstorm and flood disasters [6–8]. Their results show that the occurrence of rainstorms and floods will affect social and economic development in the current year, as well as have a significant impact on economic development in later periods [6–8]. These studies mainly focus on calculating the direct and indirect economic losses related to flood disasters. However, in China, 70% of flood-related economic losses are due to flash flood disasters [9]; nevertheless, research on the relationship between the economic losses due to flood disasters and flash flood disasters is relatively rare.

Considering the weak foundation of flash flood disaster prevention and control in China [10], the Ministry of Water Resources officially started implementing the special construction of flash flood disaster prevention and control infrastructure in 2010 [11]. During the first stage of the construction for flash flood disaster prevention and control, from 2010 to 2012, a total of 11.7 billion Yuan was invested in the construction of a non-engineering system for flash flood disaster prevention and control in 2058 counties of 29 provincial-level administrative regions. During the second stage, from 2013 to 2015, a total of 14.3 billion Yuan was invested nationwide to investigate and evaluate flash flood disasters, construct non-engineering measures for the prevention and control of flash floods, and to prevent flash flood gulley erosion. A total of 9.2 billion Yuan was invested in the third phase of the program, which lasted from 2016 to 2020. The main focus of this stage was optimizing and improving the non-engineering measures for the prevention and control of flash flood disasters, utilizing the results of flash flood disaster investigations and evaluations, and continuing to carry out the construction of disaster mass monitoring and mass preventing of flash floods [12,13].

Based on the above, this study will fully consider the dual nature and social attributes of the disaster-causing factors of floods, by employing the Kaya identity and a logarithmic mean Divisia index (LMDI) approach to quantitatively measure the driving effects of inter-annual changes in economic loss related to flood disasters in China. The Kaya identity is used to decompose the driving factors and the LMDI method is used to determine the size of each influencing factor. We also explore the influence of the flash flood disaster prevention investment, economy, annual rainfall, and other factors on the changes in flood-related economic loss.

## 2. Methods

### 2.1. Kaya Identity

The Kaya identity was proposed by Japanese scholar Yoichi Kaya in 1989 [14] and was originally used for carbon emissions research [15,16]. After years of research expansion, it is currently widely used in the field of energy research. The Kaya identity uses a simple mathematical formula to explain the relationship between the macro overall social and economic factors and describes these using simple mathematical relationships that take into account the national level of carbon emissions associated with human production and living in four elements. As it employs a simple mathematical formula to change drivers, the Kaya identity has the advantage of strong explanatory power and is thus widely used across different fields.

The Kaya identity decomposes carbon emissions into four influencing factors, and the expression formula is as follows:

$$C = P \times \left(\frac{G}{P}\right) \times \left(\frac{E}{G}\right) \times \left(\frac{C}{E}\right) \tag{1}$$

where $G$ = gross domestic product (GDP); $E$ = energy consumption; $G/P$ = per capita GDP; $E/G$ = energy intensity; and $C/E$ = carbon intensity in energy consumption.

The advantage of the Kaya identity, compared with other models for studying driving factors of carbon emissions, lies in the fact that researchers can expand the Kaya identity according to their own research needs and add other influencing factors to study its influence on the change in research objects. Recently, many researchers have applied this method to different research fields [15,16].

To better describe the effects of different factors on flood-related economic losses, we use the Kaya identity to expand the decomposition of flood-related economic loss.

We define the variable C as the economic losses from flood disasters in province i. Specifically, the economic loss due to floods in each province is decomposed into a product of five factors: the resident population or the demographic effect of province i (*P*); the per capita GDP or the economic effect of province i (*G/P*), representing current operational economic status; the flash flood disaster prevention and control of investment–GDP ratio or the flash flood disaster control effect of province i (*W/G*), representing the investment intensity in disaster prevention; the rainfall–flash flood disaster prevention and control of investment ratio or the capital efficiency effect of province i (*R/W*), representing the intensity of rainfall faced per unit defense fund; and the economic losses of flood disasters–rainfall ratio or the loss-rainfall effect of province i (*C/R*). Then, we denote the five factors *pi*, *gi*, *wi*, *ei*, and *ai*, as shown in the following expression.

$$C = P \times \left(\frac{G}{P}\right) \times \left(\frac{W}{G}\right) \times \left(\frac{R}{W}\right) \times \left(\frac{C}{R}\right) = p \cdot g \cdot w \cdot e \cdot a \tag{2}$$

## 2.2. Logarithmic Mean Divisia Index

The LMDI was first proposed by Professor Ang from Singapore [17–19]. This method was at first applied in carbon emissions research, mainly to analyze energy intensity change. The LMDI is a common factor decomposition model in global research and gives perfect decomposition [20]; this means that the results do not contain an unexplained residual term, which simplifies the result interpretation.

According to the LMDI decomposition model, the total effect $\Delta C$ on the change value of flood-related economic loss in the base period and the year t is called the total effect, representing demographic effect ($P_{effect}$), economic effect ($g_{effect}$), flash flood disaster control effect ($w_{effect}$), capital efficiency effect ($e_{effect}$), and loss-rainfall effect ($a_{effect}$). When the calculated effect value is positive, this index has an incremental effect on the direct flood-related economic loss and will lead to an increase in disaster loss. Conversely, when the effect value is negative, the index has a reduction effect on the direct flood-related economic loss, which can reduce the economic loss. The five driving factor relationships can be expressed as:

$$\Delta C = C_t - C_0 = p_{effect} + g_{effect} + w_{effect} + e_{effect} + a_{effect} \tag{3}$$

$$p_{effect} = \frac{C_t - C_0}{\ln C_t - \ln C_0} \ln\left(\frac{p_t}{p_0}\right)$$

$$g_{effect} = \frac{C_t - C_0}{\ln C_t - \ln C_0} \ln\left(\frac{g_t}{g_0}\right)$$

$$w_{effect} = \frac{C_t - C_0}{\ln C_t - \ln C_0} \ln\left(\frac{w_t}{w_0}\right)$$

$$e_{effect} = \frac{C_t - C_0}{\ln C_t - \ln C_0} \ln\left(\frac{e_t}{e_0}\right)$$

$$a_{effect} = \frac{C_t - C_0}{\ln C_t - \ln C_0} \ln\left(\frac{a_t}{a_0}\right)$$

where $C_0$ = flood-related economic loss in the base period; $C_t$ = flood-related economic loss in the year *t*.

## 2.3. Study Object and Data Sources

In this study, various sources were used to collect the data of 29 provincial areas in China, from 2010 to 2020. These areas included 21 provinces, 3 province-level megacities (Beijing, Tianjin, and Chongqing), and 5 autonomous regions. Due to the lack of relevant statistical data, Taiwan and two special administrative regions—Hong Kong and

Macao—were excluded from the study. Furthermore, as Shanghai and Jiangsu Province are not included in the flash flood disaster prevention and control project area, to ensure the consistency of data integrity of all provinces and cities, Shanghai and Jiangsu were excluded from this study.

The key data sources were the China Statistical Yearbook 2010–2020, obtained from the website of the National Bureau of Statistics of the People's Republic of China, and the China Water Resources Bulletin and the Bulletin on Flood and Drought Disasters in China 2010–2020, obtained from the website of the Ministry of Water Resources of the People's Republic of China.

The list of counties involved in flash flood disaster prevention and control was obtained from the National Flash Flood Prevention and Control Planning. The funding data of the prevention and control project were derived from the implementation plans of national flash flood disaster prevention and control projects in China. Flash flood disaster prevention and control projects are implemented by the county as a unit, the main task of which is to establish flash flood disaster monitoring and an early warning platform for flash flood disasters at the county level. This is accomplished through real-time dynamic monitoring of the water and rain situation to achieve flash flood disaster prevention and control work. In addition, as the funds for flash flood disaster prevention and control projects were distributed in stages, the public data on the projects' construction funds in each county were not available. Therefore, this study adopts the method of average allocation relating to the data of flash flood disaster prevention and control funds.

## 3. Results and Discussion

### 3.1. Temporal and Spatial Distribution of Flood-Related Economic Losses in China

According to the statistical data from the China Flood and Drought Disaster Prevention Bulletin, the average annual flood disaster loss in China from 2010 to 2020 was 237.326 billion Yuan (Figure 1). The flood-related economic loss in 2010 was the highest in the study period, at 374.543 billion Yuan, while the loss in 2011 was the lowest, at 130.127 billion Yuan. In terms of the spatial distribution of flood-related economic losses, flood disasters in China show a trend of being high in the south and low in the north (Figure 2). From 2010 to 2020, Sichuan, Guangdong, Hunan, Zhejiang, and Jiangxi were the top five provinces, highest to lowest, in terms of annual average economic losses from floods [21]. These five provinces are in the southern part of China, where rainfall is abundant and rivers are widespread, making these areas prone to flash floods. Sichuan Province had the highest average annual economic loss, at 22.130 billion Yuan. According to the statistical data and existing research [22,23], flood-related economic losses and the frequency of floods have both shown an increasing trend in Sichuan Province since 2014. Moreover, the frequency of flash flood disasters in Sichuan Province accounted for about 30% of the total number of floods in China [24]. Ningxia, Tianjin, Qinghai, Xizang, and Xinjiang had lower flood-related economic losses. These provinces are in the northern part of China, which has low rainfall and an arid climate, and are not prone to flash floods.

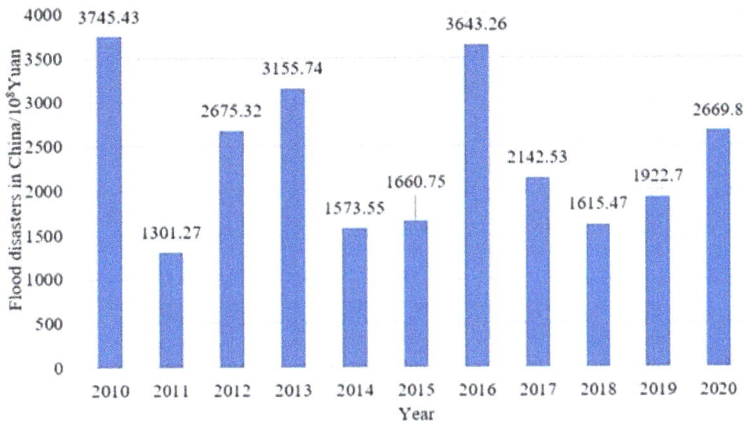


**Figure 1.** China's flood-related economic losses from 2010 to 2020. Data from the China Flood and Drought Disaster Prevention Bulletin, 2010–2020.

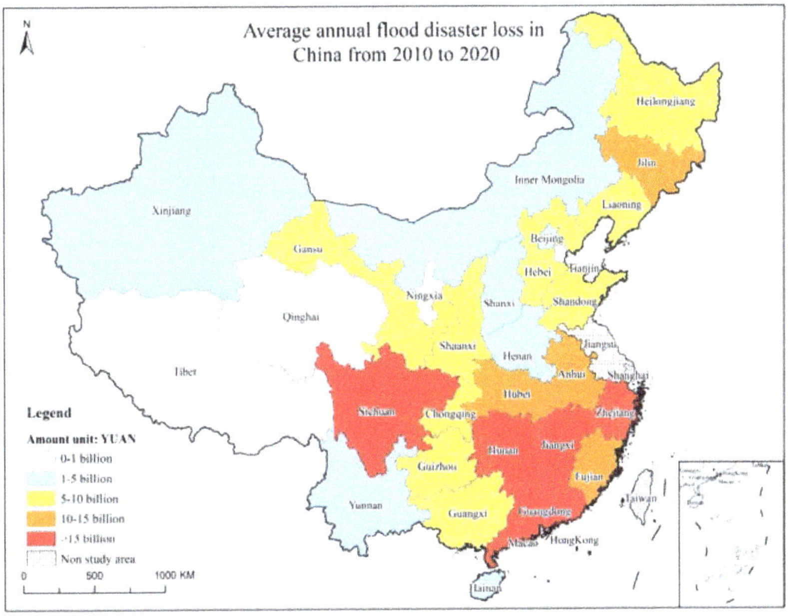


**Figure 2.** China's flood-related economic losses from 2010 to 2020.

### 3.2. Analysis of the Driving Factors of Flood-Related Economic Loss

Based on the LMDI decomposition model, this study analyzed the main driving factors affecting the direct economic losses related to flood disasters in China from 2010 to 2020. The demographic effect, economic effect, flash flood disaster control effect, disaster prevention pressure effect, and loss-rainfall effect on flood-related economic loss in 29 provinces of China were quantitatively analyzed (Tables 1 and 2). The results clearly show the reduction effect of the implemented flash flood disaster control measures. As the construction of flash flood disaster prevention and control continue, the amount of money invested in the flash flood disaster prevention and control work to reduce the economic loss plays an important role. The economic effect was characterized by obvious incremental effect. China is in the later stage of industrialization and the middle stage of urbanization [25]. Therefore, investment in urban infrastructure construction is increasing greatly [26], and the growth of regional economic density will also bring greater challenges in flood control work. It is particularly necessary to focus on the frequent occurrence of urban waterlogging disasters in recent years, which has become a common problem in major cities [27]. The demographic effect was characterized by an incremental effect; however, the effect value is relatively low, compared with the economic effect, which is caused by the low population growth rate in China. The loss-rainfall effect was characterized by a reduction effect; this effect reflects the relationship between the flood-related economic loss and the annual rainfall intensity, indicating that disaster prevention ability is strengthened under the same rainfall intensity. On the whole, the capital efficiency effect showed an incremental effect, indicating that there is still at present great pressure on capital investment to cope with the prevention and control work of heavy rainfall.

**Table 1.** Effect decomposition of China's flood-related economic losses (in Yuan).

| Year | Loss-Rainfall Effect | Capital Efficiency Effect | Flash Flood Disaster Control Effect | Economic Effect | Demographic Effect | Total |
|---|---|---|---|---|---|---|
| 2010–2011 | −2033.789 | −410.3707 | −378.5802 | 364.3994 | 14.1807 | −2444.16 |
| 2011–2012 | 1056.044 | 318.0058 | −200.9019 | 186.739 | 14.1629 | 1374.05 |
| 2012–2013 | 659.6843 | −761.9825 | 320.2077 | 245.3544 | 17.1559 | 480.4199 |
| 2013–2014 | −1494.108 | −88.0821 | −190.0587 | 174.8112 | 15.2475 | −1582.19 |
| 2014–2015 | −9.8524 | 95.3566 | −117.6909 | 111.419 | 7.9674 | 87.1997 |
| 2015–2016 | 1731.183 | 2655.194 | −2611.375 | 191.0343 | 16.4746 | 1982.51 |
| 2016–2017 | −1236.258 | −264.4714 | −302.4272 | 286.6552 | 15.7719 | −1500.73 |
| 2017–2018 | −576.1079 | 49.0477 | −156.3971 | 149.3445 | 7.0526 | −527.0602 |
| 2018–2019 | 389.8009 | −82.5708 | −171.6208 | 165.767 | 5.8539 | 307.23 |
| 2019–2020 | 561.9533 | 185.1468 | −46.1579 | 42.8678 | 3.2902 | 747.1002 |
| Effect average | −95.145 | 169.5273 | −385.5002 | 191.8392 | 11.7158 | |
| Effect standard deviation | 1208.93 | 927.548 | 803.9399 | 90.1191 | 5.1031 | |
| Effect coefficient of variation | −12.7062 | 5.4714 | −2.0854 | 0.4698 | 0.4356 | |

**Table 2.** Effect decomposition of flood-related economic losses in China (in Yuan).

| Region | Loss-Rainfall Effect | Capital Efficiency Effect | Flash Flood Disaster Control Effect | Economic Effect | Demographic Effect | Total |
|---|---|---|---|---|---|---|
| Beijing | 0.3007 | −0.4130 | −1.3098 | 1.1466 | 0.2754 | 0.0000 |
| Tianjin | 0.6224 | −0.7810 | −0.0749 | 0.1827 | 0.0449 | −0.0059 |
| Hebei | 0.0623 | 6.3323 | −10.7380 | 3.4588 | 0.2147 | −0.6699 |
| Shanxi | −1.4803 | 3.3480 | −3.7116 | 1.3850 | −0.0441 | −0.5030 |
| Inner Mongolia | −3.2358 | 2.0408 | −1.3984 | 2.0413 | −0.0950 | −0.6471 |
| Liaoning | −20.0179 | −8.3127 | −2.5122 | 4.9932 | −0.1004 | −25.9500 |
| Jilin | −43.0281 | −5.9829 | −3.0722 | 2.9943 | −0.7361 | −49.8250 |
| Heilongjiang | −0.3310 | −0.3150 | 1.2548 | 0.9660 | −1.0277 | 0.5471 |
| Zhejiang | 3.1282 | 3.1035 | −20.9594 | 9.7150 | 2.6387 | −2.3740 |
| Anhui | 44.4181 | 25.9407 | −26.8928 | 8.0031 | 0.2319 | 51.7010 |
| Fujian | −17.0980 | 25.9043 | −38.1183 | 9.5228 | 1.1102 | −18.6790 |
| Jiangxi | −10.6887 | 1.7504 | −19.2507 | 12.2608 | 0.1662 | −15.7620 |
| Shandong | −3.7440 | −5.1765 | −0.8274 | 1.9150 | 0.3940 | −7.4389 |
| Henan | −13.1402 | 4.1320 | −5.9747 | 1.7968 | 0.1081 | −13.0780 |
| Hubei | 1.1752 | 29.3494 | −35.1599 | 10.3178 | 0.0956 | 5.7781 |
| Hunan | −6.0440 | 9.1161 | −25.6694 | 12.7677 | 0.1605 | −9.6691 |
| Guangdong | −5.1864 | 6.0111 | −27.7194 | 12.9568 | 3.4269 | −10.5110 |
| Guangxi | 3.2246 | 4.4470 | −8.9474 | 4.9755 | 0.6034 | 4.3031 |
| Hainan | −9.3221 | 0.9359 | −7.9598 | 4.0317 | 0.6672 | −11.6471 |

**Table 2.** *Cont.*

| Region | Loss-Rainfall Effect | Capital Efficiency Effect | Flash Flood Disaster Control Effect | Economic Effect | Demographic Effect | Total |
|---|---|---|---|---|---|---|
| Chongqing | 8.8715 | 6.8277 | −13.4133 | 7.2321 | 0.7579 | 10.2759 |
| Sichuan | −2.3962 | −0.9183 | −19.4274 | 19.5975 | 0.5814 | −2.5630 |
| Guizhou | 2.8052 | 8.5286 | −13.3576 | 5.7722 | 0.4985 | 4.2469 |
| Yunnan | 2.2289 | 5.0178 | −9.4595 | 4.7431 | 0.1117 | 2.6420 |
| Tibet | −0.5672 | 0.7986 | −1.6222 | 0.7790 | 0.1528 | −0.4590 |
| Shaanxi | −18.7190 | 1.4214 | −7.2012 | 6.6924 | 0.3425 | −17.4639 |
| Gansu | 3.8549 | −0.8067 | −3.1115 | 5.4300 | −0.1697 | 5.1970 |
| Qinghai | −0.3791 | 0.3224 | −0.6531 | 0.3436 | 0.0222 | −0.3440 |
| Ningxia | −0.1595 | 0.5062 | −0.7422 | 0.2460 | 0.0485 | −0.1010 |
| Xinjiang | −3.3441 | 2.9019 | −4.3370 | 1.1898 | 0.2564 | −3.3330 |
| National total | −95.1450 | 169.5273 | −385.5002 | 191.8392 | 11.7158 | −107.5629 |

To simplify and analyze the influence of the driving effects on the time series changes of flood-related economic losses more directly, we created a spatial clustering to illustrate the spatial variation in the driving effects, based on an ISODATA clustering model [28]. In this model, the clustering is specific to each type of drive effect analysis.

### 3.2.1. Demographic Effect

The demographic effect was the weakest in terms of change in flood-related economic loss, with an effect value of 1.17158 billion Yuan. As China enters a stage of low fertility [29], the population growth rate is relatively slow, with the population in some provinces showing a downward trend. This is the main reason for the low demographic effect value.

One region in which demographic effect was high is Guangdong (Figure 3). During the study period, the demographic effect value in the province was 342.69 million Yuan, showing the most obvious incremental effect among all provinces. According to statistical data, Guangdong is the most populous province in China, and during the study period, its population growth showed a steady upward trend, with an increase of 21.83 million people from 2010 to 2020. The increase in population also puts higher demands on the flood prevention work.

The regions with medium levels of demographic effect are Zhejiang, Fujian, and Chongqing (Figure 3). Zhejiang and Fujian are in the southeast coastal area of China, while Chongqing is in the southwest. These three regions have high population density, and all of them are flood prone provinces. Statistics show that the population of these three provinces are increasing and therefore the demographic effect was also incremental. However, compared with the high demographic effect area, these areas' effect values are lower.

The regions with low levels of demographic effect include the following 25 provinces: Beijing, Tianjin, Hebei, Shanxi, Inner Mongolia, Liaoning, Jilin, Heilongjiang, Anhui, Jiangxi, Shandong, Henan, Hubei, Hunan, Guangxi, Hainan, Sichuan, Guizhou, Yunnan, Tibet, Shaanxi, Gansu, Qinghai, Ningxia, and Xinjiang (Figure 3). Among them, the demographic effect values of Shanxi, Inner Mongolia, Liaoning, Jilin, Heilongjiang, and Gansu showed a decreasing effect, as the population number of these six provinces showed a decreasing trend during the study period; therefore, the population pressure in terms of flood disaster prevention also showed a decreasing trend. The other 19 provinces had average demographic effects of around 25.61 million Yuan. Although the characteristics of positive effect would lead to increased flood-related economic losses, overall, they account for a relatively small effect.

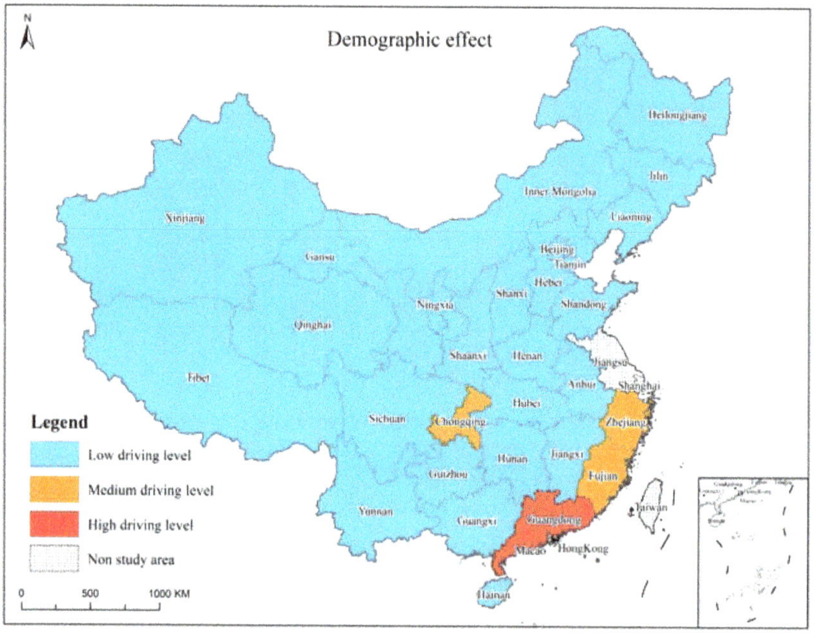


**Figure 3.** Demographic effect driving distribution.

### 3.2.2. Economic Effect

During the study period, the economic effect value of flood-related economic loss was 19.18392 billion Yuan, making it the most obvious incremental driving effect. From 2010 to 2020, China's national economy has maintained an average annual growth rate of more than 7% (Figure 4). The country's rapid economic growth also poses a growing challenge in terms of flood disaster prevention. The economic effects of flood-related economic loss in all provinces are also incremental.

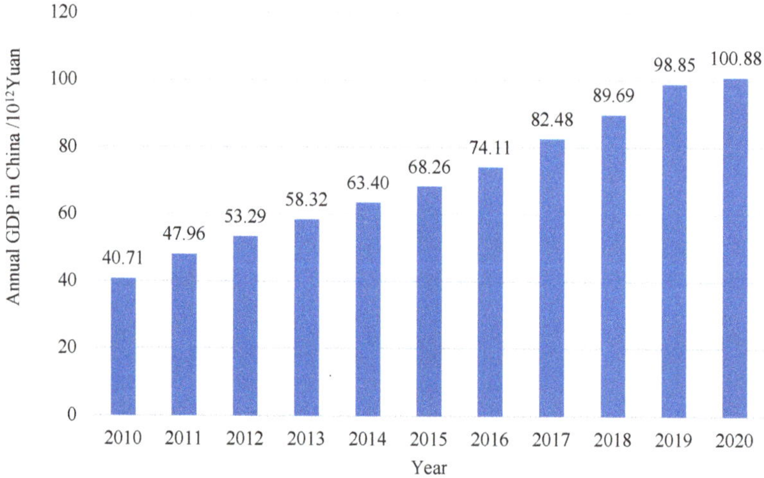


**Figure 4.** China's GDP from 2010 to 2020. Data from the China Statistical Yearbook 2010–2020.

Jiangxi and Sichuan showed the highest driving effect of flood-related economic losses, with the average effect of these two provinces during the study period as high as 1.45506 billion Yuan (Figure 5). The two provinces had similar GDPs per capita and are both prone to flash floods and flood disasters. During the study period, the average annual economic losses caused by flood disasters in Jiangxi and Sichuan ranked second and fifth, respectively, among the 29 provinces. With the continuous economic development, the per capita GDP in both Jiangxi and Sichuan has also increased significantly, which brings new challenges in terms of flood disaster prevention work; therefore, the economic effect was at a high driving level.

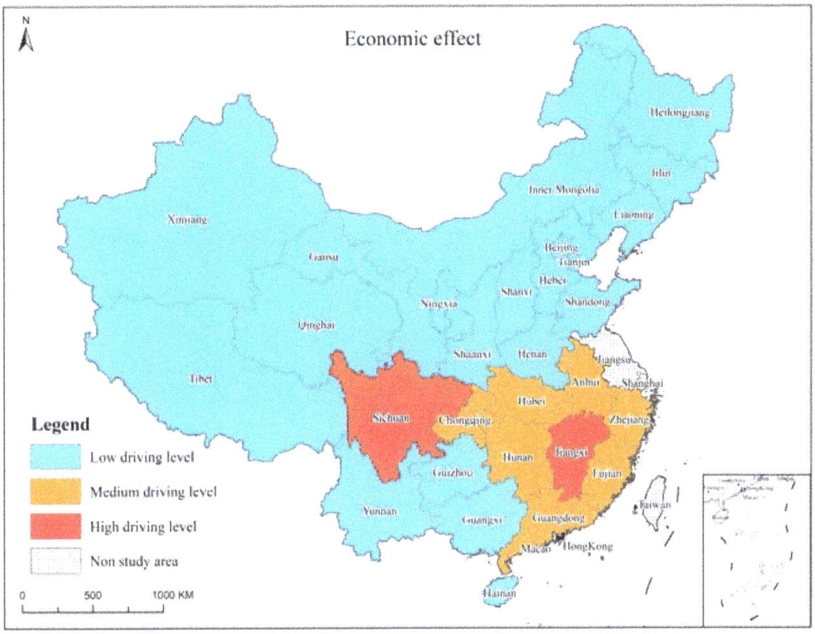


**Figure 5.** Economic effect driving distribution.

The provinces with medium driving levels of economic effect were Zhejiang, Anhui, Fujian, Hubei, Hunan, Guangdong, and Chongqing (Figure 5). According to statistical data, the economy of these seven provinces showed a general upward trend during the study period; moreover, the per capita GDPs of Zhejiang and Fujian ranked first and second in the country, with both provinces' per capita GDP exceeding 100,000 Yuan. In terms of spatial distribution, most of these provinces are in Southeast and South Central China, where annual rainfall is abundant. Therefore, these provinces are also prone to flood and flash flood disasters. Zhejiang, Fujian, and Guangdong are also typhoon-prone areas, and the flood prevention undertaking is particularly heavy in these provinces. The average annual economic losses caused by floods in these seven provinces were all among the top nine. Even though the continuous economic growth in these provinces will cause an increase in the difficulty of flood disaster prevention and control, the effect value was slightly lower than that of the high driving effect provinces.

There were 20 provinces that had low economic driving effects: Beijing, Tianjin, Hebei, Shanxi, Inner Mongolia, Liaoning, Jilin, Heilongjiang, Shandong, Henan, Guangxi, Hainan, Guizhou, Yunnan, Tibet, Shaanxi, Gansu, Qinghai, Ningxia, and Xinjiang (Figure 5). Among them, the per capita GDPs of Beijing and Tianjin in 2020 exceeded 100,000 Yuan, and the per capita GDPs of Inner Mongolia and Shandong exceeded 70,000 Yuan. These four provinces are in the northern region and their flood-related economic loss in recent years is significantly lower than that of provinces and cities in the south. As their effect value is relatively small, they have a low economic driving effect. Most of the other 16 provinces are in Northern China, where economic losses from floods are relatively low; however, the per capita GDPs of these regions are significantly lower than those of regions with high and medium economic driving effects.

### 3.2.3. Flash Flood Disaster Control Effect

The flash flood disaster control effect showed an obvious reduction effect and had the highest absolute value among all five effects. This shows that the implementation of flash flood disaster prevention and control projects has greatly reduced flood-related economic losses in China. The continuous development of flash flood disaster prevention and control plays a positive role in reducing flood-related economic losses in China.

The provinces with high levels of flash flood disaster control effects were Shanxi, Anhui, Fujian, and Hubei (Figure 6). During the study period, the average effect of flash flood disaster prevention and control in these four provinces was −3.37905 billion Yuan, and the development of flash flood disaster work greatly reduced disaster-related loss in these regions. Shanxi, located in Northern China, is a loess-covered mountain plateau, with mountains and hills accounting for more than 80% of the province's total area [30]. Flash flood disaster prevention has always been a focus point and a difficulty in terms of Shanxi flood control. Anhui and Hubei are in the central and southern parts of China, while Fujian is in the southeast coastal area, an area that experiences more serious floods [21].

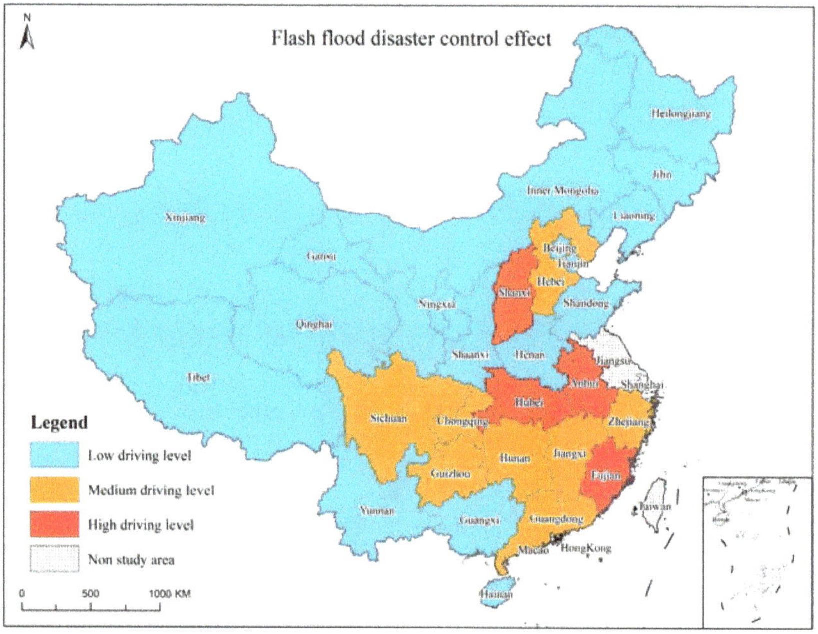


**Figure 6.** Flash flood disaster control effect driving distribution.

Provinces with a medium level of flash flood disaster control effect include Hebei, Zhejiang, Jiangxi, Hunan, Guangdong, Chongqing, Sichuan, and Guizhou (Figure 6); these provinces are mainly located in Southeast and Central China. The flash flood disaster prevention and control effect in these provinces has a reduction effect. Hebei is located at the eastern foot of Taihang Mountain, and mountainous areas account for 52.7% of the total area of the province [31]. Flash floods caused by local heavy rains occur frequently. Zhejiang, Jiangxi, Hunan, Guangdong, Chongqing, Sichuan, and Guizhou are in the south of China, an area prone to flash flood disasters [21]. Through the flash flood disaster prevention and control projects, these provinces have established a sound flash flood disaster prevention system, which has played a significant role in disaster prevention and mitigation, reducing casualties, and improving the information level of grassroots water conservancy; it also had a positive impact on reducing flood-related economic losses [9–13].

The regions with low levels of flash flood disaster control effect include Beijing, Tianjin, Inner Mongolia, Liaoning, Jilin, Heilongjiang, Shandong, Henan, Guangxi, Hainan, Yunnan, Tibet, Shaanxi, Gansu, Ningxia, and Xinjiang (Figure 6). Heilongjiang was the only province with an incremental effect value of flash flood prevention; however, the effect value was small. Although Guangxi, Hainan, and Yunnan are in Southern China—an area prone to flash flood disasters—the average annual economic losses related to floods from 2010 to 2020 were relatively low; therefore, they are at a low driving level. Most of the other 12 provinces are in the northern part of China, and the flood-related economic loss was lower than that of the provinces with high or medium driving force.

### 3.2.4. Capital Efficiency Effect

The capital efficiency effect of flood-related economic loss reflects the relationship between regional flash flood disaster input and annual rainfall. The effect values of most provinces were incremental. Due to large variations in annual precipitation in each region and the relatively stable investment of flash flood disaster prevention funds, the inter-annual variation in the effect value of each province is also apparent.

The provinces with a high driving level of capital efficiency effect were Shanxi, Anhui, Fujian, and Hubei, and the capital efficiency effect was incremental (Figure 7). From 2010 to 2020, the average annual rainfall in Shanxi, Anhui, Hubei, and Fujian was 540 mm, 1272 mm, 1182 mm, and 1778 mm, respectively. Flash floods are the most critical type of flood disaster in these provinces, and heavy rainfall is the most important factor of flash flood disasters. The inter-annual variation in rainfall in these four provinces has a strong impact on the flood-related economic losses in the region, when the investment in flash flood disaster prevention and control projects has little inter-annual variation.

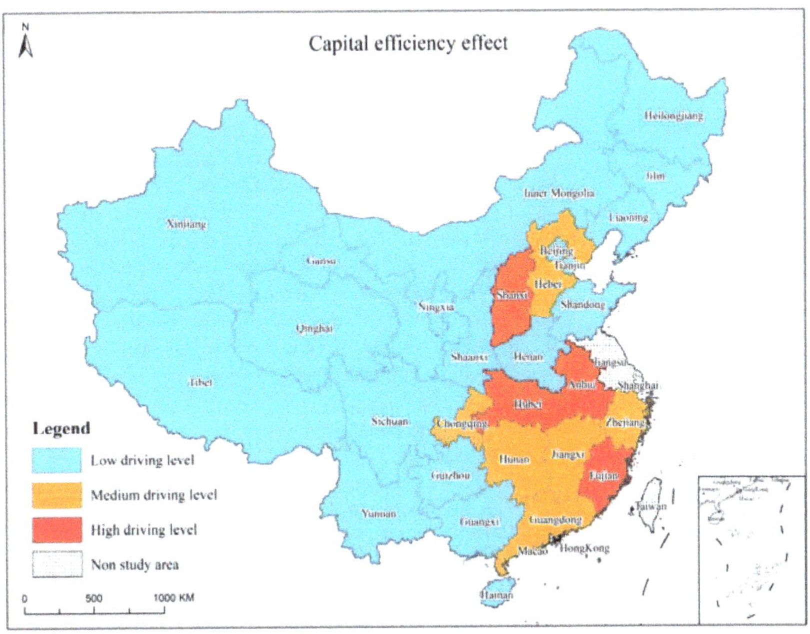


**Figure 7.** Capital efficiency effect driving distribution.

The regions with medium driving levels of capital efficiency effect were Hebei, Zhejiang, Jiangxi, Hunan, Guangdong, and Chongqing (Figure 7). From 2010 to 2020, the average annual rainfall in Hebei, Zhejiang, Jiangxi, Hunan, Guangdong, and Chongqing was 513 mm, 1795 mm, 1861 mm, 1782 mm, 1874 mm, and 1164 mm, respectively. These provinces are also regions where flash flood disasters occur more frequently, and the inter-annual variations in precipitation had a strong impact on the regional flood-related economic loss; however, the effect value was lower than in the high driving level regions.

The regions with a low driving level of capital efficiency effect were Beijing, Tianjin, Inner Mongolia, Liaoning, Jilin, Heilongjiang, Shandong, Henan, Guangxi, Hainan, Sichuan, Guizhou, Yunnan, Tibet, Shaanxi, Gansu, Qinghai, Ningxia, and Xinjiang (Figure 7). Hainan, Guizhou, Yunnan, Sichuan, and Guangxi are in the south of China with abundant precipitation; however, the capital efficiency effect value of these five provinces was obviously smaller than that of strong and medium driving regions, indicating that the inter-annual variation in rainfall had a weak impact on flood-related economic loss. Other provinces are mainly located in the northern part of China, with lower rainfall than the southern region; therefore, the flood-related economic loss in these provinces is relatively low.

### 3.2.5. Loss-Rainfall Effect

The loss-rainfall effect reflects the relationship between flood-related economic losses and annual rainfall in a region. If this effect is characterized by an incremental effect, it indicates that the economic loss of flood disaster in the region presents an increasing trend under the same rainfall intensity. On the contrary, if this effect is characterized by a reduction effect, it indicates that under the same rainfall intensity in the region, the economic loss of flood disaster shows a decreasing trend.

The loss-rainfall effect showed a clear reduction, with an average value of −9.5145 billion Yuan in China, from 2010 to 2020. During the study period, the loss-rainfall effects of most provinces were reduced, indicating that the flood-related economic loss in most provinces and cities is gradually decreasing in terms of unit rainfall intensity.

The provinces with a high driving level of loss-rainfall effect were Shanxi, Anhui, Fujian, and Hubei, and the loss-rainfall effect was incremental (Figure 8). Among these regions, Hebei and Hubei are special. Hebei suffered severe flash floods in 2016, with a single event economic loss of 50.217 billion Yuan. In the same year, the Yangtze River flood disaster occurred in Hubei Province, causing massive economic losses [24]. Although the average loss-rainfall effect of the two provinces was low during the study period, the flood-related economic loss in a single year was extremely serious, leading to the high driving level of the loss-rainfall effect. The province with the highest incremental effect value of loss-rainfall effect was Anhui Province. According to the flood-related economic loss data over the years, the economic loss caused by flood disasters in Anhui Province in 2016 and 2020 reached 500.65 billion Yuan and 600.7 billion Yuan, respectively. The flooding of the Yangtze River in 2016 and the Huai River in July 2020 caused severe economic losses in Anhui Province, leading to a high driving level of loss-rainfall effect in Anhui. The loss-rainfall effect in the other seven provinces showed an obvious reduction. The flood-related economic losses in these provinces under the same rain intensity showed a gradual decreasing trend, indicating that the defense level against flood disasters in these provinces has been greatly improved.

The regions with a medium driving level of loss-rainfall effect included Zhejiang, Guangxi, and Sichuan, and the loss-rainfall effect was incremental (Figure 8). In Zhejiang and Guangxi particularly, typhoons and flash floods are the main types of flood disasters [21,32,33]. Typhoons and flash floods are two disasters that happen very suddenly and without warning. The economic loss caused by typhoons is often especially unpredictable, which is the main reason for the effect value of these two provinces being incremental. Sichuan is in the southwest of China, and the frequent occurrence of regional floods in the province causes great economic losses, which is the main reason for the incremental effect of the loss-rainfall effect. The absolute value of the effect in these two provinces and Sichuan was lower than that in the high driving effect area.

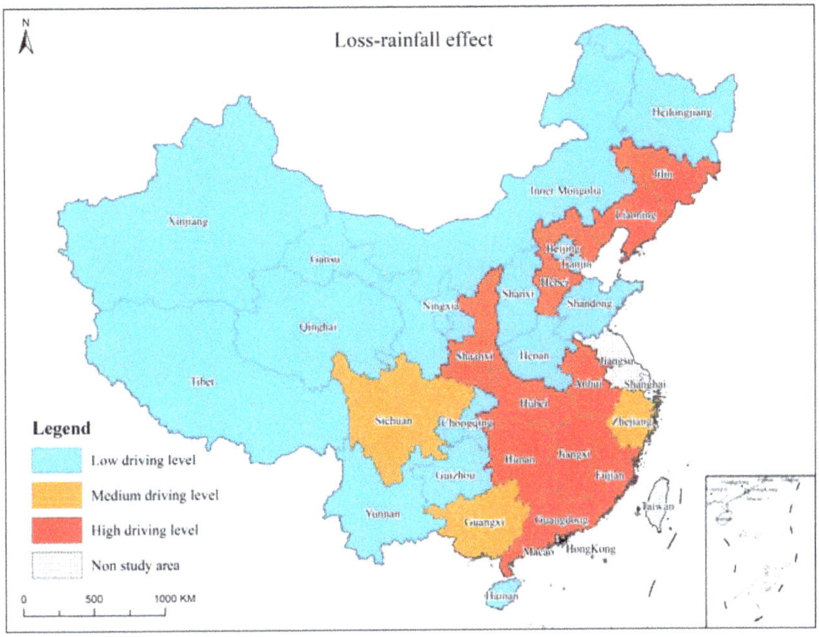


**Figure 8.** Loss-rainfall effect driving distribution.

The regions with a low driving level of loss-rainfall effect include Beijing, Tianjin, Shanxi, Inner Mongolia, Shandong, Henan, Hainan, Chongqing, Guizhou, Yunnan, Tibet, Gansu, Qinghai, Ningxia, and Xinjiang (Figure 8). Among them, in Beijing, Tianjin, Chongqing, Guizhou, Yunnan, and Gansu provinces, the loss-rainfall effect was incremental; moreover, the incremental effect value of Chongqing was higher. However, the flood-related economic losses in these provinces were relatively low in most years; therefore, they are classified as a low driving level. In Shanxi, Inner Mongolia, Shandong, Henan, Hainan, Tibet, Qinghai, Ningxia, and Xinjiang, the loss-rainfall effect showed a reduction. Furthermore, even though the absolute value of the effect was higher in Henan and Hainan provinces, the flood-related economic loss in these two provinces was relatively low during the study period.

## 4. Conclusions

This study analyzed the driving factors of flood-related economic losses in China from 2010 to 2020, using statistics obtained from the National Bureau of Statistics and the Ministry of Water Resources of China. The Kaya identity and LMDI method were used to establish a factor decomposition model to determine what affects economic losses caused by flood disasters. Five effects were isolated, measured, and analyzed: demographic effect, economic effect, flash flood disaster control effect, capital efficiency effect, and loss-rainfall effect.

Using the LMDI method, we deconstructed the driving factors affecting flood-related economic loss; the results show that changes in flood-related economic loss are the result of these five factors. The flash flood disaster control effect showed the most obvious reduction effect. This indicates that the implementation of flash flood disaster prevention and control projects has greatly reduced economic losses caused by flood disasters in China. The continuous development of flash flood disaster prevention and control plays a positive role in reducing the flood-related economic losses in China. The reduction in the loss-rainfall effect is the inevitable result of the control measures for flash flood disasters. Updating monitoring and warning equipment and the construction of water conservancy informatization have significantly improved the country's flood disaster prevention abilities, effectively reducing flood-related economic losses under the same rainfall intensity.

The demographic effect, economic effect, and capital efficiency effect were shown to be incremental. China is in the later stage of industrialization and the middle stage of urbanization, and the growth of regional economies will also bring greater challenges to flood control work. Regions with higher capital efficiency effects are concentrated in the south of China, where the annual rainfall is generally higher. When the inter-annual variation in flash flood disaster prevention funds is small, the inter-annual variation in precipitation will have a great influence on the change in the effect value. Although China is a low fertility country, the population of most provinces shows an increasing trend, which will inevitably pose more challenges to flood control. Therefore, the demographic effect also shows the characteristics of incremental effect.

Economic development is the foundation of social, scientific, and technological progress, and scientific and technological progress is necessary for the improvement of water conservancy informatization. Considering the GDP statistics of the study period, it is clear that the Chinese economy will maintain a steady growth trend in the future. It also shows that the economic effect in the future will still present the characteristics of the incremental effect. Considering these statistics along with the results of this study, we see that the flash flood disaster control effect shows obvious reduction effect characteristics. Under the condition of future economic growth, the continuous development of the special construction of mountain flood disaster prevention and control will play a positive role in reducing economic loss related to flood disasters.

**Author Contributions:** Conceptualization, Z.Z. and Q.L.; methodology, C.L.; software, Q.M.; validation, L.D.; formal analysis, Z.Z.; investigation, Q.L.; resources, Y.C.; data curation, Y.C.; writing—original draft preparation, Z.Z.; writing—review and editing, Z.Z. and Q.L; visualization, Q.M.; supervision, Q.L.; project administration, L.D.; funding acquisition, Q.L and C.L. All authors have read and agreed to the published version of the manuscript.

**Funding:** The research presented in this paper has been carried out as part of the National Key Research Program (No. 2019YFC1510603), Key Research Program of Guangxi Province (2019AB20003), and Hunan Water Conservancy Science and Technology Project (XSKJ2019081-17).

**Institutional Review Board Statement:** Not applicable.

**Informed Consent Statement:** Not applicable.

**Data Availability Statement:** Not applicable.

**Acknowledgments:** The authors thank the China Institute of Water Resources and Hydropower Research.

**Conflicts of Interest:** The authors declare no conflict of interest.

## References

1. Centre for Research on the Epidemiology of Disasters; United Nations Office for Disaster Risk Reduction. The Human Cost of Disasters: An Overview of the Last 20 Years (2000–2019) [EB/OL]. 2020. *UN Reports*. Available online: https://www.undrr.org/publication/human-cost-disasters-overview-last-20-years-2000-2019 (accessed on 30 May 2022).
2. Wen, Q.; Huo, Z.; Zhou, Y.; Che, Q.; Xiao, J.; Huang, D. Integrated risk assessment of flood disasters in southern China. *Chin. J. Ecol.* **2015**, *34*, 2900–2906.
3. Wu, J.; Li, N.; Wen, Y.; Hu, A.; Cui, W.; Zhang, P. Economic impact of natural disaster and indirect economic loss estimation methods. *Prog. Geogr.* **2009**, *28*, 877–885.
4. Jiang, L. Analysis on the characteristics of flood disaster damage in China in recent 76 years. *Adv. Meteorol. Sci. Technol.* **2018**, *8*, 11–16.
5. Jiang, L.; Qiu, Z. The assessment of indirect economic loss of urban flood disaster: A case study of Beijing. *Mod. Urban Res.* **2014**, *7*, 7–13.

6. Pauw, K.; Thurlow, J.; Bachu, M.; van Seventer, D.E. The economic costs of extreme weather events: A hydrometeorological CGE analysis for Malawi. *Environ. Dev. Econ.* **2011**, *16*, 177–198. [CrossRef]
7. Joshi, S.R.; Vielle, M.; Babonneau, F.; Edwards, N.R.; Holden, P.B. Physical and economic consequences of sea-level rise: A coupled GIS and CGE analysis under uncertainties. *Environ. Resour. Econ.* **2016**, *65*, 813–839. [CrossRef]
8. Borgomeo, E.; Vadheim, B.; Woldeyes, F.B.; Alamirew, T.; Tamru, S.; Charles, K.J.; Kebede, S.; Walker, O. The distributional and multi-sectoral impacts of rainfall shocks: Evidence from computable general equilibrium modelling for the Awash Basin, Ethiopia. *Ecol. Econ.* **2018**, *146*, 621–632. [CrossRef]
9. He, B.S.; Ma, M.H.; Li, Q.; Liu, L.; Wang, X.H. Current situation and characteristics of flash flood prevention in China. *China Rural Water Hydropower* **2021**, *5*, 133–138, 144.
10. Sun, D.; Liu, C.; He, B. Research progress on flash flood disaster control theory and technology. *China Flood Drought Manag.* **2022**, *32*, 24–33.
11. Zhang, X.; Li, C.; Sun, D. Research on national flash flood disasters risk zoning based on survey and evaluation of big data. *China Flood Drought Manag.* **2020**, *30*, 61–63.
12. Zhang, Z. Mountain torrent disaster prevention and control measures and their effects. *Water Resour. Hydropower Eng.* **2016**, *47*, 1–5, 11.
13. Guo, L.; He, B. Construction and achievements of flash flood disaster prevention systems in China. *China Flood Drought Manag.* **2019**, *29*, 16–19, 29.
14. Kaya, Y. *Impact of Carbon Dioxide Emission Control on GNP Growth: Interpretation of Proposed Scenarios*; IPCC Energy and Industry Subgroup, Response Strategies Working Group: Paris, France, 1989.
15. Zhang, Z.X.; Sun, C.Z. Driving effect measurements and spatial-temporal variation of the per capita gray water ecological footprint in China. *Acta Ecol. Sin.* **2018**, *38*, 4596–4608.
16. Sun, C.-Z.; Xie, W.; Zou, W. Contribution ratio measurement of water use efficiency driving effects and spatial driving type in China. *Sci. Geogr. Sin.* **2011**, *10*, 1213–1220.
17. Ang, B.W.; Zhang, F.Q. Inter-regional comparisons of energy-related $CO_2$ emissions using the decomposition technique. *Energy* **1999**, *24*, 297–305. [CrossRef]
18. Ang, B.W. Decomposition analysis for policymaking in energy: Which is the preferred method? *Energy Policy* **2004**, *32*, 1131–1139. [CrossRef]
19. Ang, B.W. The LMDI approach to decomposition analysis: A practical guide. *Energy Policy* **2005**, *33*, 867–871. [CrossRef]
20. Hatzigeorgiou, E.; Polatidis, H.; Haralambopoulos, D. $CO_2$ emissions in Greece for 1990–2002: A decomposition analysis and comparison of results using the Arithmetic Mean Divisia Index and Logarithmic Mean Divisia Index techniques. *Energy* **2008**, *33*, 492–499. [CrossRef]
21. Zhao, J.; Fan, B. Characteristics of temporal and spatial distribution of flash flood in China. *China Water Resour.* **2006**, *13*, 45–47.
22. Gao, J.; Wang, M.; Guo, S.; Wang, L.; Ye, B. Study and temporal distribution of rainfall disasters and analysis of its rainfall-related characteristics in Sichuan Province. *Meteorol. Sci. Technol.* **2022**, *50*, 224–233.
23. Ye, B.; Feng, H.; Liu, Z.; Zhang, X.; Luo, H. Risk assessment of Sichuan flash flooding based on logistic model. *J. Chengdu Univ. Inf. Technol.* **2020**, *35*, 573–578.
24. Tu, Y.; Wu, Z.; He, B. Analysis on the characteristics of flash flood disasters in China from 2011 to 2019. *China Flood Drought Manag.* **2020**, *30*, 22–25.
25. Gu, Q.; Cheng, Z.; Xin, X.U. Industrialization Basically Accomplished: Concept Milestone and Prospect. *Ind. Econ. Rev.* **2015**, *1*, 1–12.
26. Chu, J.; Liu, Y.; Lin, X.; Tong, D. Spatial mutual effects between municipal infrastructure investment and economic development. *Acta Sci. Nat. Univ. Pekin.* **2021**, *57*, 291–298.
27. Mugume, S.N.; Gomez, D.E.; Fu, G.; Farmani, R.; Butler, D. A global analysis approach for investigating structural resilience in urban drainage systems. *Water Res.* **2015**, *81*, 15–26. [CrossRef]
28. Raviv, G.; Fishbain, B.; Shapira, A. Analyzing risk factors in crane-related near-miss and accident reports. *Saf. Sci.* **2017**, *91*, 192–205. [CrossRef]
29. Liu, J.; Xing, C.; Zhang, Q. House price, fertility rates and reproductive intentions. *China Econ. Rev.* **2020**, *62*, 101496. [CrossRef]
30. Shanxi Provincial Department of Water Resources. Notice on the Best Efforts to do Flash Flood Disaster Prevention Work. *Shanxi Water Resour.* **2021**, *37*, 11.
31. The People's Government of Hebei Province. Overview of Hebei Province. Available online: http://www.hebei.gov.cn/hebei/14462058/14462085/14471224/index.html (accessed on 2 June 2022).
32. Ni, Z.; Wang, P.; Huang, X.; Cao, X. Contrast analysis of effects of typhoon "Lekima" and "Rananim" on precipitation and wind in Zhejiang Province. *J. Meteorol. Sci.* **2021**, *41*, 172–182.
33. Lu, Y.; Liu, H.; Li, G. Risk analysis of the extreme disaster risk caused by typhoon affecting Guangxi. *Mar. Forecast.* **2021**, *38*, 31–41.

*water*

MDPI

*Article*

# Analysis on the Disaster Mechanism of "8.12" Flash Flood in Liulin River Basin

**Sijia Hao [1,2], Wenchuan Wang [1], Qiang Ma [2,*], Changzhi Li [2], Lei Wen [3], Jiyang Tian [2] and Changjun Liu [2]**

1 College of Water Resources, North China University of Water Resources and Electric Power, Zhengzhou 450046, China; haosimaomao@163.com (S.H.); wangwen1621@163.com (W.W.)
2 China Institute of Water Resources and Hydropower Research, Beijing 100038, China; lichangzhi@iwhr.com (C.L.); tianjy@iwhr.com (J.T.); lcj2005@iwhr.com (C.L.)
3 College of Hydrology and Water Resources, Hohai University, Nanjing 210098, China; wenlei6037@hhu.edu.cn
* Correspondence: maqiang@iwhr.com

**Abstract:** Hubei province is located in the center of China with 56% total area characterized with mountainous area. Thus, flash flood caused by extreme rainfall has become one of the significant obstacles that highly affect the social and economic development of the province. In order to scientifically understand the mechanism of flash flood disasters and provide technological support to the local flood prevention and control work, the IWHR designed and developed a new distributed hydrological model named China-FFMS that can simulate the evolution of natural disasters and make an assessment by setting the flood water sources in line with the flow discharge. The FFMS was further applied to simulate the 8.12 flash flood disaster that occurred in the Liulin county of Hubei province on 12 August ("8.12") and fed by the data collected from the national flash flood disaster investigation and assessment. The calculated peak flow was 666.22 $m^3/s$ with an error of +13% compared with postdisaster investigation data (589 $m^3/s$). The results showed that using a multisourced modelling approach, e.g., mixing spatiotemporal variables and sources, to simulate the flash flood process was able to accurately reproduce the flood process and the consistence of the flow discharge, thereby explaining the underlying reason of the disaster formation and evolution. Regarding the case of the Liulin county, the main factor leading to the disaster was the overlapped peak flow where the Dunne flood peak of three different tributaries from the upper reach met together at the same time. Moreover, the peak flow of the Lianhua river at the downstream of Liulin County also arrived at the same time as the upstream peak, which obstructed the flood progress and increased the damage of the disaster. According to the analysis, several suggestions and recommendations are proposed such as the improvement of the forecast and early warning system of the upstream areas, the optimization of the current flood defense plan, and the enhancement of the residents' awareness of flash flood disasters.

**Keywords:** flash-flood modelling system; disaster mechanism; runoff generation component; disaster amplification effect

**Citation:** Hao, S.; Wang, W.; Ma, Q.; Li, C.; Wen, L.; Tian, J.; Liu, C. Analysis on the Disaster Mechanism of "8.12" Flash Flood in Liulin River Basin. *Water* **2022**, *14*, 2017. https://doi.org/10.3390/w14132017

Academic Editor: Luis Garrote

Received: 11 May 2022
Accepted: 21 June 2022
Published: 24 June 2022



## 1. Introduction

Since the 21st century, climate change has led to an increase of the frequency and intensity of heavy rainfall in mountainous areas where the demography is expanding due to counter urbanization and the amount of population under risk rises. Motivated by this, the research on flash floods has attracted an increasing interest in both China and western countries [1–4]. For example, on 12 August 2021, a severe flash flood occurred in Liulin Town, Suixian County, Suizhou City, Hubei Province, affecting more than 12,000 people and destroying 474 hectares of crops and over 160 houses. It seriously undermined local socioeconomic development with a direct economic loss of about 226 million yuan [5,6]. Studying the process and mechanism of the "8.12" flash flood in the Liulin river basin can provide a positive reference for flood control and disaster reduction in the local and other similar areas prone to flash floods.

A flash flood is highlighted by its special characteristics of highly unexpected occurrence, destructive impact, and short duration. Additionally, it is usually accompanied by secondary disasters such as landslides and debris flows; therefore, it is likely to change the landform of the affected areas. However, disaster formation monitoring and postdisaster investigation and analysis are difficult [7–9]. There have been a lot of studies on the simulation of flash floods and the analysis of disaster mechanism [10–14]. For example, Yang et al. [15] investigated the disaster mechanism of the 5·16 flash floods in the Houshan Mountains of Sanming City, Fujian Province by using the flood calculation method, and analyzed the effect of flash flood magnified by culvert blockage and slope excavation. Carle et al. [16] compared different early warning systems based on aggregated rainfall and a distributed hydrological model for regional flash flood control and found that a distributed simulation has certain advantages in the analysis of flash floods associated with a rapid water level rise. Based on the measured rainfall data, Sun et al. [17] used the kinematic-wave geomorphological instantaneous unit hydrograph model to analyze the 8.16 flash flood in the Zhongdu river basin, Sichuan Province, and concluded that the disaster was caused by regional heavy rainfall combined with human activities. Braud et al. [18] investigated the 2002 flash flood in southern France by using distributed hydrological models and identified rainfall, soil saturation, and surface soil thickness as major factors of the runoff of flash floods. They suggested that a detailed runoff description is essential to increase the accuracy of the model simulations. Based on an examination of high temporal and spatial resolution data of 25 typical flash floods across Europe, Marchi et al. [19] revealed that flash floods in Europe are seasonal and significantly affected by initial soil moisture.

To summarize, a distributed hydrological model has become the main technical tool to study the mechanism of flash floods. However, most types of this model fail to describe the details of the physical mechanism of runoff generation in river basins, let alone the analysis of the changes in runoff generation and the disaster mechanism of flash floods. This paper adopts the China Flash Flood Modelling System (FFMS) independently developed by the China Institute of Water Resources and Hydropower Research (IWHR) to analyze the "8.12" flash flood in Liulin Town, based on national flash flood investigation results and postdisaster field survey data. A distributed flash flood model with small watersheds of 0.16–1.6 km$^2$ (0.10–0.99 mi$^2$) as the minimum calculation unit was constructed for the detailed analysis and simulation, including disaster formation and runoff changes in the process, so as to analyze the amplification effect and reveal the disaster mechanism, thereby providing technical support for future flash flood prevention and control in the region.

## 2. Research Materials

### 2.1. Study Area

Liulin Town with an administrative area of 197.49 km$^2$ (76.25 mi$^2$) is located in the south of Suixian County, Suizhou City, Hubei Province. The topography of the Liulin basin is dominated by mountains and hills and mainly covered by forests and cultivated land. The average annual rainfall is recorded as 1100 mm (43.3 in) and centralized in the flood season [20]. The section of watershed controlled by Liulin Town covers about 24.29 km$^2$ (9.38 mi$^2$) that belongs to the Langhe River basin. The main river channel is about 8.47 km (5.26 mi) with a steep gradient of 8‰, where 19 tributaries including Jinyin, Taiping and Baishuwan coalesce. In the fan-shaped upper river basin of Liulin Town, the main settlements are surrounded by mountains on three sides and located at the bottom of a river valley in a good hydrological condition that is relatively closed, wide, and gentle. However, the flood discharge conditions in the lower reach are poor due to the narrow channel. As a result, there is a high probability for an intensive rainfall during the flood season to lead to flash floods in Liulin Town (Figure 1).

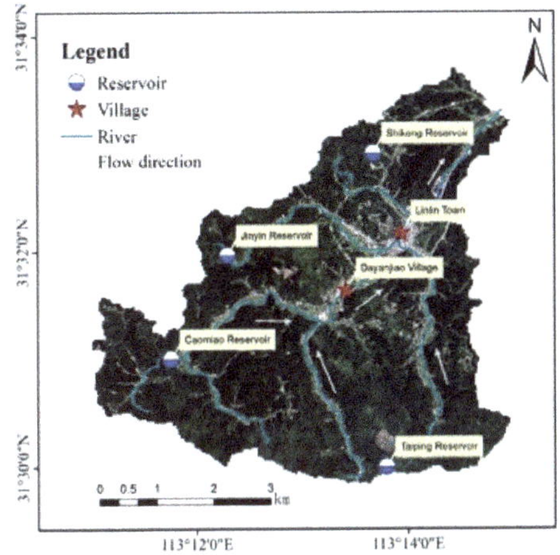


**Figure 1.** Distribution map of water systems and reservoirs in Liulin Town.

## 2.2. Data Collection

The rainfall data were measured at several reservoir sites around Liulin Town, and the total precipitation in Liulin Town from 21:00 on 11 August to 9:00 on 12 August was 503 mm, where the accumulated precipitation from 4:00 to 7:00 was 373.7 mm (1.23 ft) and the precipitation intensity at 5:00 and 6:00 was 105 mm/hour (14.7 in/h) and 104 mm/hour, respectively, both exceeding the historical record of local measurements [6]. Figure 2 shows the spatial distribution of the accumulated rainfall and hourly rainfall in the drainage area of Liulin Town from 21:00, 11 August to 12:00, 12 August.

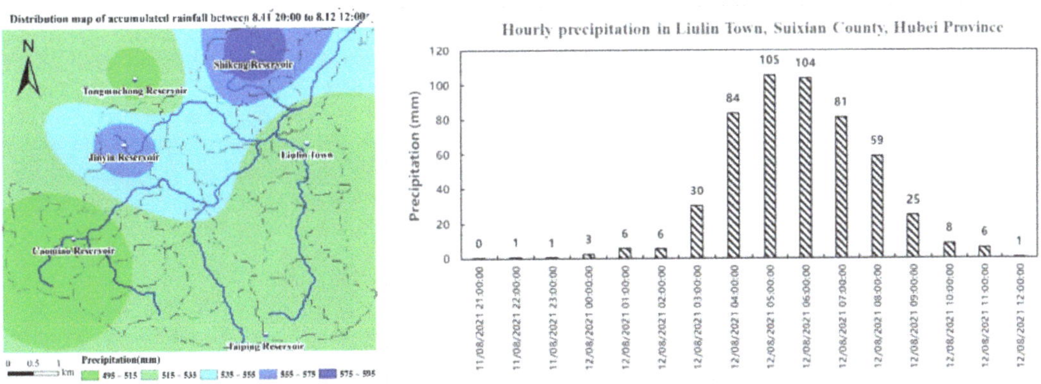


**Figure 2.** Spatial distribution map of accumulated rainfall and hourly rainfall in Liulin Town, Hubei Province.

There are four rainfall monitoring stations in the drainage area of Liulin Town, which are all small-sized reservoirs, (Grade II) namely Jinyin, Caomiao, Taiping, and Shikeng. Before the occurrence of the "8.12" flash flood, the reservoirs were kept 0.06–1.60 m (0.20–5.25 ft) below the flood limit water level. Then, when the heavy rainfall started, the water levels at Jinyin, Caomiao, and Shikeng all began to rise rapidly, with the flood peak exceeding the historical record high. However, the four reservoirs remained operational and followed the flood-season reservoir schedule and operation plan during the flash flood in Liulin Town. Figure 3 shows the changes of rainfall and water level monitored in each reservoir during the flash flood.

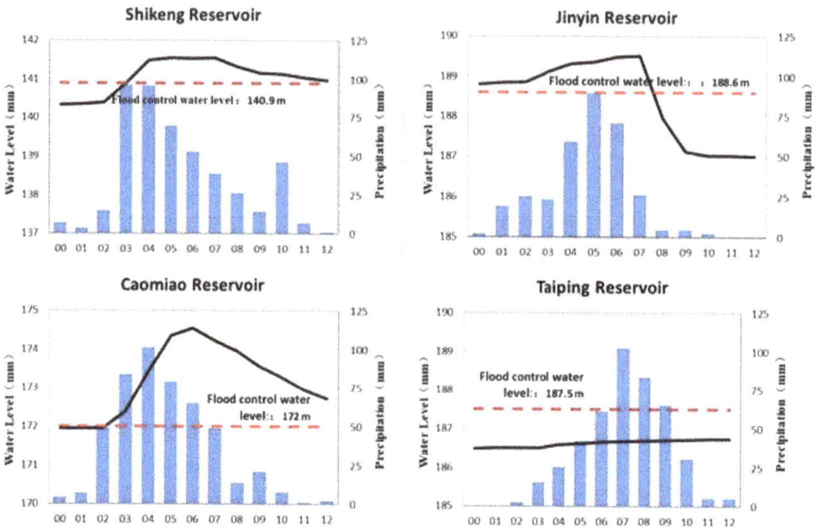


**Figure 3.** Changes of rainfall and water level in Shijinyin, Caomiao, Taiping, and Shikeng reservoirs.

## 3. Simulation Analysis

### 3.1. Hydrological Model

The FFMS system developed by IWHR was employed to simulate the "8.12" flash flood disaster. It is a new-generation flood simulation piece of software for small watersheds that integrates the advantages of C++, Fortran, Java, and other computer languages. The design aims at modularization, parameterization, intelligent, visualization and automation based on the national flash flood investigation results [21,22]. FFMS has the characteristic functions of an automatic division of small watersheds, automatic extraction of parameters, modular modeling, and automatic determination of the runoff model based on the attributes of underlying surface of small watersheds. The software architecture and functional modules are as shown in Figure 4. At present, FFMS has already integrated different hydrological models available for use, such as HEC-HMS, PRMS, spatiotemporal variable–source mixed runoff model (SVSMR), and the Xin'anjiang model.

### 3.2. Simulation Method

In view of the characteristics of flash floods in China, IWHR proposed the SVSMR model for simulating the process of flash floods [23,24]. Through the refined division of catchments with different underlying surface conditions within watersheds, a vertically, horizontally, and temporally mixed runoff model was constructed for simulations and calculations at multiple levels such as data sources, method sources, and runoff sources [25–27]. The method for computing infiltration in soil mechanics, which discretized soil moisture in the downward movement of wetting front [28–30], was introduced into distributed hydrological simulations. This enabled a detailed description of instantaneous changes of the infiltration capacity of surface soil in the vadose zone during flash floods, thereby realizing the accurate simulation of the runoff mechanism of flash floods. The detailed model simulation method can be found in reference [15].

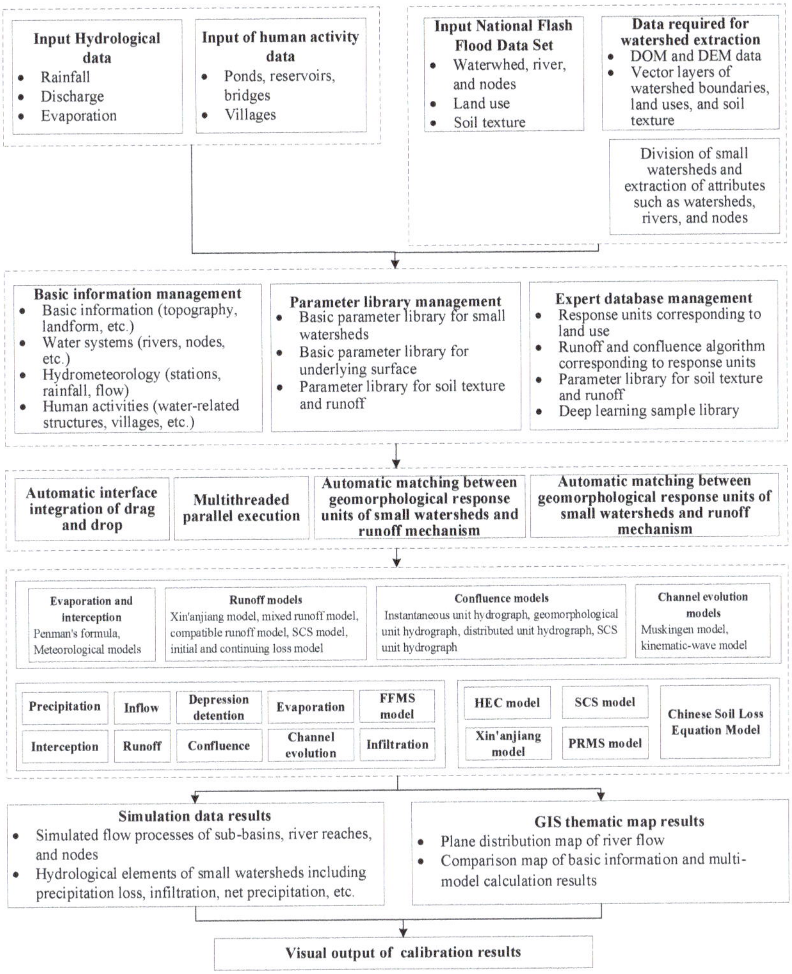


**Figure 4.** FFMS software architecture and functional modules.

### 3.3. Modeling Process

Based on basic geographic information data in the National Flash Flood Data Set, such as DEM and DLG data (1:50,000) and DOM data with a resolution of 2.5 m (8.2 ft), the drainage area upstream of the Liulin Town section were divided into 28 small watersheds each with a size of 0.16–1.6 km$^2$ (0.10–0.99 mi$^2$) as the minimum calculation unit for the model simulation (Figure 5). Meanwhile, information about 75 basic attributes of the small watersheds was extracted, including spatial attributes, land uses, and soil types.

The monitored data series of the rainfall of four reservoirs around Liulin Town were the input of the model. The areal precipitation of each calculation unit was then automatically calculated through an inverse distance-weighted interpolation by the FFMS rainfall module. The SVSMR method was used for the runoff simulation while the simplified triangular unit hydrograph method was used for the slope confluence simulation, the linear reservoir method for the baseflow simulation, and the kinematic-wave method for the flood evolution simulation. The modeling interface is presented in Figure 6, where the table displays the basic attributes of each calculation unit extracted by the model automatically.

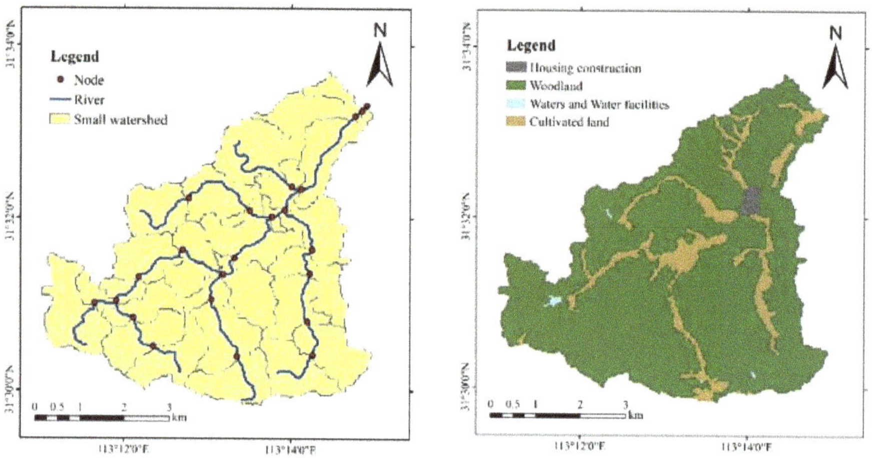


**Figure 5.** Calculation unit division in Liulin Town and land use attribute data extraction.

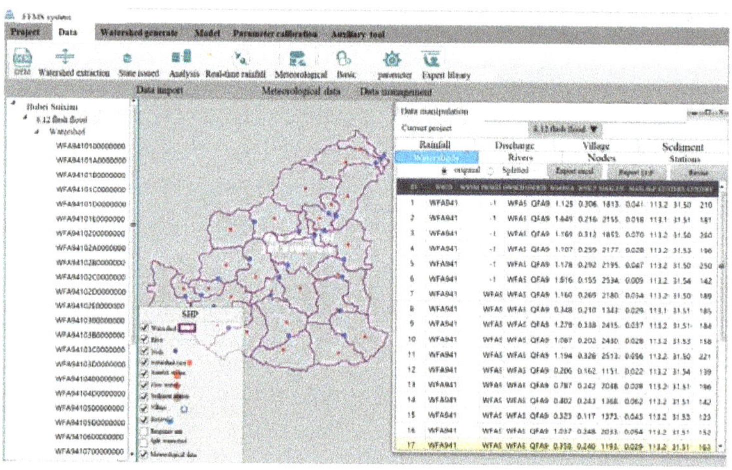


**Figure 6.** FFMS modeling process.

### 3.4. Parameter Determination

The main model parameters involved in the simulation of Liulin flash flood was divided into two categories according to their physical characteristics. One consisted of the basic attribute parameters that characterize the actual topography of calculation units, such as area, longest confluence path, and gradient. The other consisted of the runoff and confluence parameters that characterize the underlying surface of calculation units, such as maximum infiltration coefficient of surface soil, (linear/nonlinear) coefficient of preferential flow, and (linear/nonlinear) coefficient of subsurface flow. For the first category, values were obtained by automatically extracting the input data without calibration while for the second category of parameters, the model automatically determined and assigned the values based on the basic attributes of calculation units extracted from the expert database derived from the national flash flood investigation results. In the actual simulation process, the calibration and correction can be carried out for some parameters according to the simulation effect. However, this case used parameter values recommended by the expert database (Table 1) because a simulation analysis had already been conducted ahead of the post disaster survey data collection. The simulation results are compared with measured data in later verification.

**Table 1.** Parameter value.

| Symbol | Parameter | Unit | Recommended Value |
|---|---|---|---|
| $k_s$ | Saturated hydraulic conductivity | m/s | $2.78 \times 10^{-6}$ |
| $p_{erp}$ | Surface ratio of preferential flow | - | 0.82 |
| $k_{i1}$ | Linear coefficient of subsurface flow | - | 0.88 |
| $k_{p1}$ | Linear coefficient of preferential flow | - | 0.58 |
| $k_{i2}$ | Nonlinear coefficient of subsurface flow | - | 2.00 |
| $k_{p2}$ | Nonlinear coefficient of preferential flow | - | 0.00 |
| $k_g$ | Baseflow coefficient | - | 0.29 |
| $W_{fd}$ | Average depression storage | mm | 0.37 |
| $p_{erip}$ | Impervious surface ratio | - | 0.00 |

## 4. Analysis on the Disaster Mechanism

### *4.1. Analysis of Flash Flood*

The model used default parameters to perform distributed hydrological simulations for hourly flash-flood changes from 21:00 to 12:00 on 12 August 2021 in the Liulin basin. The peak flow was calculated to be 666.22 $m^3/s$ (23,527.3 $ft^3/s$). Flood peak traveled for one to two hours, and flow confluence took about two hours. The runoff coefficient was around 0.78, and the modulus of flood peak was 27.11. The simulated time of flood peak was basically consistent with the actual time by comparing with post disaster investigation data (Figure 7), with an error of +13%. Since the simulation results were obtained earlier than the investigation results, the simulation method proved to be reliable and the recommended parameter values were reasonable. Therefore, the simulation results could explain the flash flood process and support the mechanism analysis.

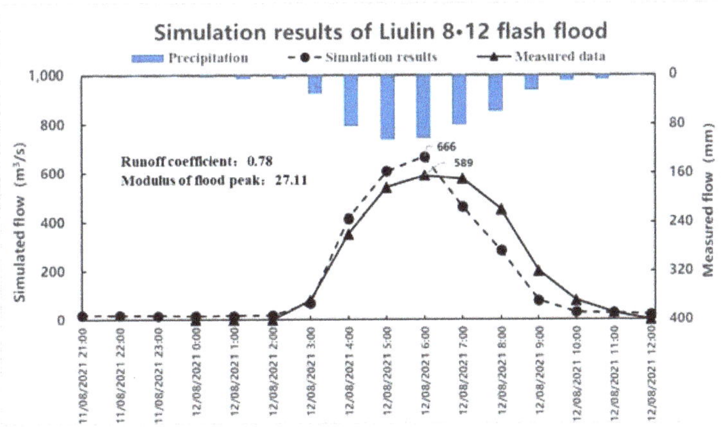


**Figure 7.** Simulation results.

By using the SVSMR model, the changes in runoff components can be examined to pinpoint the causes of the flash flood in Liulin Town. As shown in Figure 8a, in the initial stage of the "8.12" flash flood in Liulin Town, the hourly areal rainfall in the watershed was small and surface runoff was limited as a large proportion of rainfall infiltrated into the unsaturated surface soil. The subsurface flow from the upstream calculation units contributed most to the total runoff, but due to the slow confluence, it had less impact on the downstream Liulin Town. At 1:00 on 12 August, the accumulated areal rainfall of the basin exceeded 10.1 mm (0.40 in) and a runoff was generated in some calculation units where soil became saturated. The total runoff was still generally small, so the flow of the main rivers in the basin did not increase remarkably (Figure 8b). After 3:00 on 12 August, with the rapid increase of areal rainfall in the basin, a Hortonian runoff became the main mechanism of each runoff calculation unit in the watershed and was produced from almost 85% of the total rainfall before the flood peak came. Due to a relatively large channel gradient in the upper reaches, the speed of the apparent surface runoff was rapid after confluence (Figure 8c,d). As a result, the flow at the Liulin Town section surged by 500% from 3:00 to 4:00 and reached its peak of around 600 $m^3/s$ (21,188.8 $ft^3/s$) at 6:00.

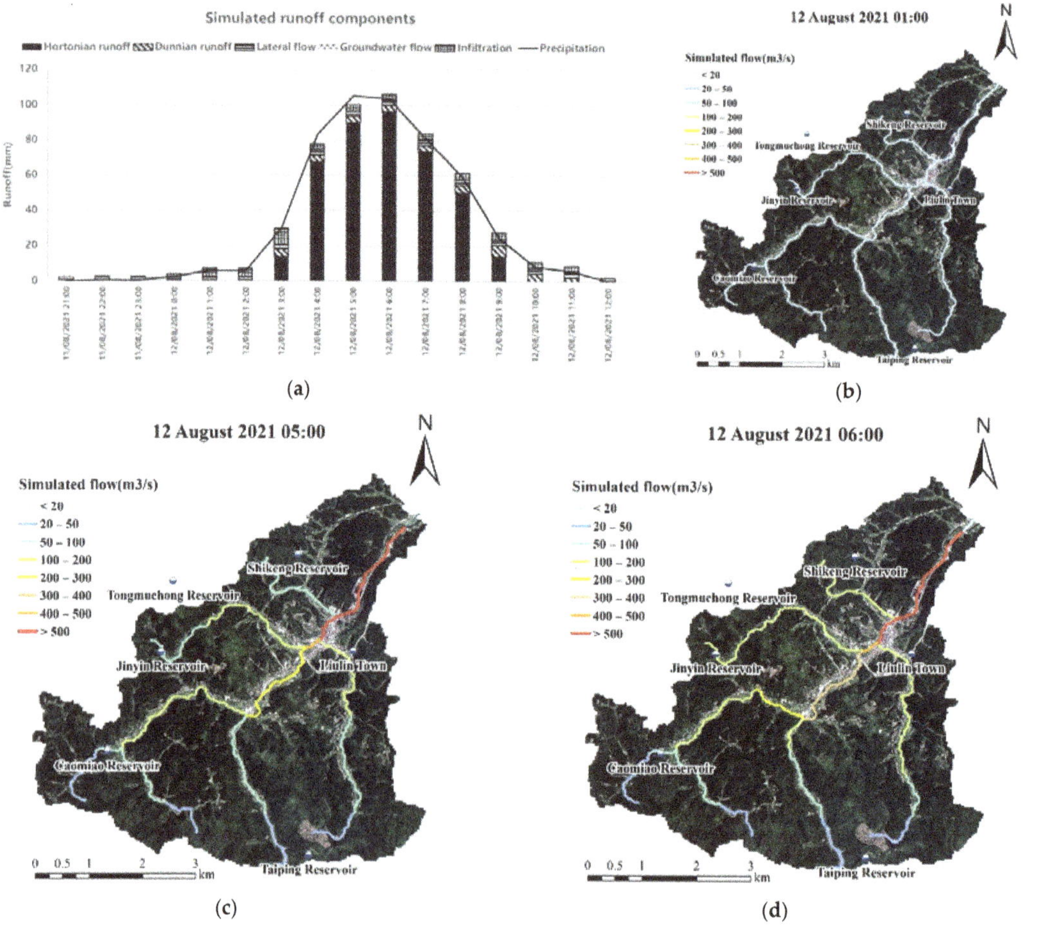


**Figure 8.** Distributed simulation of runoff composition and runoff process. (**a**) Runoff components in Liulin Town. (**b**) Simulated flow at 1:00. (**c**) Simulated flow at 5:00. (**d**) Simulated flow at 6:00.

The flash flood caused heavy casualties [6] due to several factors: an emergency evacuation that was not well organized at night; the elderly and female population that accounted for a large proportion and had relatively weak self-rescue capability; a blocked evacuation route, e.g., people could not move to a safe place in time as houses with external stairs were kept closed subject to the high water pressure of rapidly rising floods.

### 4.2. Analysis of the Amplification Effect

Based on the FFMS simulation results, the flow of the Langhe River and its three tributaries, i.e., the Shikeng Reservoir, the Jinyin Reservoir, and the Taiping River, was analyzed, and further, flood hydrographs were generated for each (Figure 9). The results showed that the peak flow of the tributaries all arrived at the section near Liulin Town at 6:00 on 12 August. An exception was the Caomiao Reservoir on the trunk of Langhe River, which is far away from Liulin Town; its peak flow appeared at 5:00 on 12 August. Such simultaneous arrival and overlapping of peak flows of the mainstream and tributaries magnified the effect of the flash flood near Liulin Town. Among them, the Taiping River contributed the largest peak flow of 205.33 $m^3/s$ (7251.2 $ft^3/s$) while the Shikeng Reservoir contributed the smallest of 104.09 $m^3/s$. However, the confluence of the Shikeng Reservoir at the downstream of Liulin Town generated a backwater effect that obstructed flood water discharge, which magnified the impact of the flash flood to a certain extent.

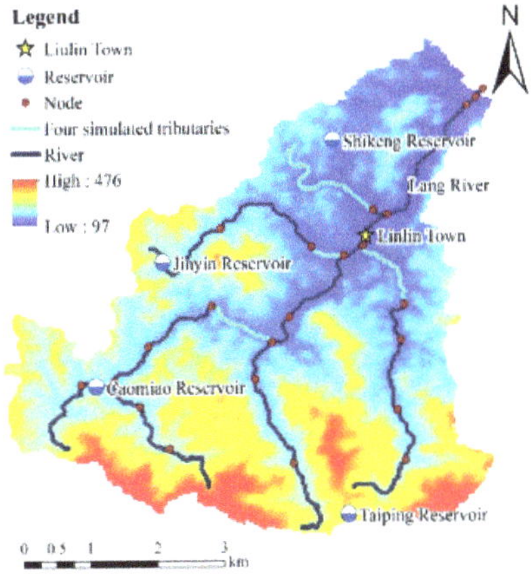


(a)

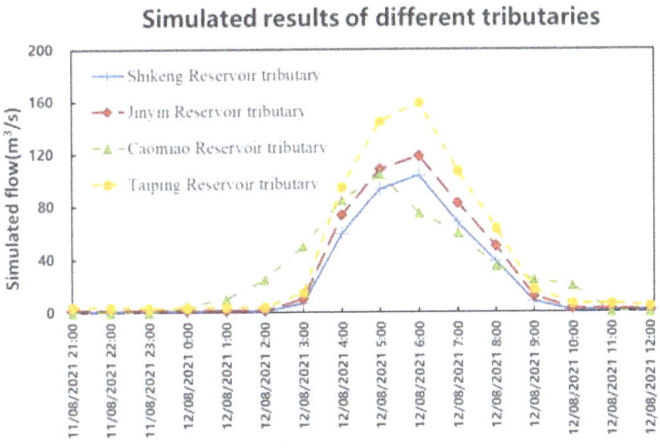


(b)

**Figure 9.** Distribution map (**a**); and simulation results of four tributaries of Lang River (**b**).

According to the FFMS simulation results, the 24 h cumulative rainfall during the "8.12" flash flood in Liulin Town was 1.6 times that of the local designed 100-year flood. The water depth corresponding to the peak flow at the extracted cross-sections was calculated using the Manning formula. As shown in Figure 10, the calculated water depth at each node of the "8.12" flash flood was higher than that of the 100-year flood, which means that the frequency of the flash flood in this case was at least once in more than 100 years.

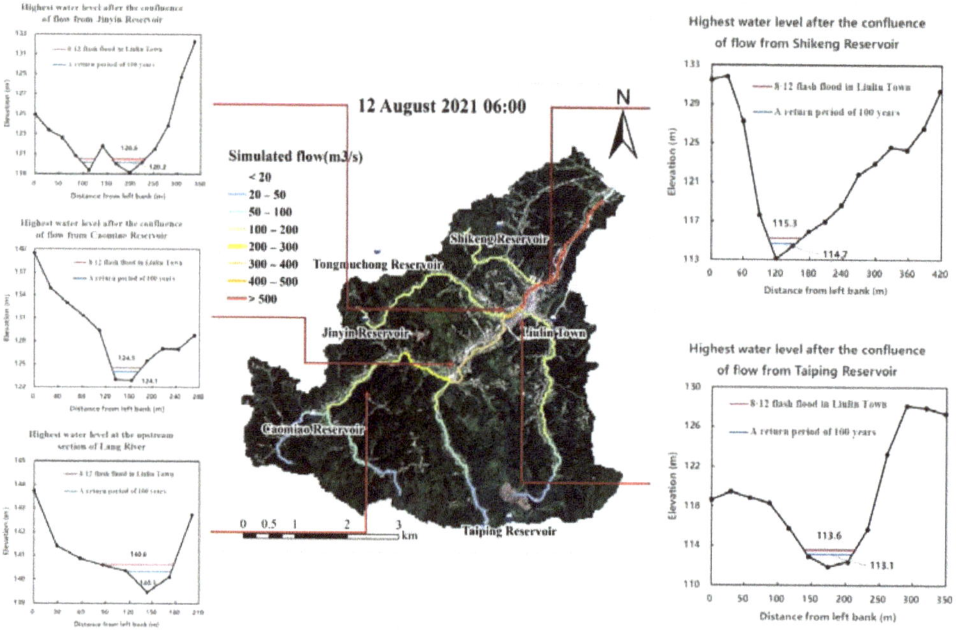


**Figure 10.** Simulated water levels of various reservoirs.

Regarding the charts for flood runoff calculation of Hubei Province and statistical parameter atlas for floods in Hubei Province in Figure 10, it was deduced that the design peak flow corresponding to the design flood with a return period of 200 years at the Liulin Town section was 634 $m^3/s$ (7251.2 $ft^3/s$), compared with 620 $m^3/s$ (21,895.1 $ft^3/s$) using the hydrological comparison method based on the design data of the nearby Baiguohe Reservoir. Taking into account the field investigations and simulation results, the 8.12 flash flood in Liulin Town was considered to correspond to a return period of 200 years or longer [6].

## 5. Conclusions

This study adopted the self-developed distributed flash flood simulation model (FFMS) to analyze the "8.12" flash flood in Liulin Town in terms of runoff generation and runoff components of each tributary in the basin, which revealed the formation mechanism and underlying reasons leading to this disaster. The main conclusions are as follows:

(1) The "8.12" flash flood in Liulin Town was mainly caused by the surge of a Hortonian runoff generated by a record extreme rainstorm in this area. The peak flow of four upstream tributaries arrived at the Liulin Town section at almost the same time and coupled with the backwater effect from the downstream Lianhua River, which magnified the effect of the flash flood and resulted in heavy casualties in Liulin Town.

(2) The capability of evacuation in response to the Liulin flash flood was restricted by factors such as the difficulty for arranging emergency evacuation at night, a large proportion of elderly and female population, and the external layout of stairs for most resident houses.

(3) Referring to the topographical features and historical flash flood events in the basin of Liulin Town, several recommendations are proposed accordingly: (1) improve the rainfall forecasting system for drainage areas of upstream tributaries; (2) build a flash flood forecasting and early warning system based on distributed models; (3) further refine the existing flash flood defense plan, and (4) increase the evacuation exercises to respond to flash floods and enhance the public awareness of extreme disaster prevention.

**Author Contributions:** C.L. (Changzhi Li) and Q.M.: Conceptualization, Methodology. W.W.: Writing—review & editing. S.H.: Program implementation, data curation, Writing—original draft preparation. L.W.: Software. J.T.: Investigation. C.L. (Changjun Liu): Project administration. All authors have read and agreed to the published version of the manuscript.

**Funding:** This research was funded by National Key Research Program (2019YFC1510603), Key Research Program of Guangxi Province (2019AB20003) and Hunan Water Conservancy Science and Technology Project (XSKJ2019081-17).

**Institutional Review Board Statement:** Not applicable.

**Informed Consent Statement:** Not applicable.

**Data Availability Statement:** All authors made sure that all data and materials support published claims and comply with field standards.

**Conflicts of Interest:** The authors declare no conflict of interest.

## References

1. Guo, L.; Ding, L.; Sun, D.; Liu, C.J.; He, B.S.; Liu, R.H. Key Techniques of Flash Flood Disaster Prevention in China. *J. Hydraul. Eng.* **2018**, *49*, 1123–1136.
2. Weng, C.; Hu, X.; Wang, Y.; Tan, W.; Li, J. A Study on the Zoning of Mountain Flood Disaster Prevention and Control in Hubei Province. *Hubei Water Power* **2006**, *002*, 12–15.
3. Zhang, Y.; Zou, Y.; Wang, X.; Zhang, P.; Xu, W. Flash Flood Warning Assessment Considering Temporal Differences. *Adv. Eng. Sci.* **2021**, *53*, 10–18.
4. Wang, X.; Yang, P.; Sun, T.; Zexing, X. Study on Division Early Warning of Flash Flood Disaster Caused by Rainstorm in Mountainous Small Watersheds. *Adv. Eng. Sci.* **2021**, *53*, 29–38.
5. Xu, S.; Jiang, Y.; Jiang, H.; Li, Q.; Xie, Q.; Lie, Y.; Zhou, J.; Wang, F.; Meng, L. Investigation and Reflection on "2021.8.12" Flood Disaster in Liulin Town, Sui County, Hubei Province. *China Flood Drought Manag.* **2022**, *32*, 54–58.
6. Chang Jiang Water Resource Commission of the Ministry of Water Resources. *Investigation and Analysis Report on "8.12" Major Mountain Flood Disaster Event in Liulin Town, Suixian County, Suizhou City, Hubei Province*; Ministry of Water Resources: Beijing, China, 2021.
7. He, B.; Ma, M.; Li, Q.; Liu, L.; Wang, X.-H. Current Situation and Characteristics of Flash Flood Prevention in China. *China Rural. Water Hydropower* **2021**, *5*, 133–138.
8. Ding, L.; Guo, L.; Liu, C.; He, B. Technical Progress of Flash Flood Disasters Prevention and Control Systems in China. *China Flood Drought Manag.* **2020**, *30*, 11–17.
9. Zhang, P.; Ding, W.; Wang, X. Research Framework and Anticipated Results of the Key Technology and Integrated Demonstration of Mountain Torrent Disaster Monitoring and Early Warning. *Adv. Eng. Sci.* **2018**, *50*, 1–11.
10. Liu, C.; Nie, R.; Liu, X.; Xu, W. Research Conception and Achievement Prospect of Key Technologies for Forecast and Early Warning of Flash Flood and Sediment Disasters in Mountainous Rainstorm. *Adv. Eng. Sci.* **2020**, *52*, 1–8.
11. Yang, P.; Xu, Z.; Yan, X.; Wang, X. Comparative Study on Methods of Early Warning Index of Flash Flood Disaster Induced by Rainstorm. *Adv. Eng. Sci.* **2020**, *52*, 157–165.
12. Pan, J. Comparative Study of Simplified and Full Hydrodynamic Models for Flash Floods. *Adv. Eng. Sci.* **2012**, *44*, 77–82.
13. Li, P.; Huang, S.; Li, D.; Huang, Q. Analysis of Flood Control Function of Three Gorges Reservoir and Flood Disaster in Poyang Lake Basin. *J. Xi'an Univ. Technol.* **2020**, *36*, 486–493.
14. Chen, Y.; Qin, J.; Wang, H. Liuxihe Model-based Method for Flood Forecast of Medium and Small-sized River. *Water Resour. Hydropower Eng.* **2017**, *48*, 12–19.
15. Yang, L.; Ye, L.; Yu, B.; Wu, Z.; Deng, J.; Zhao, B.; Liu, Q.; Chen, W. Analysis of "5.16" Rainstorm and Flood and Flash Flood Disaster Prevention in the Back Mountain of Sanming City, Fujian Province. *J. China Hydrol.* **2021**, *41*, 95–100.
16. Corral, C.; Berenguer, M.; Sempere-Torres, D.; Poletti, L.; Silvestro, F.; Rebora, N. Comparison of Two Early Warning Systems for Regional Flash Flood Hazard Forecasting. *J. Hydrol.* **2019**, *572*, 603–619. [CrossRef]
17. Sun, T.; Yang, P.; Xu, Z.; Wang, Y.; Wang, X. Analysis on The Disaster Mechanism of "8·16" Flash Flood in Zhongdu River basin. *Adv. Eng. Sci.* **2021**, *53*, 132–138.

18. Braud, I.; Roux, H.; Anquetin, S.; Maubourguet, M.M.; Manus, C.; Viallet, P.; Dartus, D. The Use of Distributed Hydrological Models for The Gard 2002 Flash-Flood Event: Analysis of Associated Hydrological Processes. *J. Hydrol.* **2010**, *394*, 162–181. [CrossRef]
19. Marchi, L.; Borga, M.; Preciso, E.; Gaume, E. Characterisation of Selected Extreme Flash Floods in Europe and Implications for Flood Risk Management. *J. Hydrol.* **2010**, *394*, 118–133. [CrossRef]
20. Cai, J. Research on Risk Zoning of Mountain Torrents Disaster in Hubei Province Based on GIS. Ph.D. Thesis, Huazhong University of Science and Technology, Wuhan, China, 2017.
21. Liu, C.; Wen, L.; Zhou, J.; Zhao, X.T.; Guo, L.; Wei, Y.Q. Comparative Analysis of Hydrological and Hydrodynamic Calculation Method for Flash Flood in Small Watershed. *J. China Inst. Water Resour. Hydropower Res.* **2019**, *17*, 262–270+278.
22. Viviroli, D.; Zappa, M.; Gurtz, J.; Weingartner, R. An Introduction to the Hydrological Modelling System PREVAH and its Pre- and Post-Processing-Tools. *Environ. Model. Softw.* **2009**, *24*, 1209–1222. [CrossRef]
23. Liu, C.; Zhou, J.; Wen, L.; Ma, Q.; Guo, L.; Ding, L.Q.; Sun, D.Y. Research on Spatio Temporally-Mixed Runoff Model and Parameter Regionalization for Small and Medium-Sized Catchments. *J. China Inst. Water Resour. Hydropower Res.* **2021**, *19*, 99–114.
24. Liu, C. A New Generation of Hydrological Model Driven by Artificial Intelligence and Big Data and its Application in Flood Forecasting and Early Warning. *China Flood Drought Manag.* **2019**, *29*, 11–22.
25. Hao, S.; Ma, Q.; Zhai, X.; Lyu, G.; Fan, S.; Wang, W.; Liu, C. A New Machine Learning Approach for parameter regionalization of Flash Flood Modelling in Henan Province, China. In Proceedings of the International Conference on Energy, Power and Environmental System Engineering (ICEPESE2021), Shanghai, China, 4–5 July 2021; p. 02010. [CrossRef]
26. Tian, J.; Fan, S.; Wang, W.; Liu, C.J.; Zhang, Q.Y.; Zhao, X.T.; Ma, S. Comparison between FFMS Model and HEC-HMS Model Applied in Rainstorm and Flood Simulation. *J. Yangtze River Sci. Res. Inst.* **2021**, *38*, 59–65.
27. Zhang, X.; Zhou, J.; Wen, L. Application of Spatio-Temporal Variable Source Mixed Runoff Model to Flood Simulation of Small Watersheds: A Case Study of Four Small Watersheds in Sichuan and Gansu Province. *J. Water Resour. Water Eng.* **2021**, *32*, 80–90.
28. Talbot, C.A.; Ogden, F.L. A Method for Computing Infiltration and Redistribution in a Discretized Moisture Content Domain. *Water Resour. Res.* **2008**, *44*, W08453. [CrossRef]
29. Lai, W.; Ogden, F.L.; Steinke, R.C.; Talbot, C.A. An Efficient and Guaranteed Stable Numerical Method for Continuous Modeling of Infiltration and Redistribution with a Shallow Dynamic Water Table. *Water Resour. Res.* **2015**, *51*, 1514–1528. [CrossRef]
30. Bao, W.; Zhao, L. Application of Linearized Calibration Method for Vertically Mixed Runoff Model Parameters. *J. Hydrol. Eng.* **2014**, *19*, 85–91. [CrossRef]

*water*

MDPI

*Article*

# Numerical Investigation on a Flash Flood Disaster in Streams with Confluence and Bifurcation

**Qingyuan Yang [1,2], Xiekang Wang [3,*], Yi Sun [4], Wengang Duan [1] and Shan Xie [5]**

1 Department of Hydraulics, Changjiang River Scientific Research Institute, Wuhan 410030, China; yangqingyuan@mail.crsri.cn (Q.Y.); duanwengang@mail.crsri.cn (W.D.)
2 MWR Key Laboratory of River Regulation and Flood Control in The Middle and Lower, Reaches of Changjiang River, Wuhan 410030, China
3 State Key Laboratory of Hydraulics and Mountain River Engineering, Sichuan University, Chengdu 610065, China
4 Department of Sediment Research, Yellow River Institute of Hydraulic Research, Zhengzhou 450003, China; sunyi031@163.com
5 Department of Planning, Wuhan Water Science Research Institute, Wuhan 410030, China; shanxie_bj@163.com
* Correspondence: wangxiekang@scu.edu.cn

**Abstract:** On 20 August 2019, a flash flood occurred in Sanjiang Town, Sichuan, China, and caused great damage to people living there. The town lies at the junction of five streams, with streams A, B, and C combining at the town and further dividing into streams D and E. The slope of streams A, B, and C is about 3~5%, while the slope of streams D and E is around 0.3%. The Sanjiang Town actually lies in the transition from supercritical slope to subcritical slope. During the flood, huge sediments were released to streams A, B, and C, and further transported to stream E. Due to the rapid change of velocity, only few sediments deposited at the supercritical slope parts of the stream, while plenty of them sedimented at the streams with subcritical slope. In order to simulate the flood with a hydrodynamic model, a field investigation was carried out to collect high DEM (digital elevation model) data, flood marks, sediment grading, etc., after the flood. The discharge curve of the flood was also obtained by the hydrometric station near Sanjiang Town. For the inlet sediment concentrations of streams A, B, and C, we made a series of assumptions and utilized the case which best fits the flood marks to set the inlet sediment concentration. Based on these data, we adopted a depth-averaged two-dimensional hydrodynamic model coupled with a sediment transport model to simulate the flash flood accident. The results revealed that the flash flood enlargement in confluence streams is mainly induced by the inflows, and the flash flood enlargement in bifurcation streams is largely affected by the sediment deposition. The bifurcation of flows can decrease the peak discharge of each branch, but may increase the flooded area near the streams. Flow in the supercritical slope runs at a very fast velocity, and seldom deposits sediment in the steep channel. Meanwhile, most sediment is transported to the streams with flat hydraulic slopes. Due to the functioning of the reservoir, the transition region from supercritical slope to subcritical slope has a much larger probability of being submerged during the flood.

**Keywords:** flash flood; bifurcation; confluence; shallow-water models

**Citation:** Yang, Q.; Wang, X.; Sun, Y.; Duan, W.; Xie, S. Numerical Investigation on a Flash Flood Disaster in Streams with Confluence and Bifurcation. *Water* **2022**, *14*, 1646. https://doi.org/10.3390/w14101646

Academic Editor: Marco Franchini

Received: 28 April 2022
Accepted: 19 May 2022
Published: 21 May 2022



## 1. Introduction

Flash floods are usually induced by rapidly gathered rain. The flash flood runs down the hill and moves quickly, affects a large region, and has little time for early warning. In mountainous areas, people usually reside in flat areas along the river bank, and facilities such as villages, roads, railway, etc., are all in these areas. The flash flood often leads to great damage to people living nearby.

In order to reduce the threat of flash floods, pre-warning systems are set by many countries. The key point of a warning system is creating a standard for the identification of disasters, and using the standard to classify the possible impact area of flash flood inundation. These warning systems should respond in a rather short time. In the past, the computation resources were limited, and the modeling tools for flash flood propagation were simplified to save in calculation time. These systems usually adopted some simple models, such as statistical models [1] and simplified hydrodynamic models [2,3] focusing on water levels, discharge, rainfall, etc., but the propagation process of flow was seldom mentioned [4–6]. The warning standard obtained by these models was rough and inaccurate for residential areas with complex channels and buildings. With the development of computers and in order to study the warning standard of flash floods in detail, many researchers have studied flash floods with depth-averaged two-dimensional hydrodynamic models, which are powerful tools to capture flow dynamic behavior and save computational time for a spatially large-scale flow domain [7,8]. These models solve the full governing equations, including the rainfall and infiltration sections. The discretization equations using the Godunov method [9] can accurately capture the shock wave and deal with the sharp change of the bed form. The calculation time is acceptable, and this is a very promising too [7,10–15].

Furthermore, flash flood enlargement, where the peak water level or peak discharge is increased or decreased in areas with rapid bed form obstruction [16] will largely impact the flash flood warning standard. Many studies are mainly focused on the natural issues of the flash floods, such as sediment transport, landslide dams [17], confluence of rivers, etc. Yang et al. [18] studied the impact of sediment transport on the peak level and peak discharge of flash floods with numerical models, and Chen et al [4] investigated the effect of landslide dams on the flash flood peak level with a large-scale physical model. Wang et al. [19] made progress in the flash flood propagation in the river confluence zone with both numerical and experimental techniques. Chen et al [20] analyzed the flash flood wave propagation in the confluence of open channels. Hackney et al. [21] made a dent in discharge variation of the flash flood in the river confluence region with field investigations.

Generally speaking, flash floods in mountainous areas occur in channels with very steep slopes, and huge sediments can be transported far away due to the high-speed flow. However, if there is a reservoir setting on the stream, the slope will be slowed by the highly raised water level. A transition region from the supercritical slope to the subcritical slope will be produced by the sedimentation. Flash floods in streams with confluence and bifurcation, especially in the transition region from supercritical slope to subcritical slope, are seldom studied in detail. In this study, we use a depth-average two-dimensional hydrodynamic model to conduct a numerical inversion of a flash flood that occurred on 20 August 2019 in Sanjiang Town, Sichuan, China.

## 2. Study Area

The study area covers Sanjiang Town, which lies in Sichuan Province of China. Figure 1 shows the satellite image of Sanjiang Town. The town lies at the confluence of three streams, A, B, and C. Firstly, stream B meets stream C at the upper part of the town, then they both join stream A. The distance between the two junctions is about 100 m. Right after the confluence, stream A is further divided into streams D and E. For stream D, there is a milldam near the bifurcation, and the milldam can discharge part of the flash flood through stream D. Stream E has twice the width of stream D, and releases most of the flood. Streams D and E meet again at the Sanjiang reservoir. The central bar surrounded by streams D and E is the major developing part of the town in recent years. At the end of stream E, a hydrometric station collects hydraulic data during the floods. An overflow dam lies at the end of the Sanjiang reservoir, and gates of the dam are all opened to discharge floods.

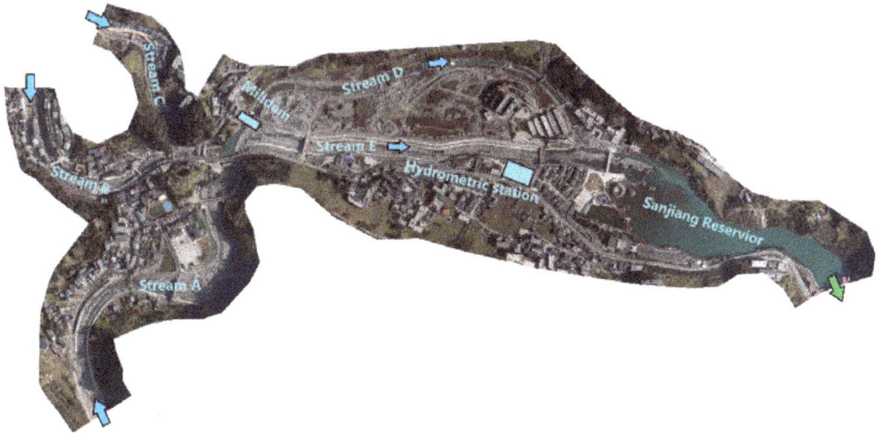


**Figure 1.** Satellite image of Sanjing Town, Sichuan, China.

## 3. Methodology

In this study, we adopted the depth-averaged, two-dimensional, shallow-water equations, coupled with sediment transport and bed variation equations [22], to simulate the flash flood and sediment transport during the disaster.

Continuity equation:

$$\frac{\partial}{\partial t}(h) + \frac{\partial}{\partial x}(hu) + \frac{\partial}{\partial y}(hv) = -\frac{\partial Z_b}{\partial t} \tag{1}$$

Momentum equation:

$$\frac{\partial}{\partial t}(hu) + \frac{\partial}{\partial x}\left(hu^2 + \frac{1}{2}gh^2\right) + \frac{\partial}{\partial y}(huv) = gh\left(S_{bx} - S_{fx}\right) + hv_t\left(\frac{\partial^2 u}{\partial x^2} + \frac{\partial^2 u}{\partial y^2}\right) - \frac{\Delta\rho g h^2}{2\rho_s\rho_m}\frac{\partial S_T}{\partial x} + \frac{\rho_0 - \rho_m}{\rho_m}u\frac{\partial Z_b}{\partial t} \tag{2}$$

$$\frac{\partial}{\partial t}(hv) + \frac{\partial}{\partial y}(huv) + \frac{\partial}{\partial y}\left(hv^2 + \frac{1}{2}gh^2\right) = gh\left(S_{by} - S_{fy}\right) + hv_t\left(\frac{\partial^2 v}{\partial x^2} + \frac{\partial^2 v}{\partial y^2}\right) - \frac{\Delta\rho g h^2}{2\rho_s\rho_m}\frac{\partial S_T}{\partial y} + \frac{\rho_0 - \rho_m}{\rho_m}v\frac{\partial Z_b}{\partial t} \tag{3}$$

Equation of bed load transport:

$$\frac{\partial}{\partial t}(hq_b) + \frac{\partial}{\partial x}(huq_b) + \frac{\partial}{\partial y}(hvq_b) = -\alpha_b\omega_b(q_b - q_{b*}) = -\rho'\frac{\Delta Z_b}{\Delta t} \tag{4}$$

Equation of bed deformation:

$$\frac{\Delta Z_b}{\Delta t} = \frac{\alpha_b\omega_b(q_b - q_{b*})}{\rho'} \tag{5}$$

in which $t$ denotes time, $h$ means water depth, $u$ and $v$ present velocity components in the x and y directions, respectively, $g$ is gravitational acceleration, $v_t$ is the turbulent viscosity coefficient, and $\Delta\rho = \rho_s - \rho_w$, in which $\rho_s$ is the sediment density, and $\rho_w$ is the water density.

$$\rho_m = (1 - S_v)\rho_w + S_v\rho_s = \left(1 - \frac{S}{\rho_s}\right)\rho_w + S_T \tag{6}$$

where $S_v$ is the volumetric sediment concentration, and $S_T$ is the total concentration of graded sediments (kg/m$^3$). In this study, only the bed load transport is calculated, so $S_T$ equals $q_b$.

$$\rho_0 = \left(1 - \frac{\rho'}{\rho_s}\right)\rho_w + \rho' \tag{7}$$

in which $\rho_0$ and $\rho'$ denote the density of saturated and dry bed material, respectively. The bed slope terms $(S_{bx}, S_{by})$ and friction slope terms $(S_{fx}, S_{fy})$ are written as $S_{bx} = -\partial Z_b/\partial x$, $S_{by} = -\partial Z_b/\partial y$, and $S_{fx} = n^2 u\sqrt{u^2+v^2}/h^{4/3}$, $S_{fy} = n^2 v\sqrt{u^2+v^2}/h^{4/3}$ in the x and y directions, respectively, where $Z_b$ is bed elevation and $n$ is Manning's roughness coefficient.

$$q_{b0} = \frac{K_b}{C_0^2}\frac{\rho_s\rho_m}{\rho_s-\rho_m}(U-U_c)\frac{U^3}{g\omega_b} \tag{8}$$

$$q_{b*} = q_{b0}/(hU) = \frac{K_b}{C_0^2}\frac{\rho_s\rho_m}{\rho_s-\rho_m}(U-U_c)\frac{U^2}{hg\omega_b} \tag{9}$$

where $q_b$ (in kg/m$^3$) is the amount of bed load in a unit volume of water, $\omega_b$ is the settling velocity of bed load, $q_{b0}$ is the transport capacity of bed load in a unit volume of water, in kg/m$^3$, and $\alpha_b$ is the non-equilibrium adaptation coefficient of bed load. $K_b$ is an empirical coefficient, $U_c$ is the incipient velocity of bed load, calculated by $U_c = 0.265\log(11h/d)\sqrt{\frac{(\rho_s-\rho_w)}{\rho_w}gd}$, $C_0$ is the dimensionless Chézy coefficient, and $q_{b*}$ is the value of bed load in a unit volume, and is obtained according to $q_{b*} = q_{b0}/(hU)$.

The governing Equations (1)–(4) can be rewritten in the form:

$$\frac{\partial U}{\partial t} + \frac{\partial E}{\partial x} + \frac{\partial G}{\partial y} = S \tag{10}$$

in which

$$U = \begin{bmatrix} h \\ hu \\ hv \\ hq_b \end{bmatrix}, E = \begin{bmatrix} h \\ hu \\ hv \\ hq_b \end{bmatrix}, G = \begin{bmatrix} h \\ hu \\ hv \\ hq_b \end{bmatrix}$$

$S$ is the source section, and is equal to the rest of the sections of each governing equation.

$$U_i^{n+1} = U_i^n + \sum(En_x + Gn_y) + S \tag{11}$$

The governing equations are solved by the method of finite volume with hybrid triangular and quadrangular meshes. The variables $(h, u, v, q_b, Z_b)$ in each mesh are updated in a time-marching manner (Equation (11)). $U_i^n$ is the variables' value at time $n$, $U_i^{n+1}$ is the variables' value at time $n$ + 1, and $En_x + Gn_y$ denotes the flux crossing the edges of the mesh. In order save the calculation time, a NVIDIA GeForce RTX 3050 GPU (graphics processing unit) was adopted to speed up the calculation. The flow chart of the calculation is shown in Figure 2.

The calculation codes used in this study have been successfully used in the former studies [18,23]. The simulation region covers most parts of Sanjiang Town, and the number of simulation meshes is about 110,000. In order to capture the flash flood propagation accurately with a moderate quantity of mesh, a non-uniform (triangular and quadrangular hybrid) mesh is adopted. The size is about 2.0 m in regions near the stream channel and bank, and about 5.0 m in regions far from the streams (Figure 3).

We collected high-resolution (5.0 m) DEM (digital elevation model) data (Figure 1) after the flash flood in 2019, and the flow rates of stream E (Qt(CS14)) at the hydrometric station of Sanjiang Town. The flow rates at the hydrometric station were gauged at an interval of 20 min. The flash flood lasted about 120 h (from 20 to 24 August in 2019). The 120 h-long flash flood can be divided into four stages, and each of them has one discharge peak (Figure 4a, stage 1, 2, 3, and 4, respectively). We deduced the discharge of streams A ($Q_A$), B ($Q_B$), and C ($Q_C$) based on their upstream catchment area (Figure 4b). For example, since the milldam in stream D was closed during the flood, the total discharge of streams A, B, and C should be equal to the discharge at CS14. Supposing the catchment areas of streams A, B, and C are $A_A$, $A_B$, and $A_C$, respectively, and the total area of them $A_T = A_A + A_B + A_C$, then $Q_A$ = Qt$A_A/A_T$, $Q_B$ = Qt$A_B/A_T$, and $Q_C$ = Qt$A_C/A_T$, respectively. All these data contribute to the numerical inversion of the flash flood.

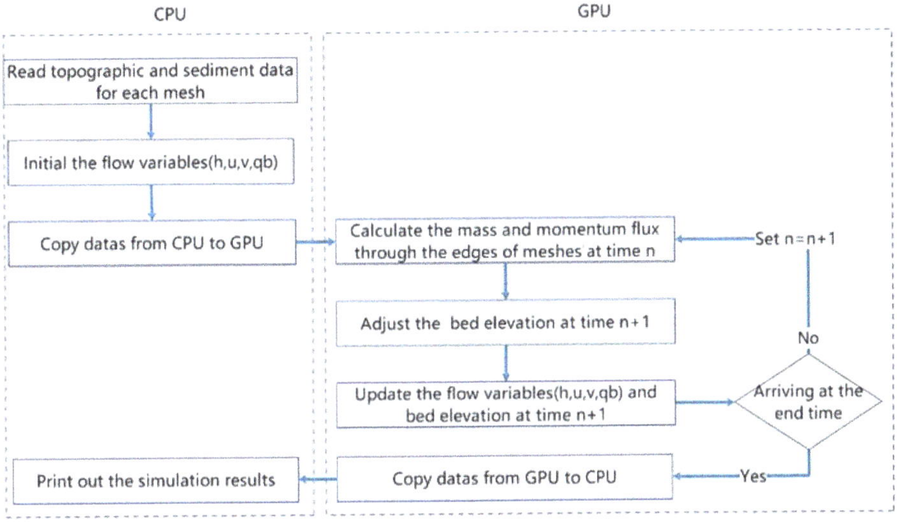


**Figure 2.** Flow chart of the calculation.

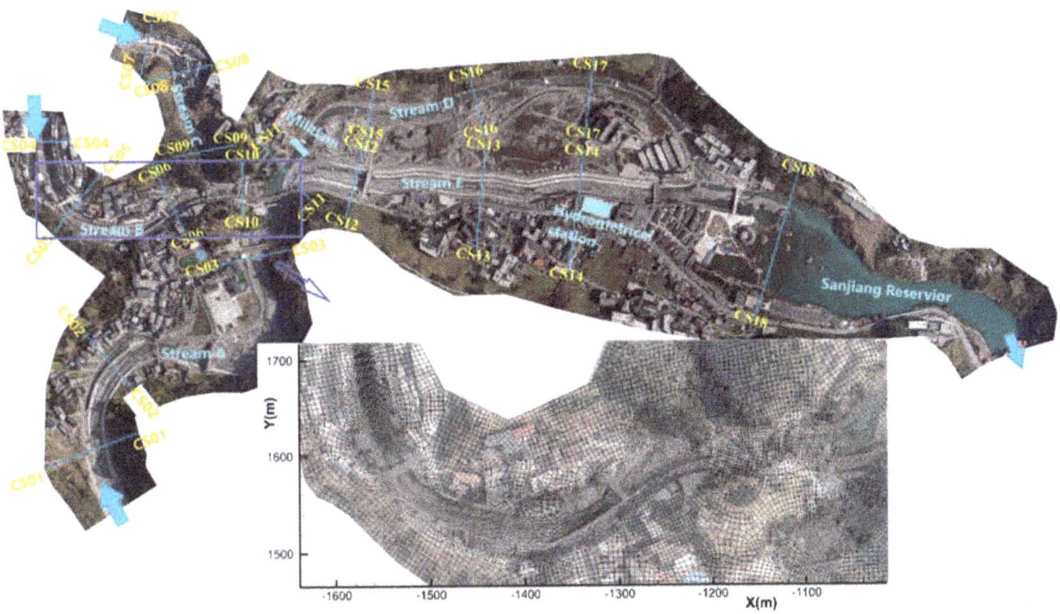


**Figure 3.** Recorded cross-sections (top) and part of the calculation meshes (bottom).

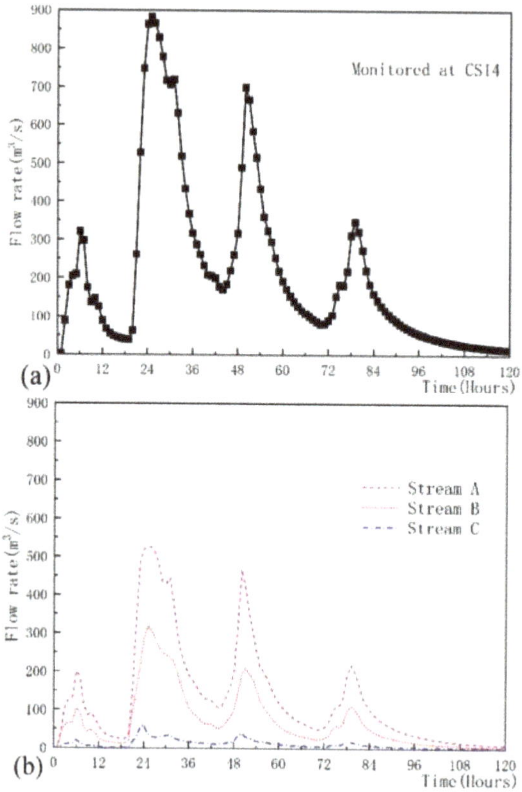


**Figure 4.** Monitored (**a**) and deduced (**b**) discharges of the flash flood.

In this study, the roughness of the bed was set with the Manning's roughness, $n$, of 0.025, which is suggested by hydraulics manuals [24] and verified by our former studies [18]. The bed elevations at each mesh were initiated by interpolating from the high-resolution terrain data, and updated by the simulation codes at each timestep based on Equation (5). In order to study the unsteady process of the flash flood, hydraulic data of 18 cross-sections (Figure 3) were saved at every timestep, and data of the full simulation region were saved at every 100 timesteps during the calculation.

The inlet discharges for streams A, B, and C were consistent with the deduced rate curves in Figure 4b, and flow rate of the outlet boundary was automatically calculated according to the empirical formula for weir flow. The formula is $Q = mnb\sqrt{2g}H^{1.5}$, in which $Q$ = discharge, $m$ = discharge coefficient, about 0.502, $n$ = number of gates opened, $b$ = width of each gate, and $H$ = the difference between water level and elevation of the weir top. Once the discharge is calculated, the velocity at the outlet is set based on the discharge.

The milldam was closed during the flash flood in 2019 (Figure 5), and released little water to stream D. The water in streams A, B, C, and E performs as the confluence flow, but the flow will move as bifurcation flow when the milldam is open. Therefore, in order to investigate the impact of the milldam on the flash flood, scenarios with both a closed and an opened milldam were simulated in this study.

**Figure 5.** Image of the milldam after the flash flood in 2019.

As the sediment characteristics were not monitored by the hydrometric station during the flash flood disaster, it is difficult to figure out the amount of bed load at inlet boundaries of streams A, B, and C. We assumed the inlet sediment concentrations with different percentages of transport capacity of bed load at the inlets. Transport capacity will change with velocity and depth. For example, if the transport capacity of bed load at the inlets at time A is 100 $kg/m^3$, Qs-0.05 means the inlet sediment concentration is set as $100 \times 0.05 = 5$ $kg/m^3$ at time A, and Qs-0 means the inlet sediment concentration is zero all the time. For the diameters of sediment, we collected some sediment of the river bed after the flash flood in 2019, and obtained a median size of 0.02 m. The flow conditions for the simulation scenarios are listed in Table 1.

**Table 1.** Flow conditions for the simulation scenarios.

| Scenarios | Inlet Discharge | Inlet Sediment Concentration | Milldam of Stream D |
|---|---|---|---|
| Case Qs-0 | Same as Figure 4b | 0 | Closed |
| Case Qs-0.05 | | $0.05 \times q_{b0}$ | Closed |
| Case Qs-0.075 | | $0.075 \times q_{b0}$ | Closed |
| Case Qs-0.1 | | $0.1 \times q_{b0}$ | Closed |
| Case Qs-0.125 | | $0.125 \times q_{b0}$ | Closed |
| Case Qs-0.1-bifurcation | | $0.1 \times q_{b0}$ | Open |

## 4. Results

### 4.1. Model Calibration

The purpose of this section is to figure out the scenario which best fits the monitored data. We analyzed the simulation results of the first five scenarios in Table 1. In these scenarios, the milldam in stream D was closed during the flood.

Figure 6a shows the simulated and monitored discharge profiles at CS14, and Figure 6b presents a detailed view of stage 3. It can be seen that the simulated discharge profile agrees well with the monitored data at CS14, except for that of case Qs-0.125 in stage 3. Figure 7 shows the distribution of unit width flux of cases Qs-0.125 and Qs-0.1. The flooded area of case Qs-0.125 was a bit larger than that of case Qs-0.1, and some flow travelled downstream through CS17 instead of CS14, which contributed to the bigger difference in Figure 6b.

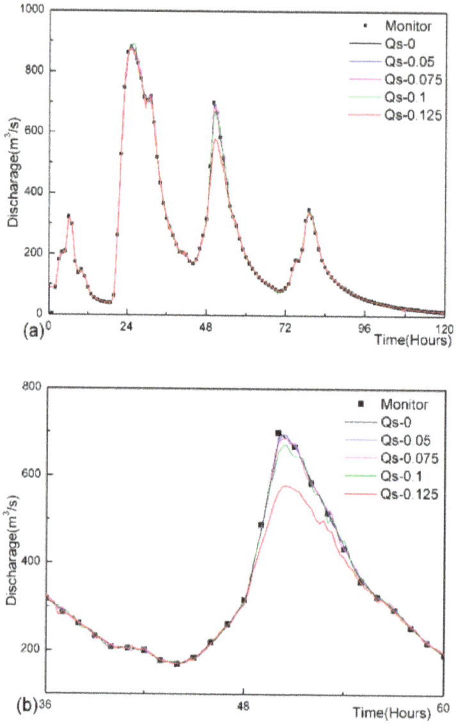


**Figure 6.** Simulated and monitored discharge profiles at CS14: (**a**) whole flood and (**b**) stage 3 of the flood.

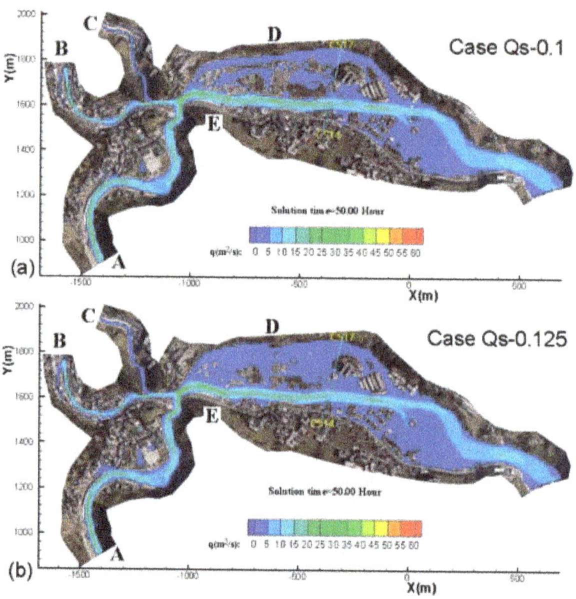


**Figure 7.** Flooded region of cases Qs-0.1 (**a**) and Qs-0.125 (**b**).

Figure 8a shows the simulated water level of each scenario. For cases Qs-0 and Qs-0.05, the maximum level occurred in stage 2 of the flood, while for cases Qs-0.075, Qs-0.1, and Qs-0.125, the maximum level occurred in stage 3. Figure 8b shows the simulated bed elevation of each scenario. The bed elevation increased rapidly in flood stages 2 and 3, and the larger the inlet sediment concentration, the more the bed elevation increased. The lifted bed elevation further contributed to the occurrence time of the maximum level being delayed from flood stage 2 to 3. The sediment deposition can change the amplitude and occurrence time of the peak level.

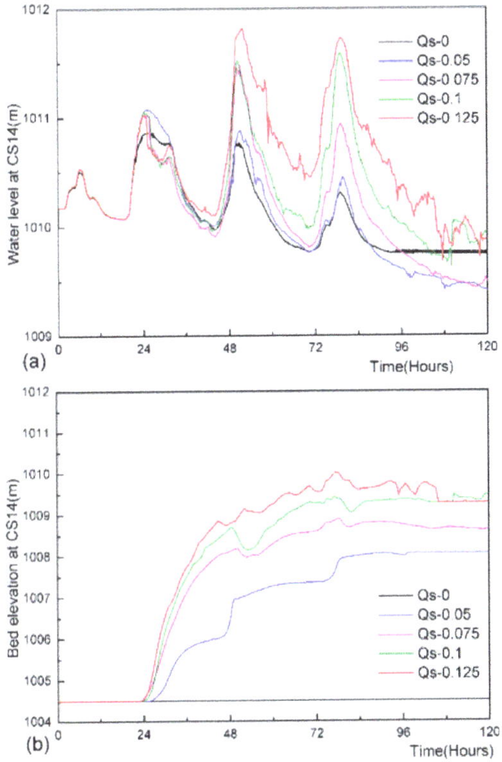


**Figure 8.** Simulated water level (**a**) and bed elevation (**b**) at CS14.

According to the analysis above, the discharge curves were nearly coincident with each other among the first four simulation scenarios. However, water level and bed elevation were much different from each other for cases with different inlet sediment concentrations. The different water level and bed elevation further impacted the flooded area during the flood. Based on the field investigation carried out right after the flash flood in 2019, the simulation results of case Qs-0.1 fit better concerning the flood marks, final bed elevation, flood area, etc. Therefore, we used the 10% of transport capacity (Qs-0.1) as the inlet sediment concentration.

*4.2. Flash Flood in Confluence Streams (Case Qs-0.1)*

As the gates of the milldam were closed in this case, all of the flow coming from streams A, B, and C went to the reservoir through stream E. Figure 9 presents the peak discharge along the streams. In the four stages of the flash flood, the peak discharges increased rapidly at CS11, and showed little change at other parts of the streams.

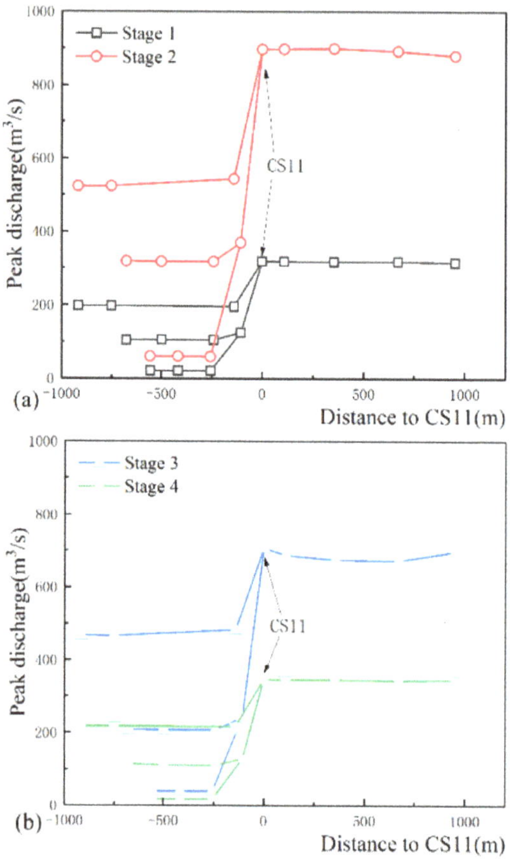


**Figure 9.** Peak discharge along the streams (Case Qs-0.1, (**a**) stage 1 and 2, (**b**) stage 3 and 4).

Figure 10a shows the peak level at each cross-section. The water level dropped sharply along the channel of streams A, B, and C, with a minimum slope of 2.5%. On the contrary, in stream E, the peak water level slowly decreased along the channel, and the slope was about 0.3%. Once again, this proves that the town lies in the transition region from a steep channel to a gentle slope.

Peak water levels were generally coincident with the flood marks obtained by field investigations, and it was difficult to compare them as the slope was steep (Figure 10a). Figure 10b presents the relative peak water level at all the cross-sections. The relative peak water level was obtained by subtracting the peak water level of each flood stage from that of stage 1, so the relative peak water levels of stage 1 were all zero. For cross-sections in streams A, B, and C, the maximum level occurred in stage 2, in which the maximum peak discharge existed. However, for cross-sections in stream E, the maximum level occurred in stage 3 of the flood. The bed elevation in stage 3 was much higher than that of stage 2, and the deposition lifted the water to a higher level with a relatively smaller discharge (Figure 8).

For a better understanding of the flow properties along the streams, we adopted the unit width flux to analyze the main flow variation. As the unit width flux, $q$, is equal to the product of depth and velocity magnitude, regions with larger $q$ denote more flow passing through. Figure 11 shows the distribution of unit width flux ($q$) at the four stages of the flood. Since the milldam was closed, the main flow came from streams A and B, and went to the reservoir only through stream E.

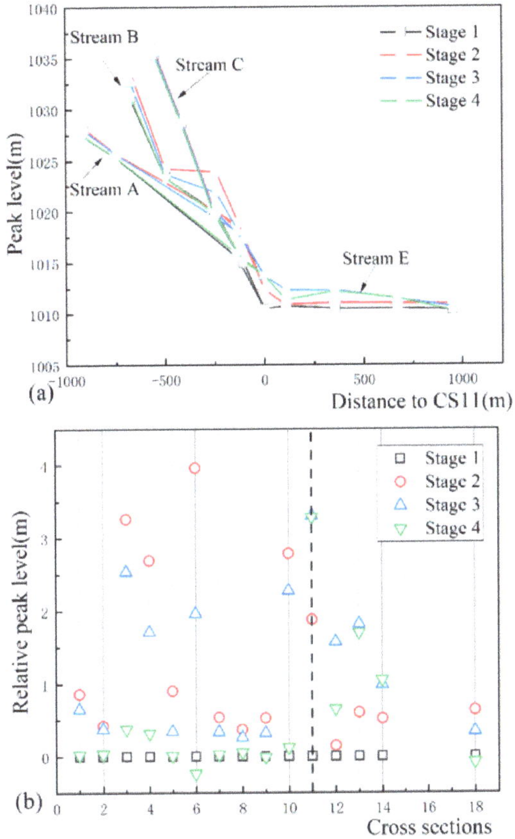


**Figure 10.** Peak water level (**a**) and relative peak water level (**b**) (Case Qs-0.1).

Siltation thickness denotes the differences between bed elevation at any time and that of the initial time. The larger the thickness, the more sediments deposited there. Figure 12 presents the distribution of siltation thickness at the four stages of the flood. Sediments deposited mainly at the confluence region of streams A and B in stages 1 and 2, while the main siltation region shifted to stream E in stages 3 and 4. Meanwhile, the siltation thickness in streams A and B became thinner in stages 3 and 4, which explains the finding in the field investigation whereby the flood marks were much higher but the elevation of sediments was much lower in stream A. In stages 1 and 2, the water level in streams A and B was higher due to the larger discharge and higher bed elevation, while in stages 3 and 4, the water level in streams A and B was lower due to the smaller discharge and lower bed elevation, due to erosion. After the flood, only the final bed elevation and flood marks could be measured.

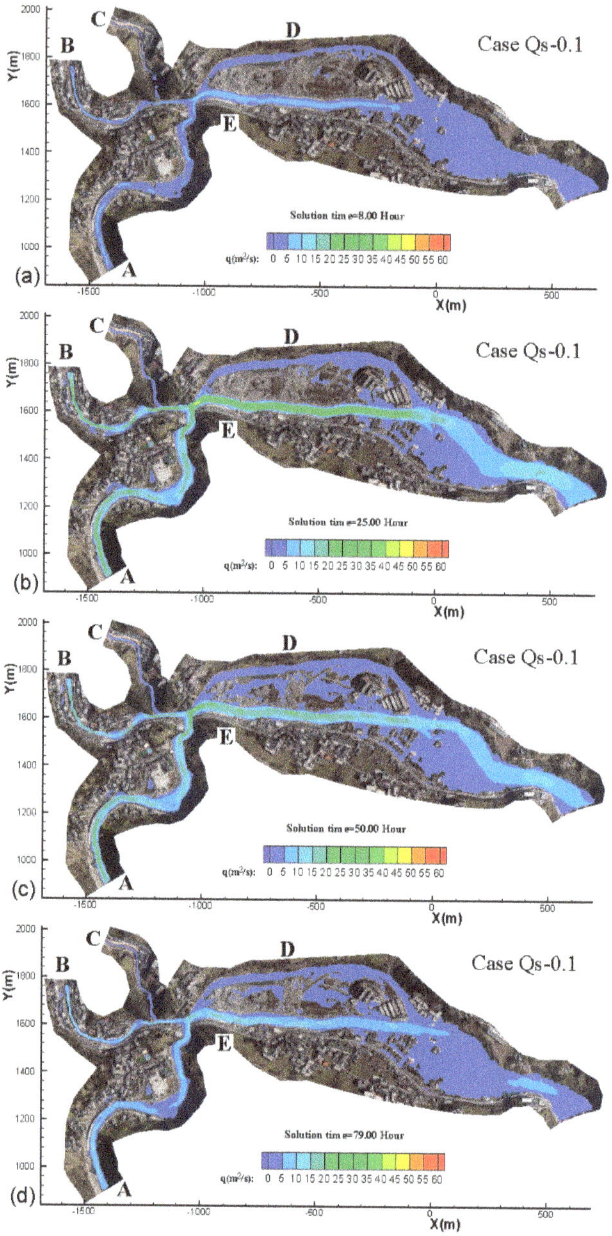


**Figure 11.** Distribution of unit width flux ($q$) at the four stages of the flood ((**a**) stage 1, (**b**) stage 2, (**c**) stage 3, (**d**) stage 4).

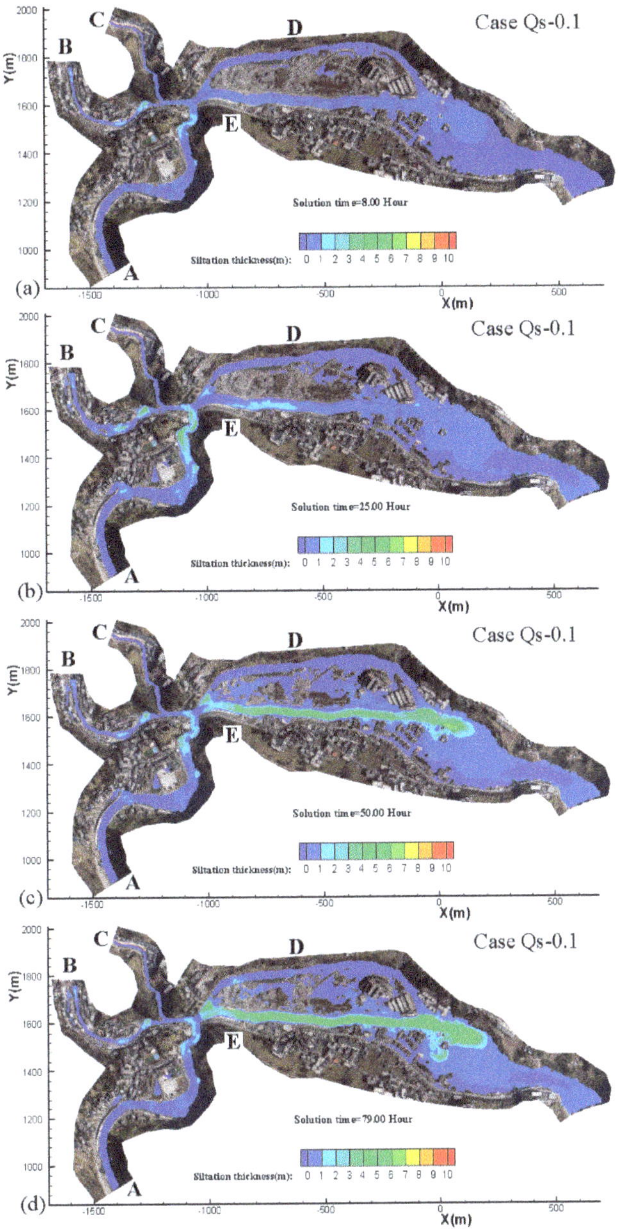


**Figure 12.** Distribution of siltation thickness at the four stages of the flood ((**a**) stage 1, (**b**) stage 2, (**c**) stage 3, (**d**) stage 4).

Figure 13 shows the distribution of the Froude number (Fr) at the four stages of the flood. The Froude number is defined as the velocity magnitude divided by water depth (sqrt(u ˆ 2 + u ˆ 2)/h). If Fr > 1, the flow is subcritical, if Fr > 1, the flow is supercritical, and if Fr = 1, the flow is critical. For streams A, B, and C, the Froude number of most regions was larger than 3 in the four flood stages, while for stream E, the flow was subcritical in stages 1 and 2, and supercritical in stages 3 and 4. There is little siltation at places with a larger Froude number. On the contrary, for stream E, the regions with a larger Froude number were nearly coincident with those of higher sediment thickness (Figure 12). The highly raised bed decreased the water depth and increased the Froude number.

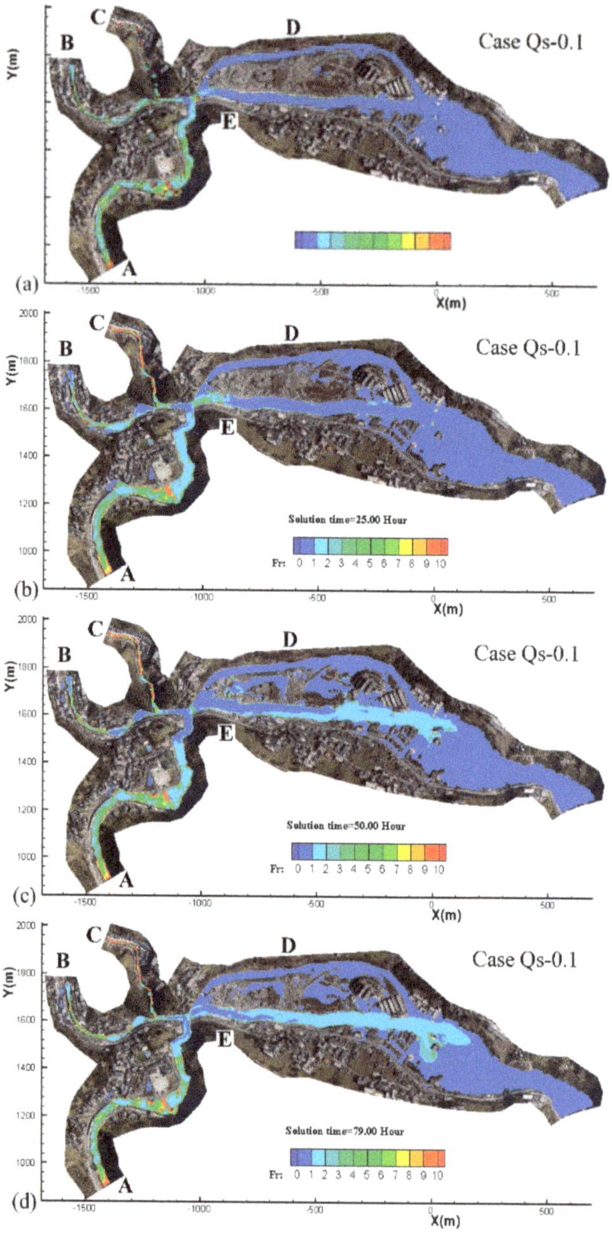


**Figure 13.** Distribution of the Froude number at the four stages of the flood ((**a**) stage 1, (**b**) stage 2, (**c**) stage 3, (**d**) stage 4).

### 4.3. Flash Flood in Streams with Bifurcation (Case Qs-0.1-Bifurcation)

As the gates of the milldam were open in this case (Qs-0.1-bifurcation), all of the flow coming from streams A, B, and C travelled to the reservoir through both stream D and stream E. Figure 14 presents the peak discharge along the streams. The peak discharge increased rapidly at CS11 and CS18 due to the combination of flows, decreased sharply after CS11, and showed little change at other parts of the stream.

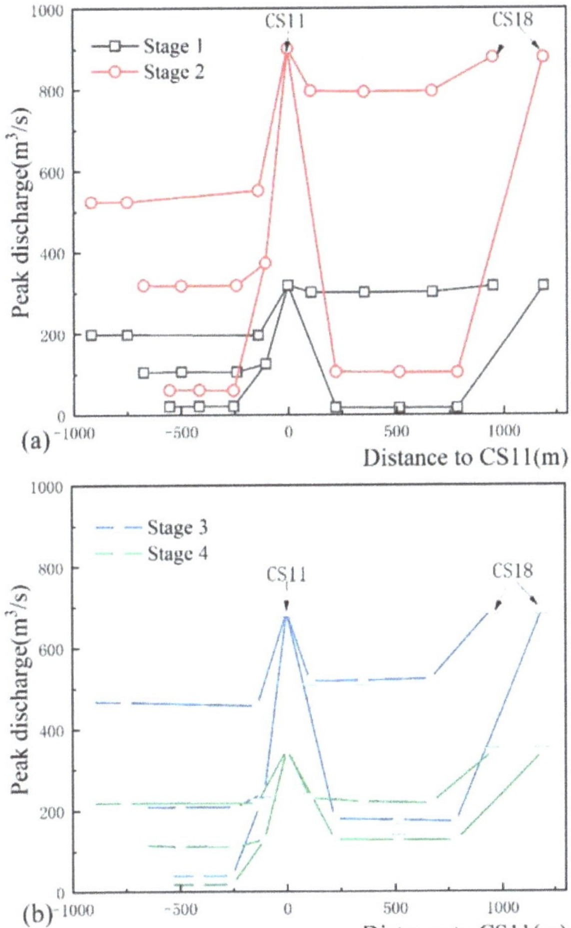


**Figure 14.** Peak discharge along the streams (Qs-0.1-bifurcation, flood (**a**) stage 1, stage 2. (**b**) stage 3, stage 4).

Figure 15 shows the peak level at each cross-section. The water level varied in a similar way to that in Figure 10. The maximum water level of streams D and E occurred in flood stage 3 (the second largest discharge stage), while the maximum water level of streams A, B, and C occurred in flood stage 2 (the largest discharge stage).

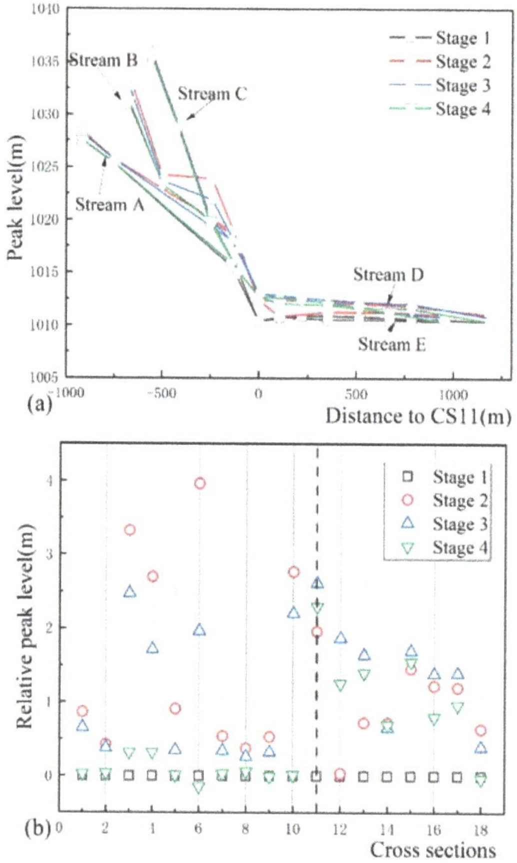


**Figure 15.** Peak water level (**a**) and relative peak water level (**b**) (Qs-0.1-bifurcation).

Figure 16 shows the distribution of unit width flux (q) at the four stages of the flood. The main flow came from streams A and B, and travelled to the reservoir through stream E in stages 1 and 2, while more and more flow travelled to the reservoir via stream D in stages 3 and 4. Figure 17 presents the flow rates and flow rate ratios of streams D and E. In stages 1 and 2 (before 30 h), the flow rate ratio of stream D was about 0.1, while in stages 3 and 4, the flow rate ratio of stream D increased to around 0.3. Figure 18 presents the distribution of siltation thickness at the four stages of the flood as the milldam was open. Sediments deposited mainly at the confluence region of streams A, B, and C in stages 1 and 2. The siltation region shifted to streams D and E in stages 3 and 4, and siltation in stream E increased much quicker than that in stream D. The uplifted bed of stream E pushed more water to stream D. For the distribution of Froude numbers in Figure 19, supercritical regions were still coincident with places of thick sediment deposition in streams D and E, and with little sediment deposits at places with a higher Froude number in streams A, B, and C.

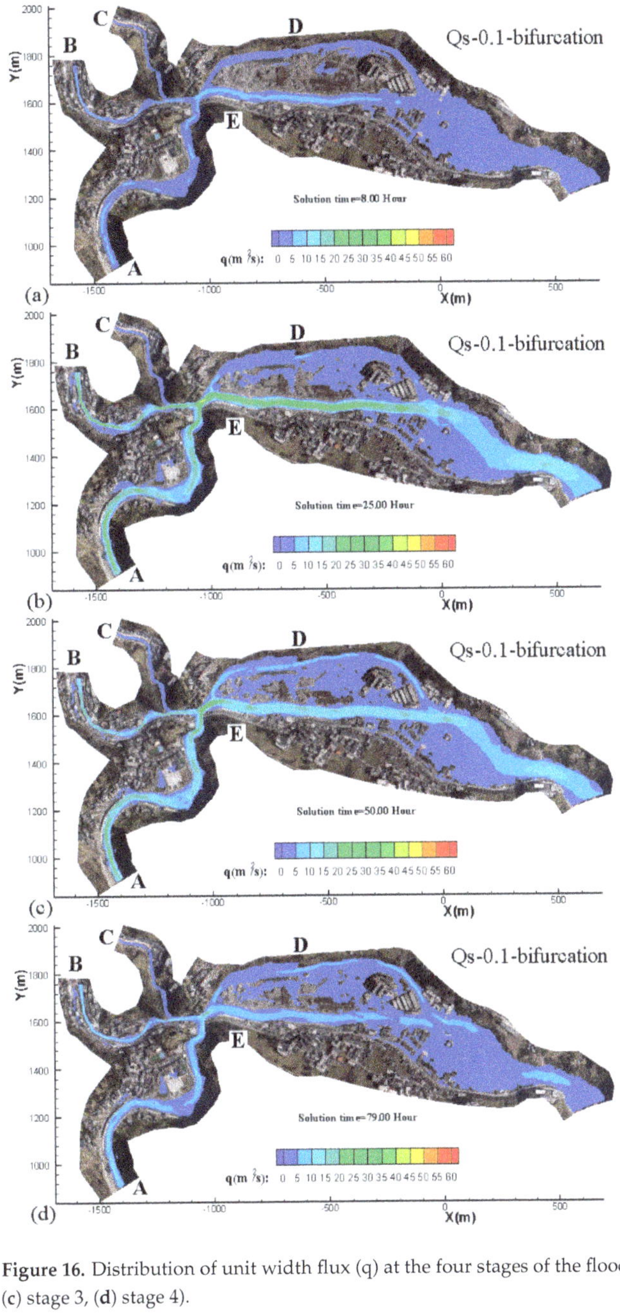


**Figure 16.** Distribution of unit width flux (q) at the four stages of the flood ((**a**) stage 1, (**b**) stage 2, (**c**) stage 3, (**d**) stage 4).

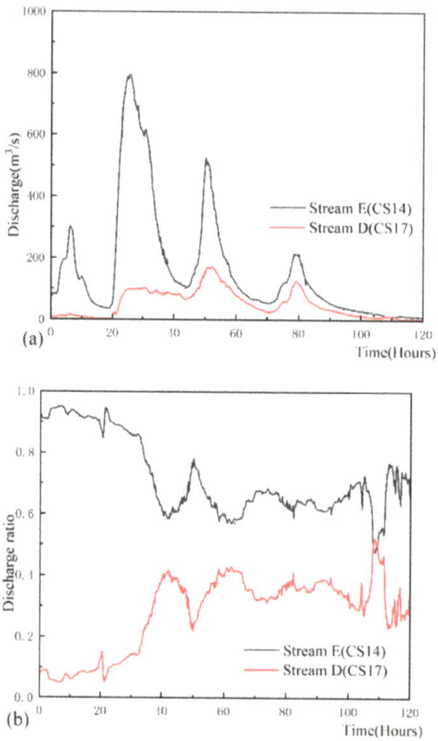


**Figure 17.** Discharge (**a**) and discharge ratios (**b**) between streams D and E.

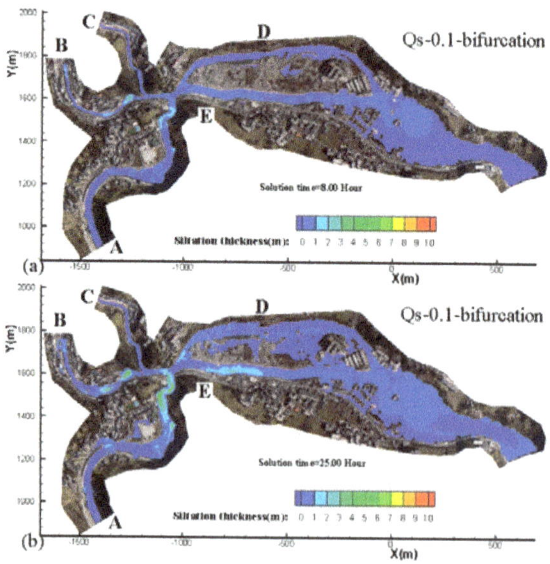


**Figure 18.** *Cont.*

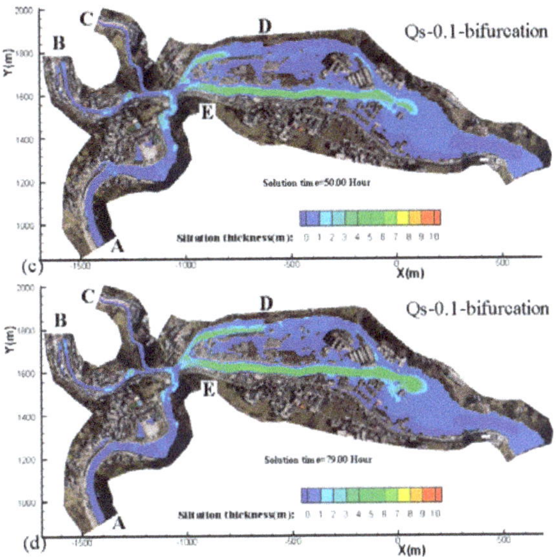


**Figure 18.** Distribution of siltation thickness at the four stages of the flood ((**a**)stage 1, (**b**) stage 2, (**c**) stage 3, (**d**) stage 4).

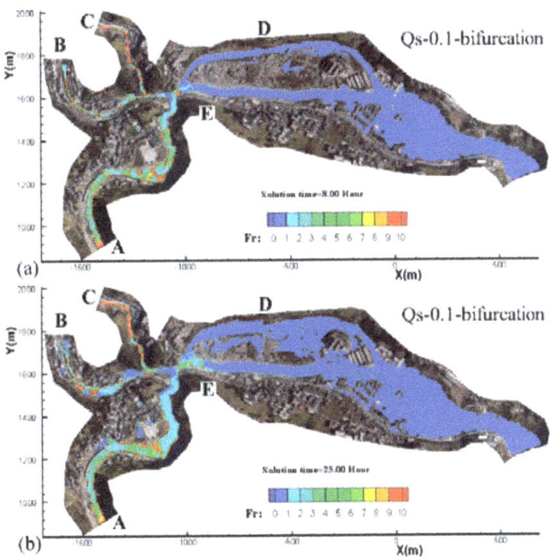


**Figure 19.** *Cont.*

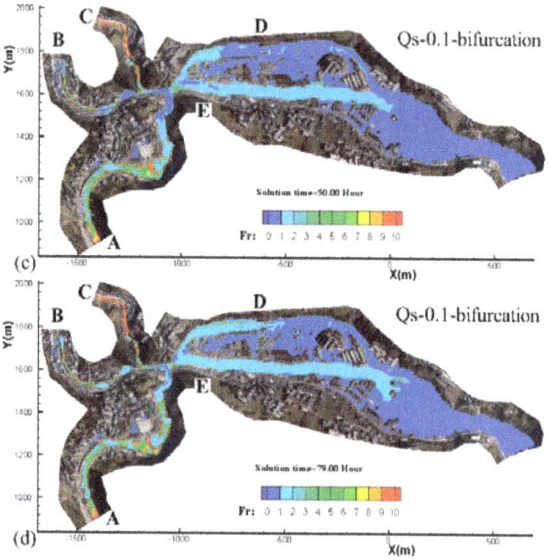


**Figure 19.** Distribution of the Froude number at the four stages of the flood ((**a**)stage 1, (**b**) stage 2, (**c**) stage 3, (**d**) stage 4).

Figure 20 shows the flooded area of cases Qs-0.1 and Qs-0.1-bifurcation. Among all the four stages of the flood, the flooded area of case Qs-0.1-bifurcation was larger than that of case Qs-0.1. The flooded area ratio between case Qs-0.1-bifurcation and case Qs-0.1 is shown in Figure 20b, and the ratio had a maximum of 1.1. From the point of view of reducing the flooded area, it is better to keep the milldam closed during the flood.

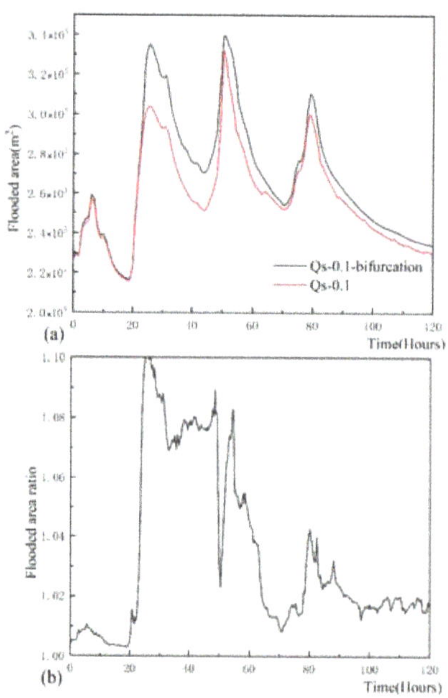


**Figure 20.** Flooded area of cases Qs-0.1 (**a**) and Qs-0.1-bifurcation (**b**).

## 5. Discussion

### 5.1. Boundary Conditions for the Numerical Inversion of Flash Flood

In this study, we adopted the two-dimensional shallow-water models coupled with sediment transport models to analyze the flash flood that occurred in Sanjiang Town in 2019. On most occasions, flash floods occur in a region with limited monitoring facilities. It is difficult to accurately set the boundaries for flash flood simulations. Usually, the elevation of the bed form, the grade curve of sediment, and flood marks can be measured after the flood disaster. However, the time-discharge curve, the inlet sediment concentration, and the outlet level are rather difficult to estimate. Luckily, we obtained the monitored discharge curves from the hydrometric station near Sanjiang Town, and deduced the discharge for streams A, B, and C by their upstream catchment areas. For the inlet boundary, we made assumptions with a series of sediment concentrations, and used the sediment concentration that best fit the field investigation data. Theoretically, the method is more accurate than those using a constant inlet sediment concentration, although it is still not certain that the sediment concentration is accurate due to the complexity of sediment transport. The method of evaluating the boundaries in this paper is helpful for the numerical analysis of other flash flood disasters, and hydrological models coupled with hydrodynamic models may be taken into account if the discharge curve cannot be directly obtained [2].

Furthermore, the quantitative simulation results are still helpful for taking measures against the flood. For example, keeping the milldam closed can minimize the flooding area.

### 5.2. Impact of Sediment on Flash Flood in Stream Confluence and Bifurcation

Flash flood wave enlargement is common in flash flood propagation, and it is usually caused by the sediment intrusion and bed form obstruction [18]. For sediment intrusion, the enlargement comes from the bulk effect, and the volume and density of fluid are both enlarged. For the bed form obstruction, the enlargement is rooted in the sharp change of the water level and velocity of the bed form. For example, the landslide dams, the change in the bed slope, etc., can increase or decrease the peak level and discharge.

The confluence and bifurcation are prototypes of streams. The flash flood enlargement in confluence streams is impacted not only by the bed form, such as the junction angle, elevation differences, etc., but also by the flow rate ratio of the streams [19]. In this study, as the flow rate ratio among streams A, B, and C was constant, the flash flood enlargement in confluence streams was mainly induced by the merging of inflows. Meanwhile, the flash flood enlargement in bifurcation streams is largely affected by the sediment deposition. The flow rate ratio between the two branches was changed by the uneven sedimentation of streams D and E. The adjustment in the flow rate ratio is similar to that in plain rivers, except that the variation in mountainous streams is much quicker [8].

The distinctive part of this study is that the study region lies in the transition region from a supercritical slope to a subcritical slope. The hydraulic slopes of streams D and E were flattened by the Sangjiang reservoir. Huge sediments deposited in streams D and E as the velocity was slowed down. From the point of view of reducing the flooded area, the position of Sangjiang reservoir was not well-planned considering the sediment transport mode, and it should be moved downstream, far away from Sanjiang Town.

## 6. Conclusions

(1) In this study, the flash flood that occurred on 20 August 2019 in Sanjiang Town was analyzed by a depth-averaged two-dimensional model. The simulation showed the flash flood process in detail, and introduced a method of evaluating the boundary conditions for the numerical inversion of flash floods.

(2) Mountainous streams with many people living nearby are usually reformed by manmade facilities, which will affect the flash flood propagation properties. In this study, the milldam in stream D could largely reduce the flooded area. The manmade facilities should be taken into account in the study of flash floods, and be properly utilized during the

flood events. Furthermore, planning of reservoirs should consider the sediment transport properties, as the hydraulic slope will be flattened by the reservoir.

(3) For streams in the transition region from supercritical slope to subcritical slope, the flash flood enlargement in confluence streams is mainly induced by the combination of inflows, and the flash flood enlargement in streams with bifurcation is largely affected by the sediment deposition. Flow in the supercritical slope runs at a very fast velocity, and seldom deposits sediment in the steep channel. In the meantime, the sediment is transported to the streams with a flat hydraulic slope and deposited there. The transition region has a much larger probability of being submerged during the flood.

**Author Contributions:** Conceptualization, Q.Y. and X.W.; investigation, X.W.; resources, W.D.; data curation, Y.S. and S.X. All authors have read and agreed to the published version of the manuscript.

**Funding:** This research was funded by the National Key R&D Program of China (Grant No. 2019YFC1510703-03), the National Natural Science Foundation of China (Grant Nos. 51639007, 51979007, 51609014, and 51809107), and the Fundamental Research Funds for Central Public Welfare Research Institutes (Grant No. CKSF2021482/SL).

**Institutional Review Board Statement:** Not applicable.

**Informed Consent Statement:** Written informed consent has been obtained from the patient(s) to publish this paper.

**Conflicts of Interest:** The authors declare no conflict of interest.

## Glossary of Terms

| | |
|---|---|
| $t$ | time |
| $h$ | water depth |
| $Z_b$ | bed elevation |
| $C_0$ | dimensionless Chézy coefficient |
| $n$ | Manning's roughness coefficient |
| $u$, $v$ | velocity components in the x and y directions |
| $g$ | gravitational acceleration |
| $v_t$ | turbulent viscosity coefficient |
| $\rho_w$ | clear water density |
| $\rho_S$ | sediment density |
| $\rho_m$ | density of water-sediment mixture |
| $\Delta\rho$ | $\rho_s - \rho_w$ |
| $\rho_0$ | density of saturated bed material |
| $\rho'$ | density of dry bed material |
| $q_b$ | amount of bed load in a unit volume of water |
| $q_{b*}$ | value of bed load in a unit volume |
| $q_{b0}$ | transport capacity of bed load in a unit volume of water |
| $\alpha_b$ | non-equilibrium adaptation coefficient of bed load |
| $\omega_b$ | setting velocity of bed load |
| $S_v$ | volumetric sediment concentration |
| $S_T$ | total concentration of graded sediments (kg/m$^3$) |
| $S_{bx}, S_{by}$ | bed slope terms in the x and y directions |
| $S_{fx}, S_{fy}$ | friction slope terms in the x and y directions |
| $K_b$ | empirical coefficient |
| $U_c$ | incipient velocity of bed load |
| $S$ | source section |
| $U$ | vector of the conserved variables |
| $E$, $G$ | convective flux vectors of the flow in the x and y directions |
| $n_x, n_y$ | components of unit normal vector in the x and y directions |

## References


1. Sayama, T.; Matsumoto, K.; Kuwano, Y.; Takara, K. Application of Backpack-Mounted Mobile Mapping System and Rainfall–Runoff–Inundation Model for Flash Flood Analysis. *Water* **2019**, *11*, 963. [CrossRef]
2. Li, W.J.; Lin, K.R.; Zhao, T.T.G.; Lan, T.; Chen, X.H.; Du, H.W.; Chen, H.Y. Risk assessment and sensitivity analysis of flash floods in ungauged basins using coupled hydrologic and hydrodynamic models. *J. Hydrol.* **2019**, *572*, 108–120. [CrossRef]
3. Prasad, R.N.; Pani, P. Geo-hydrological analysis and sub watershed prioritization for flash flood risk using weighted sum model and Snyder's synthetic unit hydrograph. *Model. Earth Syst. Environ.* **2017**, *3*, 1491–1502. [CrossRef]
4. Chen, H.Y.; Cui, P.; Zhou, G.G.D.; Zhu, X.H.; Tang, J.B. Experimental study of debris flow caused by domino failures of landslide dams. *Int. J. Sediment Res.* **2014**, *29*, 414–422. [CrossRef]
5. Zhou, G.G.; Cui, P.; Zhu, X.; Tang, J.; Chen, H.; Sun, Q. A preliminary study of the failure mechanisms of cascading landslide dams. *Int. J. Sediment Res.* **2015**, *30*, 223–234. [CrossRef]
6. Zhang, Y.; Wang, Y.; Zhang, Y.; Luan, Q.; Liu, H. Multi-scenario flash flood hazard assessment based on rainfall–runoff modeling and flood inundation modeling: A case study. *Nat. Hazards* **2020**, *105*, 967–981. [CrossRef]
7. Guan, M.F.; Carrivick, J.L.; Wright, N.G.; Sleigh, P.A.; Staines, K.E.H. Quantifying the combined effects of multiple extreme floods on river channel geometry and on flood hazards. *J. Hydrol.* **2016**, *538*, 256–268. [CrossRef]
8. Liu, T.H.; Wang, Y.K.; Wang, X.K.; Duan, H.F.; Yan, X.F. Morphological environment survey and hydrodynamic modeling of a large bifurcation-confluence complex in Yangtze River, China. *Sci. Total Environ.* **2020**, *737*, 139705. [CrossRef]
9. Toro, E.F. *Shock Capturing Methods for Free Surface Shallow Flows*; John Wiley & Sons: Chichester, UK, 2001.
10. Schippa, L.; Pavan, S. Numerical modelling of catastrophic events produced by mud or debris flows. *Int. J. Saf. Secur. Eng.* **2011**, *4*, 403–422. [CrossRef]
11. Yang, Q.Y.; Lu, W.Z.; Zhou, S.F.; Wang, X.K. Impact of dissipation and dispersion terms on simulations of open-channel confluence flow using two-dimensional depth averaged model. *Hydrol. Process.* **2014**, *28*, 3230–3240. [CrossRef]
12. Yoshioka, H.; Unami, K.; Fujihara, M. A dual finite volume method scheme for catastrophic flash floods in channel networks. *Appl. Math. Model.* **2015**, *39*, 205–229. [CrossRef]
13. Hu, X.Z.; Song, L.X. Hydrodynamic modeling of flash flood in mountain watersheds based on high-performance GPU computing. *Nat. Hazards* **2018**, *91*, 567–586. [CrossRef]
14. Bellos, V.; Papageorgaki, I.; Kourtis, I.; Vangelis, H.; Kalogiros, I.; Tsakiris, G. Reconstruction of a flash flood event using a 2D hydrodynamic model under spatial and temporal variability of storm. *Nat. Hazards* **2020**, *101*, 711–726. [CrossRef]
15. Contreras, M.T.; Escauriaza, C. Modeling the effects of sediment concentration on the propagation of flash floods in an Andean watershed. *Nat. Hazards Earth Syst.* **2020**, *20*, 221–241. [CrossRef]
16. Lorenzo-Lacruz, J.; Amengual, A.; Garcia, C.; Moran-Tejeda, E.; Homar, V.; Maimo-Far, A.; Hermoso, A.; Ramis, C.; Romero, R. Hydro-meteorological reconstruction and geomorphological impact assessment of the October 2018 catastrophic flash flood at Sant Llorenc, Mallorca (Spain). *Nat. Hazards Earth Syst. Sci.* **2019**, *19*, 2597–2617. [CrossRef]
17. Liu, W.; He, S. Dynamic simulation of a mountain disaster chain: Landslides, barrier lakes, and outburst floods. *Nat. Hazards* **2017**, *90*, 757–775. [CrossRef]
18. Yang, Q.Y.; Guan, M.F.; Peng, Y.; Chen, H.Y. Numerical investigation of flash flood dynamics due to cascading failures of natural landslide dams. *Eng. Geol.* **2020**, *276*, 105765. [CrossRef]
19. Wang, X.K.; Yan, X.F.; Duan, H.F.; Liu, X.N.; Huang, E. Experimental study on the influence of river flow confluences on the open channel stage–discharge relationship. *Hydrol. Sci. J.* **2019**, *64*, 2025–2039. [CrossRef]
20. Chen, S.; Li, Y.; Tian, Z.; Fan, Q. On Dam-Break Flow Routing in Confluent Channels. *Int. J. Environ. Res. Public Health* **2019**, *16*, 4384. [CrossRef]
21. Hackney, C.R.; Darby, S.E.; Parsons, D.R.; Leyland, J.; Aalto, R.; Nicholas, A.P.; Best, J.L. The influence of flow discharge variations on the morphodynamics of a diffluence-confluence unit on a large river. *Earth Surf. Process. Landf.* **2018**, *43*, 349–362. [CrossRef]
22. Xia, J.; Lin, B.; Falconer, R.A.; Wang, G. Modelling dam-break flows over mobile beds using a 2D coupled approach. *Adv. Water Resour.* **2010**, *33*, 171–183. [CrossRef]
23. Yang, Q.; Liu, T.; Zhai, J.; Wang, X. Numerical Investigation of a Flash Flood Process that Occurred in Zhongdu River, Sichuan, China. *Front. Earth Sci.* **2021**, *9*, 486. [CrossRef]
24. Chaudhry, M.H. *Open-Channel Hydraulics*, 2nd ed.; Springer: New York, NY, USA, 2008.

*water*

*Article*

# Assessment of an Alternative Climate Product for Hydrological Modeling: A Case Study of the Danjiang River Basin, China

**Yiwei Guo [1,2], Wenfeng Ding [1,2,*], Wentao Xu [1,2], Xiudi Zhu [3], Xiekang Wang [4] and Wenjian Tang [1,2]**

1 Changjiang River Scientific Research Institute, Changjiang Water Resource Commission, Wuhan 430010, China; guoyiwei32ww@gmail.com (Y.G.); xuwt@mail.crsri.cn (W.X.); tangwj@mail.crsri.cn (W.T.)
2 Research Center on Mountain Torrent Geological Disaster Prevention of Ministry of Water Resources, Wuhan 430010, China
3 Changjiang Water Resources Protection Institute, Wuhan 430051, China; zhuxiudi@mail.bnu.edu.cn
4 State Key Laboratory of Hydraulics and Mountain River Engineering, Sichuan University, Chengdu 610065, China; wangxiekang@scu.edu.cn
* Correspondence: dingwf@mail.crsri.cn; Tel.: +86-189-7168-9795

**Abstract:** Precipitation has been recognized as the most critical meteorological parameter in hydrological studies. Recent developments in space technology provide cost-effective alternative ground-based observations to simulate the hydrological process. Here, this paper aims to evaluate the performance of satellite-based datasets in the hydrological modeling of a sensitive area in terms of water quality and safety watershed. Three precipitation products, i.e., rain gauge observations (RO), the China Meteorological Assimilation Driving Datasets for the SWAT model (CMADS), and Tropical Rainfall Measuring Mission Multi-satellite (TRMM) products, were used to develop the Soil and Water Assessment Tool (SWAT) model to simulate the streamflow in the Danjiang River Basin (DRB). The results show that: (1) these three precipitation products have a similar performance with regard to monthly time scale compared with the daily scale; (2) CMADS and TRMM performed better than RO in the runoff simulations. CMADS is a more accurate dataset when combined with satellite-based and ground-based data; (3) the results indicate that the CMADS dataset provides reliable results on both monthly and daily scales, and CMADS is a possible alternative climate product for developing a SWAT model for the DRB. This study is expected to serve as a reference for choosing the precipitation products for watersheds similar to DRB where the rain gauge data are limited.

**Keywords:** SWAT; CMADS; TRMM; the Danjiang river basin

**Citation:** Guo, Y.; Ding, W.; Xu, W.; Zhu, X.; Wang, X.; Tang, W. Assessment of an Alternative Climate Product for Hydrological Modeling: A Case Study of the Danjiang River Basin, China. *Water* **2022**, *14*, 1105. https://doi.org/10.3390/w14071105

Academic Editor: Pankaj Kumar

Received: 14 March 2022
Accepted: 29 March 2022
Published: 30 March 2022



## 1. Introduction

Precipitation has been recognized as the most critical meteorological parameter in relation to developing hydrological models, because its spatiotemporal variability has a significant impact on hydrological behavior and water distribution [1–3]. Previous research studies have illustrated that having less precipitation information uncertainty has a sizable effect on stabilizing model parameterization and calibration [4–6]. However, there are severe limitations to describing rainfall inputs' true spatiotemporal variability of a river basin accurately, such as the rainfall pattern influenced by the complex topography and impacted by a hierarchy of regionally dominated atmospheric cycles [7,8].

Precipitation observed from a rain gauge, in general, is considered to be actual rainfall [9,10]. In most cases, point rainfall measurements are spatially interpolated to illustrate the rainfall field at a basin scale, and hence they are used as inputs in spatial-distributed hydrological models [11,12]. Field rainfall obtained from such interpolation, however, can represent the true distribution of precipitation well only if the rain gauges are deployed with reasonable density and uniform distribution [13]. Unfortunately, in most areas, especially in remote and developing areas, rain gauges are distributed irregularly and sparsely [14–17]. Consequently, the true rainfall field is poorly represented through interpolation, challenging the application of hydrological models. The accidental missing of the ground observations also exacerbate this challenge [18,19].

Recently, the feasibility of satellite-based data as alternatives for describing the temporal and spatial variability of the true rainfall field has been frequently tested. For example, Hur et al. [20] compared two high-resolution satellite rainfall datasets (TRMM 3B42 v7.0 and GSMaP v5.222) with rain gauge observations in Singapore. It was found that TRMM 3B42 v7.0 and GSMaP v5.222 both tended to overestimate the light rain and frequency but underestimate high-intensity precipitation when extreme precipitation was analyzed. Jiang et al. [21] researched a middle-latitude basin in South China, pointing out that rainfall was overall largely underestimated when using TMPA 3B42RT, Precipitation Estimation from Remote Sensing Information using Artificial Neural Network (PERSIAN), and the NOAA/Climate Precipitation Center Morphing Technique (CMORPH). Duncan et al. [22] assessed the accuracy of satellite-derived precipitation estimation (TRMM) over Nepal and found that though the precipitation of TRMM was significantly correlated with ground-based observations in all seasons, satellite precipitation estimates consistently overestimated the amount of precipitation and inaccurately detected extreme precipitation events.

The distributed hydrological model is beneficial for understanding the hydrological process [23–25]. The most habitually utilized distributed hydrological models have been appeared to effectively consolidate information from rain gauges, whereas satellite-based precipitation has been persistently moved forward and integrated into distinctive modules that assess its execution in simulating watershed streamflow [26,27]. The Soil and Water Assessment Tool (SWAT) is the most widely used distributed hydrological model among all the various hydrological models [28–31]. Huang et al.'s [32] study in the German state of Baden-Württemberg used three precipitation datasets with different time scales (daily, sub-daily, and diurnal) as inputs to drive a SWAT model to simulate the runoff, and found that there is a positive correlation between model performance and higher precipitation resolution. Yeganantham et al. [33] found that Climate Hazards Group InfraRed Rainfall with Station (CHIRPS) performed better than Climate Forecast System Reanalysis (CFSR) in simulating streamflow when using the SWAT model in ten watersheds located in the USA, Brazil, Spain, Ethiopia, and India. Hamoud et al.'s research [34] showed that the applicability of CHIRPS and TRMM 3B42 in runoff simulations were better than that of CFSR, Artificial Neural Networks–Climate Data Record (PERSIANN-CDR), and European Atmospheric Reanalysis (ERA-5) in the Highland Region of Yemen. Moreover, the performances of the satellite-based data are various in different areas. For example, Mararakanye et al.'s research in the lower Vaal River Catchment area (South Africa) [35] found that the CFSR performed well in simulating runoff by using a hydrological model, while according to Dao et al.'s study in the Cau River Basin (North Vietnam) [36], the performance of CFSR in runoff simulation was unsatisfactory. Gao et al. [37] proved that the performance of PERSIANN-CDR as an input to drive the SWAT model to simulate runoff was not suitable for the Xiang River Basin (China); however, its performance when simulating runoff was good in the Lancang River Basin (China). Like the studies above, the results simulated using the data-based SWAT model are heterogeneous and the performance of satellite-based datasets to simulate runoff should be evaluated for the specific basin.

Originating from the Q-DM, the Danjiang River Basin (DRB) is the main water source of the central route projects of the South-to-North Water Diversion Project [38]. This project is one of the most important hydraulic engineering projects in China and aims to improve the water shortage problem in northern China and improve the ecological environment along the related region. The quantity and quality of the water delivered are influenced by the erosion of the DRB [39,40]. Therefore, the DRB is considered to be a sensitive area in terms of water quality and safety with regard to the watershed. However, the uneven distribution of the meteorological stations in Q-DM makes it difficult to understand the real hydrological process. A previous study [41] used CFSR-driven SWAT models to simulate the runoff in the Bahe River Basin (Q-DM area) and found that the runoff simulated by uncorrected CFSR data were only satisfactory in this basin, while corrected data performed better. This indicates that it is necessary to verify the applicability of meteorological data in the DRB (Q-DM area).

Here, this study explores the results of the CMADS, TRMM, and rain gauge data when simulating rainfall estimation and surface runoff at monthly and daily scales in the DRB. The study aims to verify the applicability of the CMADS data and TRMM data in the DRB, and it can, therefore, serve as a reference for choosing the precipitation datasets in watersheds similar to the DRB where the ground-based rain gauge data are unavailable. With the objectives above, this study involves (1) a comparison of rainfall estimations from CMADS, TRMM 3B42 data, and rain gauge observations (Gauge) at monthly, daily, and spatial scales, (2) setting up a SWAT model with CMADS, TRMM 3B42 data, and rain gauge observations to simulate monthly and daily runoff, (3) calibrating and validating the simulated streamflow at three hydrological stations using the SWAT Calibration Uncertainties Program (SWAT-CUP) which uses the Sequential Uncertainty Fitting ver.2 (SUFI-2) algorithm, and (4) evaluating the multi-statistical performance of the simulation against the observed streamflow data. The main goal of this study is to evaluate the use of satellite-based and reanalysis precipitation products as model operation driving data, and assess whether they can drive the model in a watershed similar to the DRB where the gauge observations are limited.

## 2. Materials and Methods

### 2.1. Study Area

The largest tributary of the Hanjiang River, the Danjiang River, is a mountain river that covers a drainage area of 8887 km$^2$. The total length of its main stream is 280 km. Originating from the South Qinling Mountains and flowing into the Hanjiang River [42], the Danjiang River flows through the Shaanxi, Henan, and Hubei Provinces. It stretches between 33°04′10″ N and 34°11′09″ N and across 109°30′08″ E and 111°15′51″ E. The Danjiang River Basin (DRB) features a high-rising west and a low-lying east, with a relative elevation difference of 1915 m. The continental monsoon climate contributes to the distinct seasons of the DRB. According to the records from 1950 to 2015, the long-term annual precipitation of the DRB is 732.29 mm and the spatial distribution difference shows an increasing trend from the west to the east. Rainfall is concentrated in the period from May to October, accounting for about 80% of the annual precipitation. Moreover, the annual average temperature ranges from 7.8 °C to 13.9 °C and the annual runoff is 14.36 × 108 m$^3$.

Forestland occupies the largest area in the DRB, followed by the cropland. The yellow-brown soil and sandy loam are the dominant soil types in the DRB [38]. There are 3 hydrological stations (Majie Station upstream, Danfeng Station midstream, and Jingziguan Station downstream) and 58 ground-based rain gauges in the study area. The digital elevation model (DEM), stream network, weather stations, and hydrological stations are shown in Figure 1.

### 2.2. Hydrological Model and Data Sources

In this study, the SWAT model was used for hydrological modeling, which was developed by USDA-ARS. Because the SWAT model is designed for long-term simulations on a daily scale, it is suitable for evaluating the performance of three precipitation products. To ensure the accuracy of relative changes induced by different precipitation inputs, all input parameters, such as temperature, wind, solar radiation, and humidity, were kept the same, except precipitation. Additionally, the temperature, wind, solar radiation, and humidity inputs were simulated by the internal weather generator of SWAT.

Moreover, the target watershed is required by the SWAT model to be divided into sub-watersheds. Each sub-watershed may include one or more Hydrologic Response Units (HRUs). On the basis of the 30 m DEM and by choosing the Jingziguan Station as the outlet, the controlled watershed was delineated. The threshold to discretize the sub-watershed was based on the 2% area. Other input parameters, such as soil type and land use, were downloaded from websites (Table 1). The data of measured runoff were obtained from the Department of Hydrology of the Ministry of Water Resources of China. Additionally, the SPWA (Soil–Plant–Air–Water) software was used to analyze the soil–water characteristics of each soil type.

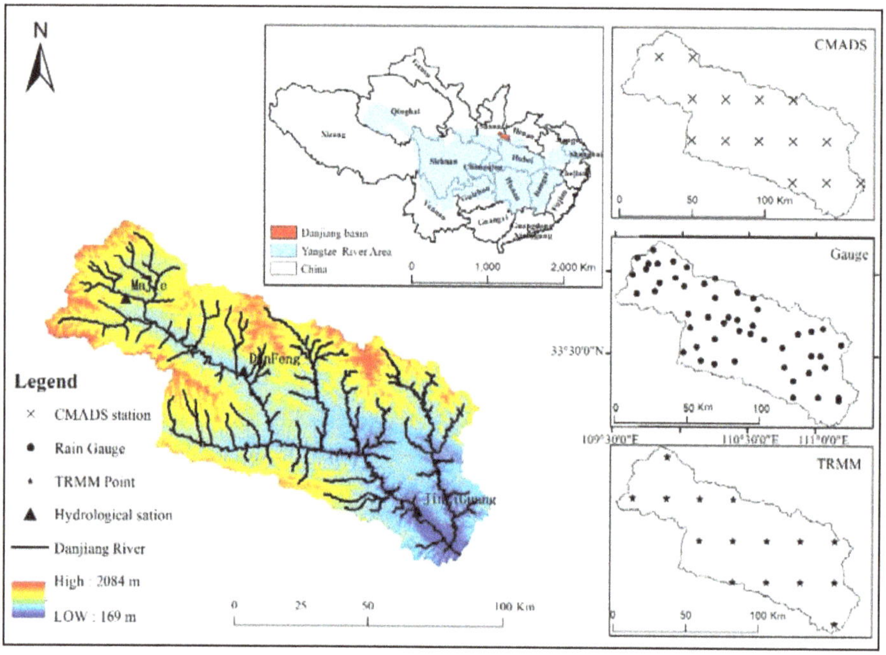


**Figure 1.** Location of the Danjiang River Basin and the distribution of hydrological stations, CMADS stations, TRMM points, and rain gauges.

**Table 1.** Summary of the input parameters.

| Parameters | Dataset | Developed Organization | Resolution | Data Source (Accessed on 1 August 2021) |
|---|---|---|---|---|
| DEM | Shuttle Radar Topography Mission (SRTM) | National Aeronautics and Space Administration (NASA) | 30 m | https://earthexplorer.usgs.gov/ |
| Land cover | 30 m-resolution Global Land Cover (GLC30) | The National Geomatics Center of China (NGCC) | 30 m | http://www.globallandcover.com/ |
| Soil type | World Soil Database (HWSD) | The Food and Agriculture Organization of the United Nations (FAO) | 1000 m | http://www.fao.org/soils-portal/soil-survey/soil-maps-and-databases/harmonized-world-soil-database-v12/en/ |

Daily rainfall data were collected from weather gauge stations and the two satellite-based and reanalysis precipitation products used were CMADS and TRMM 3B42 version 7.

Daily precipitation data obtained from the fifty-eight rain gauges in the DRB were available from the website of the Department of Hydrology of the Ministry of Water Resources of China. The rain gauge data covered from 1964 to 2015.

The dataset CMADS introduces the technology of The Space and Time Mesoscale Analysis System (STMAS) assimilation algorithm. Multiple technologies and scientific methods were used to develop CMADS [43,44]. The dataset, containing information relating to precipitation, temperature, and other variables, can be used to run hydrological models such as SWAT. The precipitation data of CMADS are merged with the hourly precipitation data collected by the China National Meteorological Information Center using the CPC MORPHing technique (CMORPH). CMADS stations provide information throughout the day from 2008 to 2016 in the areas between 0–65° N and 60–160° E. There are a total of 19 CMADS stations in the study area.

In late 1997, the TRMM satellite was launched by the National Aeronautics and Space Administration (NASA) and the Japanese Aerospace Exploration Agency (JAXA) to monitor precipitation [45]. TRMM 3B42 is one of the RMM Multi-satellite Precipitation Analysis (TMPA) products [46]. It provides daily precipitation data from 50° S to 50° N at a resolution of 0.25° spatially and temporally from 1998 to 2015 [47,48]. There is a total of 19 TRMM 3B42 pixels in the study area. Further information about TRMM and CMADS can be found in Table 2.

**Table 2.** Summary of remote-sensing/reanalysis precipitation datasets.

| Full Name | Abbreviation | Coverage | Spatiotemporal Resolution Used | Data Source (Accessed on 1 August 2021) |
|---|---|---|---|---|
| The China Meteorological Assimilation Driving Datasets for the SWAT model Version 1.1 | CMADS V1.1 | 0–65° N 60–160° E | Daily, 0.25° | http://www.cmads.org/ |
| Tropical Rainfall Measuring Mission Multi-satellite Precipitation Analysis 3B42 Version 7 | TRMM 3B42 V7 | 50° S–50° N | Daily, 0.25° | https://disc.gsfc.nasa.gov/ |

The SWAT model uses data from the station nearest to the centroid of each sub-basin to categorize precipitation data into sub-basins [47].

### 2.3. Model Calibration and Evaluation

When all parameters were entered into the SWAT model, the SWAT model ran with three precipitation products (rain gauge data, CMADS dataset, and TRMM dataset) separately at the monthly and daily scale. The watershed was divided into a total of 237 sub-catchments by the SWAT model, and these sub-watersheds were further divided into 980 HRUs on the basis of the land use, soil type, and slope classes. The simulated period was selected to be the period from 2008 to 2015 to ensure its consistency, because the available gauge data, CMADS data, and TRMM data were, respectively, collected from 1964 to 2015, 2008 to 2018, and 1998 to 2015. Here, 2008 was taken as the warm-up period.

The SUFI-2 algorithm in SWAT-CUP was used in the calibration procedure. On the basis of Duan et al.'s research [48] and the official guide, 17 parameters were selected. Considering the influence of elevation on precipitation, the precipitation lapse rate (PLAPS) was introduced [49]. Moreover, the simulated results of the Majie Station, Danfeng Station, and Jingziguan Station were calibrated together. The model was calibrated by first using the initial value ranges of each parameter and then using the suggested ranges of the previous simulation. The simulations were calibrated five times with 500 iterations each.

In this study, the coefficient of determination, Nash–Sutcliffe efficiency (NSE), and percent bias (PBIAS) were used to evaluate the accuracy of runoff modeling results. The formulas are as follows [35]:

$$R^2 = \left( \frac{\sum_{i=1}^{n} (Q_i - \overline{Q_i})(S_i - \overline{S_i})}{\sqrt{\sum_{i=1}^{n} (Q_i - \overline{Q_i})^2} \sqrt{\sum_{i=1}^{n} (S_i - \overline{S_i})^2}} \right) \tag{1}$$

$$NSE = \frac{\sum_{i=1}^{n} (S_i - Q_i)^2}{\sum_{i=1}^{n} (Q_i - \overline{Q_i})^2} \tag{2}$$

$$PBIAS = \frac{\sum_{i=1}^{n} (S_i - Q_i)}{\sum_{i=1}^{n} Q_i} \times 100\% \tag{3}$$

where $Q_i$ is the observed value, $S_i$ is the simulated value, and $\overline{Q_i}$ and $\overline{S_i}$ are the mean values of the observed and simulated values. The statistical threshold values that were used to evaluate the performance of the model are shown in Table 3.

**Table 3.** The statistical threshold values used for interpreting model performance.

| Performance Ratings | $R^2$ | NSE | PBIAS |
|---|---|---|---|
| Very Good | 0.7~1 | 0.75~1 | $<\pm10$ |
| Good | 0.6~0.7 | 0.65~0.75 | $\pm10$~$\pm15$ |
| Satisfactory | 0.5~0.6 | 0.50~0.65 | $\pm15$~$\pm25$ |
| Unsatisfactory | $\leq0.5$ | $\leq0.5$ | $\geq\pm25$ |

## 3. Results

### *3.1. Evaluation of the Three Precipitation Products*

#### 3.1.1. Monthly Scale

The comparison of the SWAT model results using the three precipitation products from 2009 to 2015 in the study area is shown in Figure 2. The dry (drought), wet (rainy), and normal years were defined on the basis of the commonly used precipitation year classification standard [50]. In this study, 2010, 2013, and 2015 were denoted as drought years and 2009, 2011, and 2012 were denoted as rainy years. It is noted from Figure 2 that the annual rainfall mainly concentrates in the period from June to August (the flood season), and the precipitation calculated by the rain gauge, CMADS, and TRMM from June to August, respectively, account for 51.53%, 54.61%, and 54.60% of the annual precipitation. Figure 2 also shows that, before 2013, the CMADS and TRMM both underestimated the rainfall severely in the flood season of the rainy years compared with actual precipitation, by 32.51% and 11.66%, respectively, and that the CMADS and TRMM overestimated the rainfall by 18.12% and 40.48%, respectively, in the flood season of the drought years. The rainfall estimated by CMADS and TRMM was similar to the estimation of Gauge in non-flood seasons. The situation has improved since 2013. The precipitation trend of CMADS and Gauge became similar after 2013 because the underestimation of CMADS's precipitation was narrowed. However, TRMM still underestimated the rainfall in the flood season of rainy years and overestimated the rainfall in the flood season of normal years after 2013, though the precipitation deviation was reduced. In addition, the Pearson correlation coefficient of Gauge and CMADS (Gauge-CMADS) and Gauge and TRMM (Gauge-TRMM) were 0.74 and 0.75, respectively, indicating that the precipitation of CMADS and TRMM were highly similar to the precipitation of the rain gauges. In other words, the rainfall in this area can be effectively represented by CMADS and TRMM.

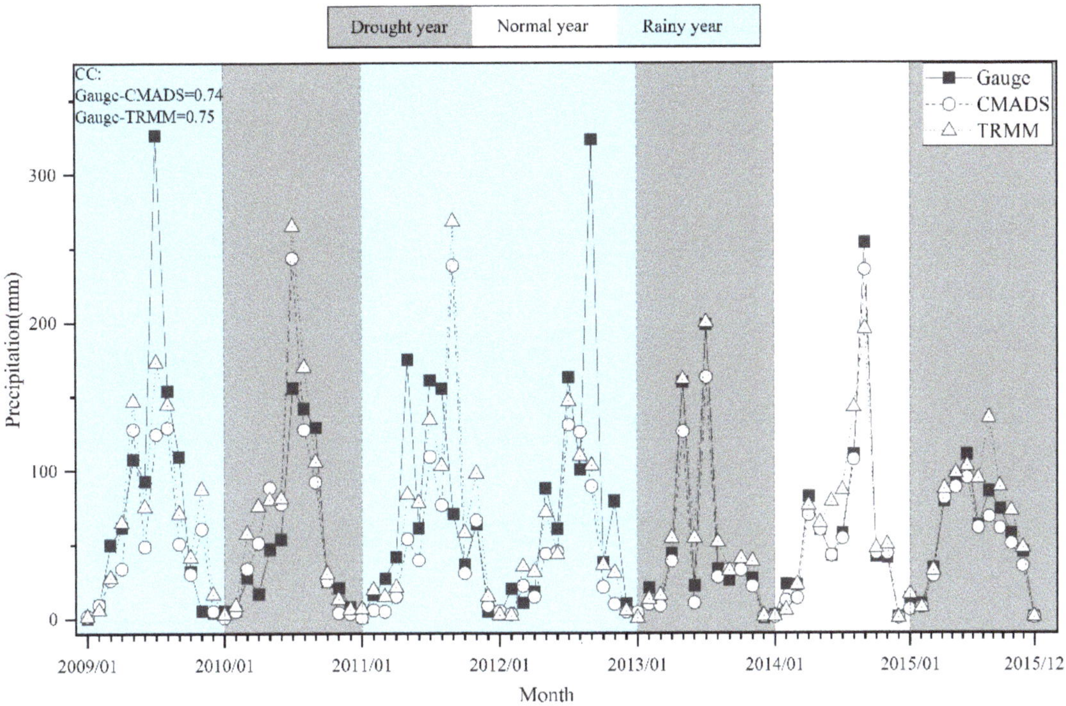


**Figure 2.** Three different precipitation records at monthly scale in the DRB (the CC value of Gauge-CMADS and Gauge–TRMM were 0.74 and 0.75, respectively; 2009, 2011, and 2012 were denoted as rainy years, 2014 was denoted as the normal year, and 2010, 2013, and 2015 were denoted as drought years. Additionally, the data in this figure were given as year and month.

It can be seen from the box plot of Figure 3 that the precipitation featured three peaks, with the peak values appearing in May, July, and September. The monthly precipitation of CMADS had the largest average and median line, while the average and median line of TRMM was the smallest. Moreover, CMADS reported the largest maximum rainfall except in October and December, and TRMM had the smallest minimum precipitation throughout the year. In addition, the PBIAS values of Gauge-CMADS and Gauge-TRMM were −18.86 and 3.20, respectively, indicating that the CMADS precipitation was underestimated compared to the Gauge precipitation, with the TRMM estimation the exact opposite. However, the total precipitation was not much different. In summary, compared with the Gauge records, CMADS tends to overestimate the rainfall, while TRMM tends to underestimate the rainfall.

### 3.1.2. Daily Scale

The intensity and frequency of precipitation are the critical parameters used to describe the characteristic of daily rainfall [51]. It is noted from Figure 4 that the angle between the CMADS model's 95%-line estimates and the horizontal axis was <45° and that the angle between the TRMM model's 95%-line estimates and the horizontal axis was <45° as well. This suggested that although the precipitation trends of CMADS and TRMM were similar to that of the Gauge records at a daily scale, the CMADS and the TRMM rainfall data tended to be underestimated when extreme rainstorms occurred. The CMADS and the TRMM, respectively, underestimated the storm rainfall (>50 mm/day) by 13.11% and 10.65%, showing that the CMADS and TRMM were less capable of accurately simulating the storm rainfall than Gauge.

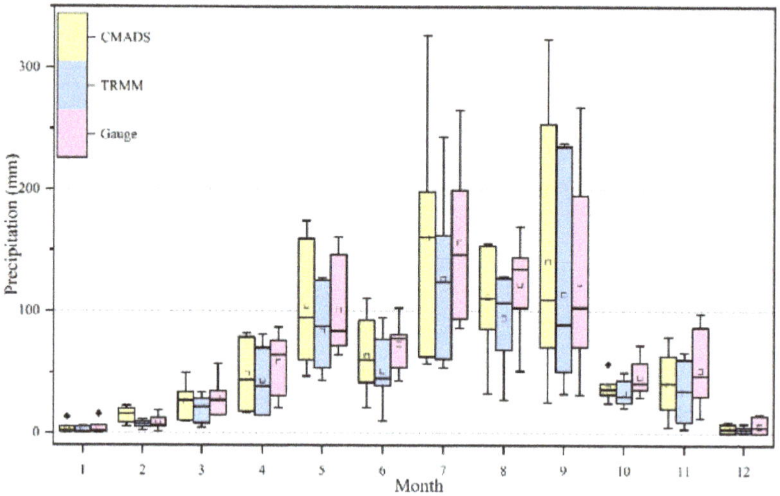


**Figure 3.** The box diagrams of three precipitation records at a monthly scale in the DRB.

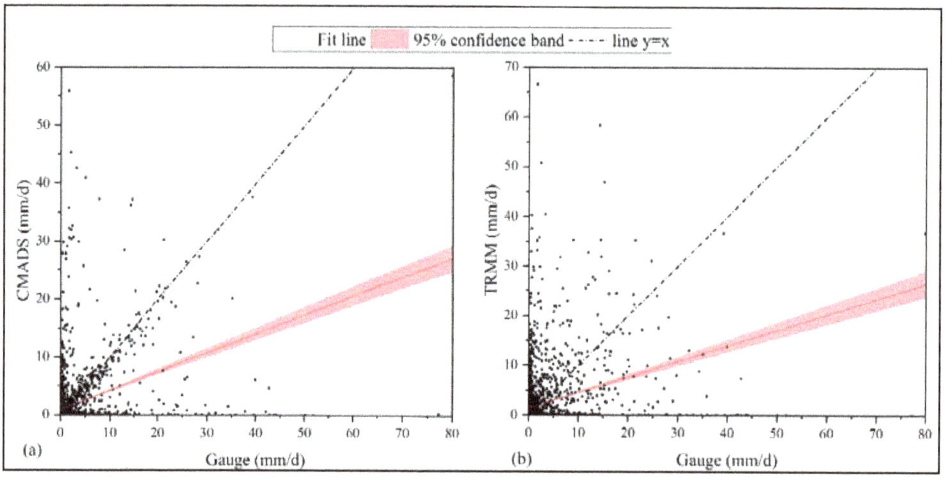


**Figure 4.** Scatterplot of the CMADS and TRMM records compared with Gauge records at a daily scale: (**a**) comparison of CMADS and Gauge and (**b**) comparisons of TRMM and Gauge. Note that the straight lines that pass through the origin are dividing lines with an angle of 45° to the x-axis, which means that the precipitation products overestimated the rainfall if the point is higher than this line.

Moreover, the Pearson correlation coefficients of Gauge-CMADS and Gauge-TRMM were 0.39 and 0.32, respectively, and the Pearson correlation coefficient of CMADS-TRMM was 0.80, indicating that there were big differences between Gauge and CMADS data and between Gauge and TRMM data, while the CMADS and TRMM data were similar.

The cumulative daily precipitation intensity frequencies of the three precipitation products are shown in Figure 5. Taking 50 mm/day as the panel line, Figure 5a was divided into Figure 5b,c to describe the frequency trend of the three precipitation products clearly. It can be noticed that the three products have the smallest difference in the events of less-than-heavy rain (30 mm/d), but the largest difference in the events of torrential rain (>50 mm/d). Additionally, the number of torrential rain events identified by TRMM was lower than that in Gauge records, and the frequency of torrential rain identified by CMADS was the lowest among the three precipitation products.

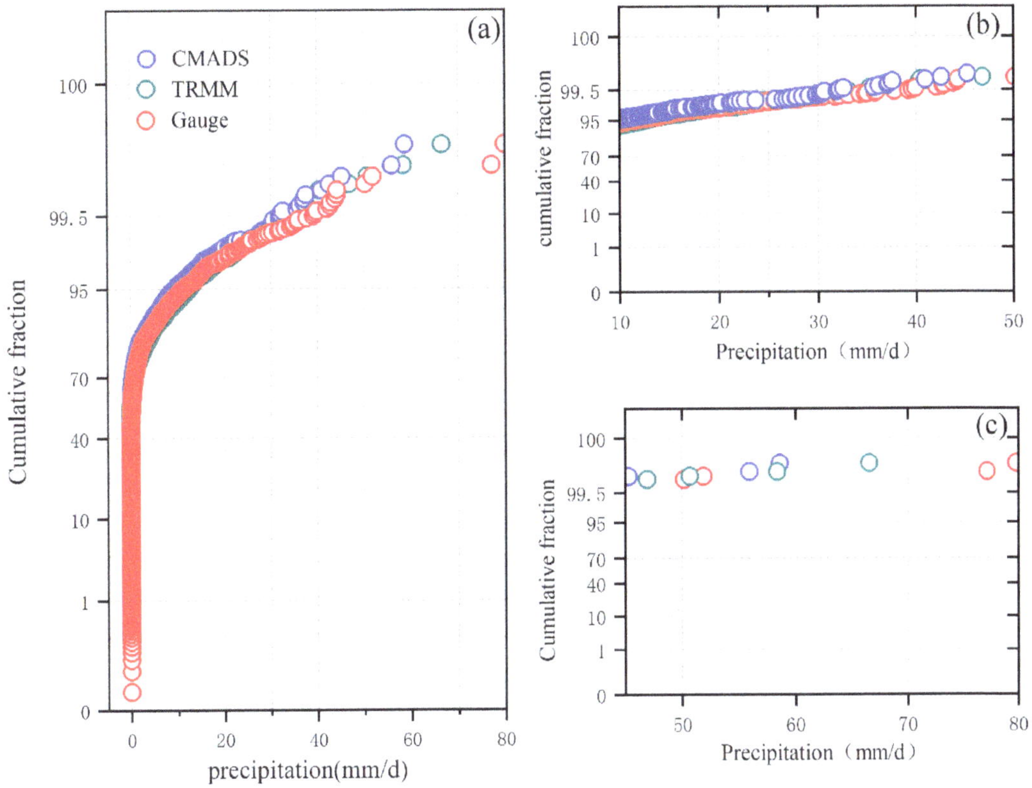


**Figure 5.** Cumulative frequencies of daily precipitation intensity for Gauge (red points), CMADS (blue points), and TRMM (green points) in the DRB: (**a**) distribution of all precipitation values, (**b**) distribution of precipitation values that are <50 mm, and (**c**) distribution of precipitation values that are ≥50 mm.

### 3.1.3. Spatial Scale

The spatial distributions suggested by the three precipitation products were almost completely different, as shown in Figure 6. CMADS suggested that the rainfall increased from the center to the surroundings, with the highest rainfall in the central north. The clear trend of precipitation suggested by TRMM was that rainfall increased from upstream to downstream, with the highest rainfall in the east. However, the rainfall of Gauge in each sub-basin varied greatly, and there was no obvious spatial distribution pattern mainly due to the distribution of the rain gauge stations. Though there are a large number of rain gauge stations (58 stations) in the study area, most of them are located in the north and the east, leaving a vast area in the central west and southeast of the basin with no rainfall stations. Meanwhile, the CMADS (15 stations) and TRMM (15 stations) grid data were collected from uniformly distributed stations, causing the different spatial distribution of rainfall. Moreover, regardless of the daily or the monthly scale, the similarity of rainfall between CMADS and Gauge was higher than that of TRMM.

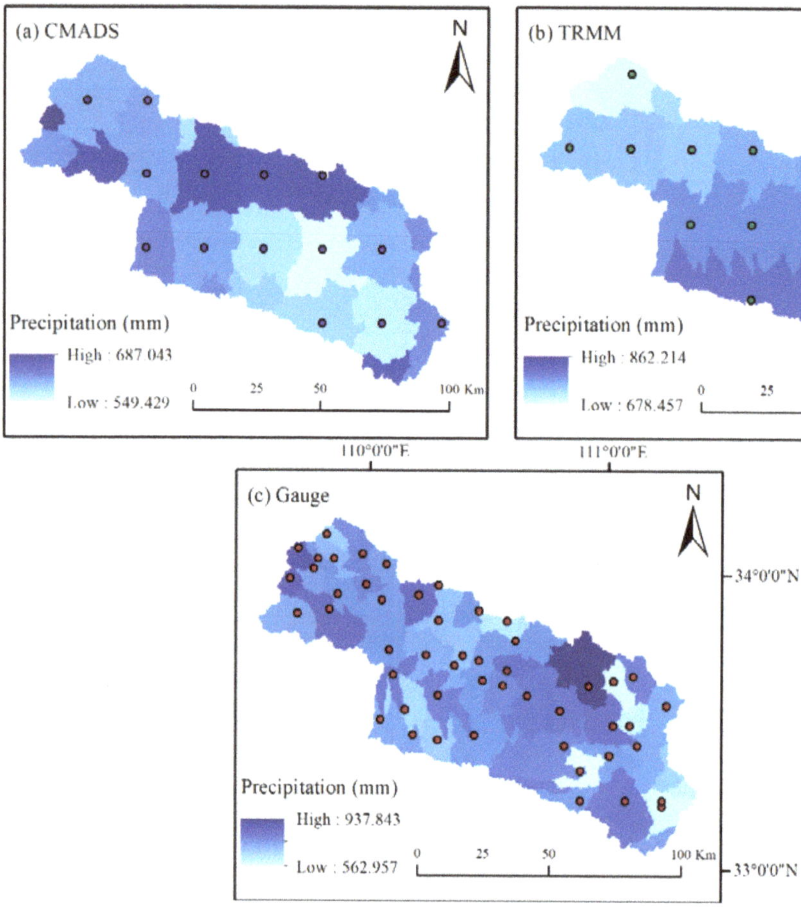


**Figure 6.** Spatial variation of precipitation at a yearly scale for all sub-basins calculated with precipitation inputs from (**a**) CMADS, (**b**) TRMM, and (**c**) Gauge.

Though the rainfall gauges are able to reflect more information than CMADS and TRMM when describing the variation of precipitation, the areal rainfall interpolated from the gauges may be distorted because the gauges' observations are point data. However, the CMADS and TRMM data are evenly distributed grid data with a resolution of 0.25° and reflect the areal rainfall. Thus, despite the better performance of the rain gauge data in describing watershed areal rainfall among three products, which products perform the best in driving SWAT model to simulate runoff is uncertain.

### *3.2. The Performance of Different Precipitation Products in Simulating Runoff*

#### 3.2.1. Pre-Calibration Model Results

Before the model was calibrated, we conducted a statistical analysis of the simulation results of runoff during the simulation period (2008–2015) of the model. As is shown in Table 4, the runoff simulation results of the three precipitation products downstream, midstream, and upstream showed different trends. The best simulation effect was upstream (Majie Station), while the worst performance was midstream (Danfeng Station). Moreover, the best simulation performance was achieved by CMADS, whose NSE was 0.74 upstream and 0.63 downstream (Jingziguan Station), while the worst simulation effect occurred in Gauge with its NSE almost all below zero (it was only above zero in the Majie Station).

Furthermore, the PBIAS values of the Gauge model were all below zero, showing that the simulation value of the Gauge model was smaller than the measured runoff. The PBIAS values of the CMADS model and TRMM model were both above zero upstream and below zero midstream and downstream, indicating that the simulation values of these two products were larger than the measured runoff in the upper stream and smaller than the actual runoff in the middle and downstream.

**Table 4.** The pre-calibration performance error amounts of the SWAT model simulated with CMADS, TRMM, and Gauge data on a monthly scale.

| Station | $R^2$ | | | NSE | | | PBIAS | | |
|---|---|---|---|---|---|---|---|---|---|
| | CMADS | Gauge | TRMM | CMADS | Gauge | TRMM | CMADS | Gauge | TRMM |
| Majie Station | 0.79 | 0.5 | 0.68 | 0.74 | 0.43 | 0.66 | 23.4 | −22.38 | 10.44 |
| Danfeng Station | 0.04 | 0.05 | 0.05 | −0.7 | −1.52 | −7.24 | −7.67 | −56.43 | −32.1 |
| Jingziguan Station | 0.7 | 0.06 | 0.63 | 0.63 | −0.82 | 0.51 | −63.35 | −164.97 | −164.7 |

The pre-calibration results of the monthly scale indicated that the CMADS and TRMM data were reliable in estimating runoffs; the $R^2$ values of CMADS-SWAT and TRMM-SWAT were 0.79 and 0.68, respectively. Though the performance of Gauge-SWAT was not as good as CMADS-SWAT, it was still a valuable data source for use in the model, for its $R^2$ was 0.50.

It is noted from Table 5 that the simulation results of the three precipitation products on a daily scale showed the same trend as that on a monthly scale in the upper, middle, and lower reaches of the basin. The simulation performance in the upstream was the best and the performance downstream was the worst. Moreover, the NSE values of the CMADS model and the TRMM model were all below 0.6 in the entire basin, meaning that their performances were unsatisfactory. Additionally, the PBIAS values of the CMADS model were all above zero, the PBIAS values of the Gauge model were all below zero, and the PBIAS values of the TRMM model were above zero in the upstream and below zero in the midstream and downstream, indicating that the simulation values of the CMADS model and Gauge model were, respectively, higher and lower than the measured runoff in the whole basin and that the simulation value of TRMM model was higher in the upper reach but lower in the middle and lower reaches.

**Table 5.** The pre-calibration performance error amounts of the SWAT model simulated with CMADS, TRMM, and Gauge data on the daily scale.

| Station | $R^2$ | | | NSE | | | PBIAS | | |
|---|---|---|---|---|---|---|---|---|---|
| | CMADS | Gauge | TRMM | CMADS | Gauge | TRMM | CMADS | Gauge | TRMM |
| Majie Station | 0.39 | 0.18 | 0.22 | 0.38 | 0.11 | 0.18 | 22.98 | −23.02 | 9.83 |
| Danfeng Station | 0.32 | 0.05 | 0.2 | 0.16 | −1.24 | −0.23 | 1.65 | −47.57 | −24.83 |
| Jingziguan Station | 0.26 | 0.02 | 0.16 | 0.25 | −0.57 | 0.03 | 60.82 | −160.86 | −143 |

All these three datasets seriously overestimate the runoff in the middle stream and downstream, according to Figure 7B,C. Moreover, the average observed streamflow was smaller than the average simulated runoff throughout the year in the upstream, except in September. The mean standard deviation of the three precipitation products computed at the monthly time scale and averaged over the 8 years considered are 0.28, 1.71, and 4.30 for the Majie Station, Danfeng Station, and Jingziguan Station, respectively. Additionally, the mean standard deviation of three products computed at the daily time scale are 0.21, 0.87, and 1.62, respectively.

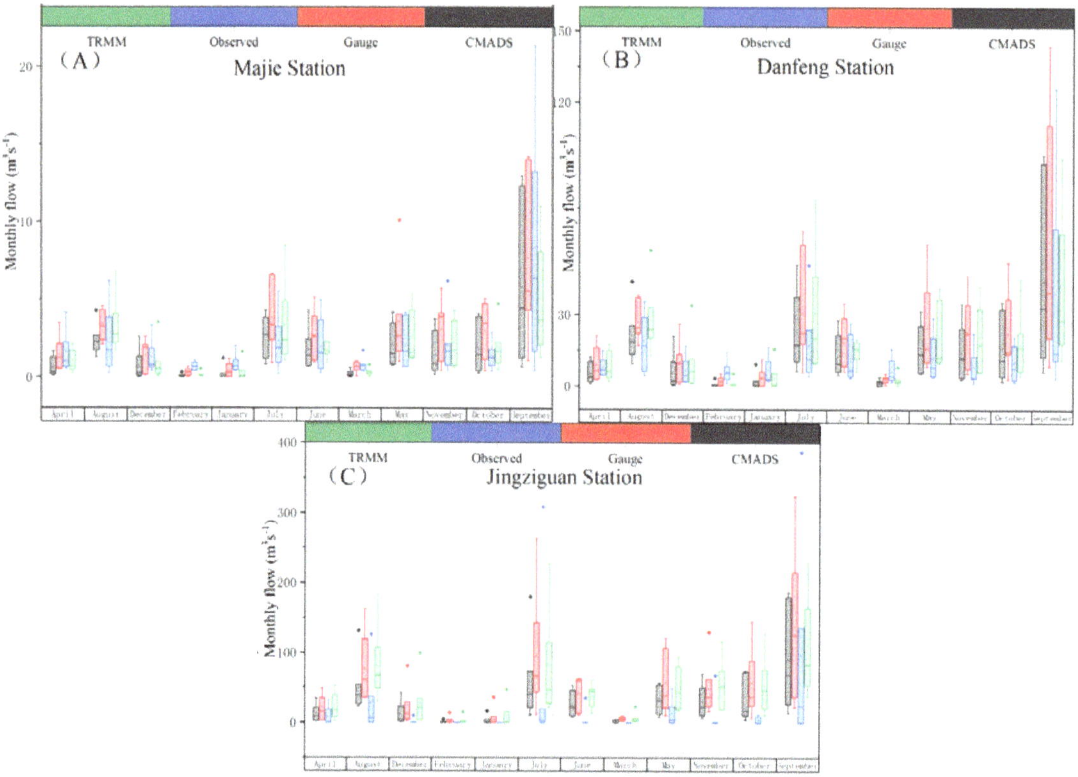


**Figure 7.** Box plot of the monthly runoff from 2006 to 2015, observed data and simulated streamflow using CMADS, TRMM, and Gauge data at (**A**) Majie Station, (**B**) Danfeng Station, and (**C**) Jingziguan Station.

### 3.2.2. Post-Calibration Model Results

Seventeen parameters related to the hydrological process were selected to calibrate the SWAT model and the range of the parameters, as well as the result of the calibration, which are shown in Table 6. Be aware that the streamflow of Majie Station, Danfeng Station, and Jingziguan Station were calibrated together.

It is noted from Figure 8 that there is a positive correlation between the simulated runoff and rainfall. Moreover, the runoff of the CMADS model had a positive correlation with the measured runoff, but the smaller flood peaks were not simulated in the year when extreme flood events occurred or flood events were relatively continuous. The runoff trend of the TRMM model was similar to the measured runoff in the upper and lower reaches, but several consecutive flood events were simulated as a larger one in the middle stream. Though the runoff of the Gauge model was positively correlated with measured runoff in the upper stream, large floods in certain months were simulated as several smaller flood events in the midstream and downstream.

It is also noted form Table 7 that the $R^2$ values of the CMADS-SWAT (SWAT model derived from CMADS) and TRMM-SWAT were above 0.8 in the whole basin, except for the $R^2$ of the TRMM-SWAT in the middle reaches, which was 0.77. The NSE values of the CMADS-SWAT and TRMM-SWAT were all close to 0.8. However, the $R^2$ and NSE values of Gauge-SWAT were all below 0.6 in the whole basin and its $R^2$ and NSE values were only greater than 0.5 in the upstream. This suggested that the performances of CMADS-SWAT and TRMM-SWAT were better, while the performance of Gauge-SWAT was unsatisfactory.

Moreover, the PBIAS values of CMADS-SWAT and TRMM-SWAT were both greater than 25% in the upper and lower reaches and the PBIAS values in the middle stream were 9.9% and 0.0%, respectively, indicating that the CMADS-SWAT and TRMM-SWAT severely overestimated the runoff in the upper reach, severely underestimated the runoff in the downstream, and slightly overestimated the runoff in the middle reach. The PBIAS values of the Gauge-SWAT were both smaller than 15% in the middle and upper reaches and the PBIAS in the downstream was −89.3, showing that the simulation values in the upper and middle reaches were higher and the simulation value was severely low in the downstream.

**Table 6.** Hydrological parameters chosen to calibrate the model (r_ and v_ mean a relative change and a replacement to the initial parameter values, respectively).

| Paraments | Description | Monthly Scale | | | Daily Scale | | |
|---|---|---|---|---|---|---|---|
| | | CMADS | Gauge | TRMM | CMADS | Gauge | TRMM |
| r__SOL_AWC().sol | Available water capacity of the soil layer (mm HzO/mm soil) | −0.06 | −0.26 | −0.06 | 0.52 | −0.76 | −0.22 |
| r__CN2.mgt | scS runoff curve number | 0.14 | −0.04 | 0.17 | −0.21 | −0.03 | 0.32 |
| v__ALPHA_BF.gw | Baseflow alpha factor (days) | 0.6 | 0.98 | 0.51 | 1.01 | 1.04 | 0.62 |
| v__GW_DELAY.gw | Groundwater delay (days) | −139.76 | 234.5 | −198.89 | −216.84 | 209.46 | 481.86 |
| r__GWQMN.gw | Threshold depth of water in the shallow aquifer required for return | −0.07 | 0.46 | −0.64 | −0.11 | 2.01 | 0.21 |
| v__CH_K2.rte | Effective hydraulic conductivity (mm/h) | −367.12 | 175.49 | −12.96 | 232.1 | 249.42 | 131.99 |
| v__CH_N2.rte | Manning's n value for main channel | 0 | 0.04 | 0.13 | 0.03 | 0.15 | 0.13 |
| v__REVAPMN.gw | Threshold depth of water in the shallow aquifer for "revamp" to occur | 81.71 | 326.5 | 421.85 | 387.87 | 481.2 | 571.41 |
| r__GW_REVAP.gw | Groundwater "revap" coefficient | −0.39 | −0.38 | −0.23 | 0.44 | −0.34 | 0.21 |
| r__OV_N.hru | Manning's "n" value for overland flow | 0.32 | 0.05 | 0.39 | 0.21 | 0.06 | −1.11 |
| r__SLSUBBSN.hru | Average slope length (m) | 0.9 | 0.45 | 1.12 | 0.84 | 0.7 | −0.52 |
| r__HRU_SLP.hru | Average slope steepness (m/m) | −0.1 | −0.38 | −0.18 | 0.02 | 0.08 | −0.47 |
| v__EPCO.hru | Plant uptake compensation factor | 0.34 | 0.59 | −0.47 | 0.07 | 0.26 | 0.43 |
| v__ESCO.hru | Soil evaporation compensation factor | 0.15 | 0.05 | 0.02 | 0.89 | 0.15 | 0.37 |
| r__SOL_BD().sol | Moist bulk density (g/cm$^3$) | 2.19 | 1.5 | 1.22 | 1.31 | 1.79 | 1.64 |
| r__SOL_K().sol | Saturated hydraulic conductivity (mm/h) | −0.55 | −0.35 | −0.56 | −0.68 | −0.75 | −0.98 |
| v__PLAPS.sub | Precipitation lapse rate (mm) | −716.86 | 690 | −368.75 | −867.07 | 615.05 | −660.12 |

In summary, the performances of the CMADS model and TRMM model in the DRB were satisfactory across the sub-basins, but the performance of the Gauge-SWAT was only satisfactory in the upstream, deviating significantly from the observed data in the middle stream and downstream.

As was shown in Figure 9, though the CMADS and TRMM inputs replicated the runoff successfully at a daily scale, some small flood peaks were not simulated and the discharge of extreme floods during the flood season was significantly underestimated. The runoff of the Gauge model was notably different from the measured records and most of the floods were not simulated by the Gauge model.

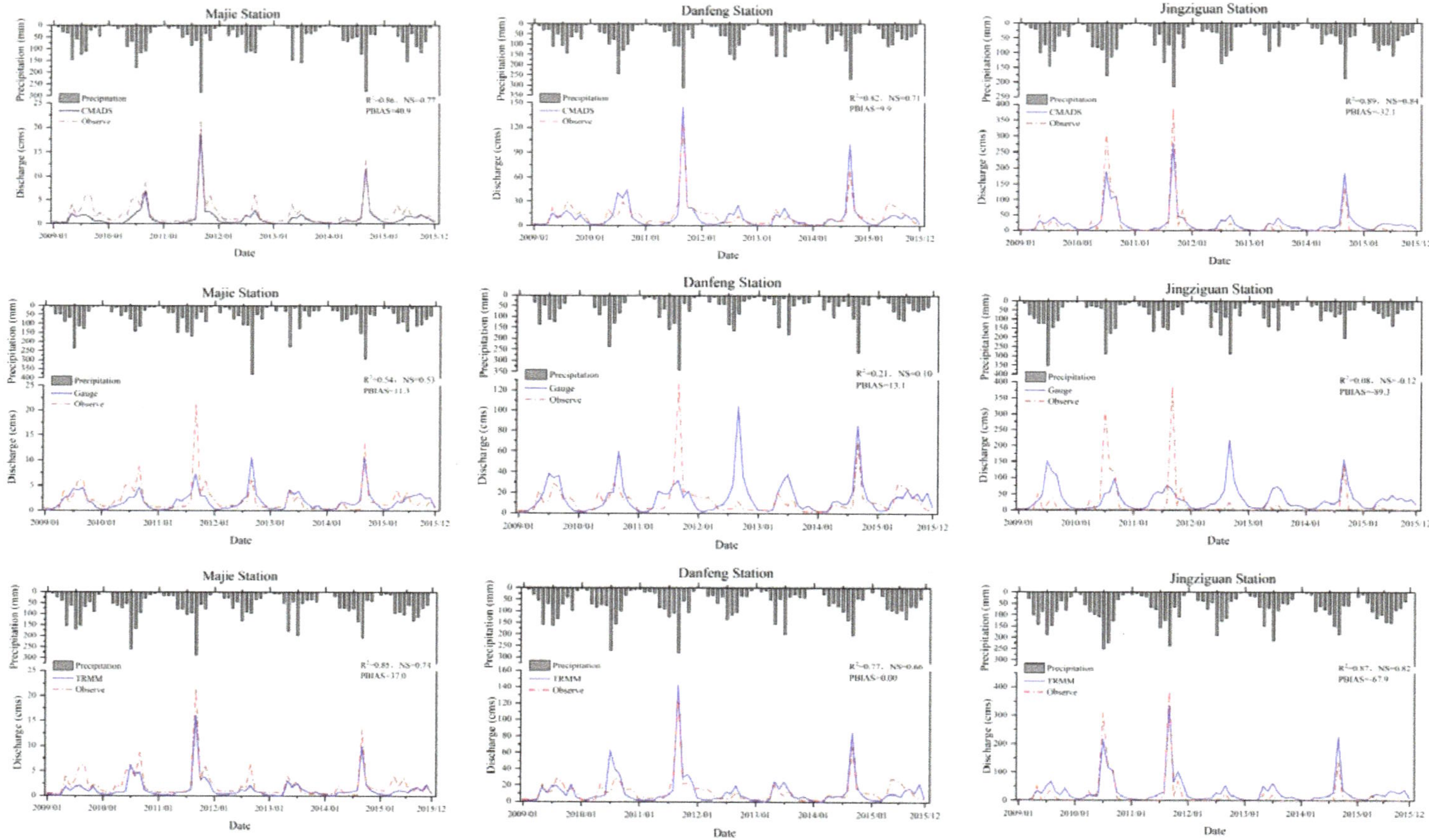


**Figure 8.** Observed and simulated discharges at Majie Station, Danfeng Station, and Jingziguan Station in the DRB at a monthly scale using inputs from Gauge, CMADS, and TRMM. Note that this figure is given as year/month.

**Table 7.** The post-calibration performance error amounts of the SWAT model simulated with CMADS, TRMM, and Gauge data on a monthly scale.

| Station | $R^2$ | | | NSE | | | PBIAS | | |
|---|---|---|---|---|---|---|---|---|---|
| | CMADS | Gauge | TRMM | CMADS | Gauge | TRMM | CMADS | Gauge | TRMM |
| Majie Station | 0.86 | 0.54 | 0.85 | 0.77 | 0.53 | 0.74 | 40.09 | 11.3 | 37 |
| Danfeng Station | 0.82 | 0.21 | 0.77 | 0.71 | 0.1 | 0.66 | 9.9 | 13.1 | 0 |
| Jingziguan Station | 0.89 | 0.08 | 0.87 | 0.84 | 0.12 | 0.82 | −32.1 | −89.3 | −67.9 |

Specifically, according to Table 8, the CMADS-SWAT was the most successful in estimating runoffs because its $R^2$ and NSE values in the upstream and middle stream were all above 0.5, while its $R^2$ and NSE values were 0.42 and 0.34 in the downstream. The $R^2$ and NSE values of the TRMM-SWAT in the upper and lower reaches were all below 0.5, with the exception of that in the midstream, which was above 0.5, showing that the TRMM-SWAT underperformed in the upstream and downstream. Different from the CMADS-SWAT and the TRMM-SWAT, the performance of the Gauge-SWAT was unsatisfactory in the whole basin, for its $R^2$ and NSE values were all below 0.5. In addition, the PBIAS values of the CMADS-SWAT and the Gauge-SWAT were above zero in the upstream, below zero in the middle stream, and below 0.5 in the downstream, indicating that the discharge values of the CMADS-SWAT and the Gauge-SWAT were overestimated in the upstream and underestimated in the middle and lower reaches, especially in the downstream. Furthermore, the PBIAS values of the TRMM-SWAT in the upper, middle, and lower reaches were 42.8%, 20.4%, and −59.7%, respectively, showing that the discharge values derived from TRMM-SWAT were too high in the upstream and too low in the downstream.

**Table 8.** The post-calibration performance error amounts of the SWAT model simulated with CMADS, TRMM, and Gauge data on a daily monthly scale.

| Station | $R^2$ | | | NSE | | | PBIAS | | |
|---|---|---|---|---|---|---|---|---|---|
| | CMADS | Gauge | TRMM | CMADS | Gauge | TRMM | CMADS | Gauge | TRMM |
| Majie Station | 0.59 | 0.35 | 0.49 | 0.51 | 0.33 | 0.45 | 20.3 | 10.6 | 42.8 |
| Danfeng Station | 0.52 | 0.1 | 0.54 | 0.52 | 0.07 | 0.52 | −4.5 | −12.9 | −20.4 |
| Jingziguan Station | 0.42 | 0.03 | 0.53 | 0.34 | 0.05 | 0.49 | −65.5 | −86.9 | −59.7 |

The results after calibration indicated that CMADS-SWAT was superior to the other two precipitation products in both monthly and daily runoff simulation with the highest $R^2$ and NSE and a similar hydrological process line to the observed runoff. However, it was unexpected that the performance of the Gauge-SWAT was the worst, although the simulated runoff and rainfall had the same trend as the other two products. Moreover, all the three products tended to overestimate the runoff in the upper and middle reaches and underestimate that in downstream at the monthly scale. When it comes to the daily scale, all these three products overestimated the streamflow in upstream and underestimated the runoff in the lower and middle streams.

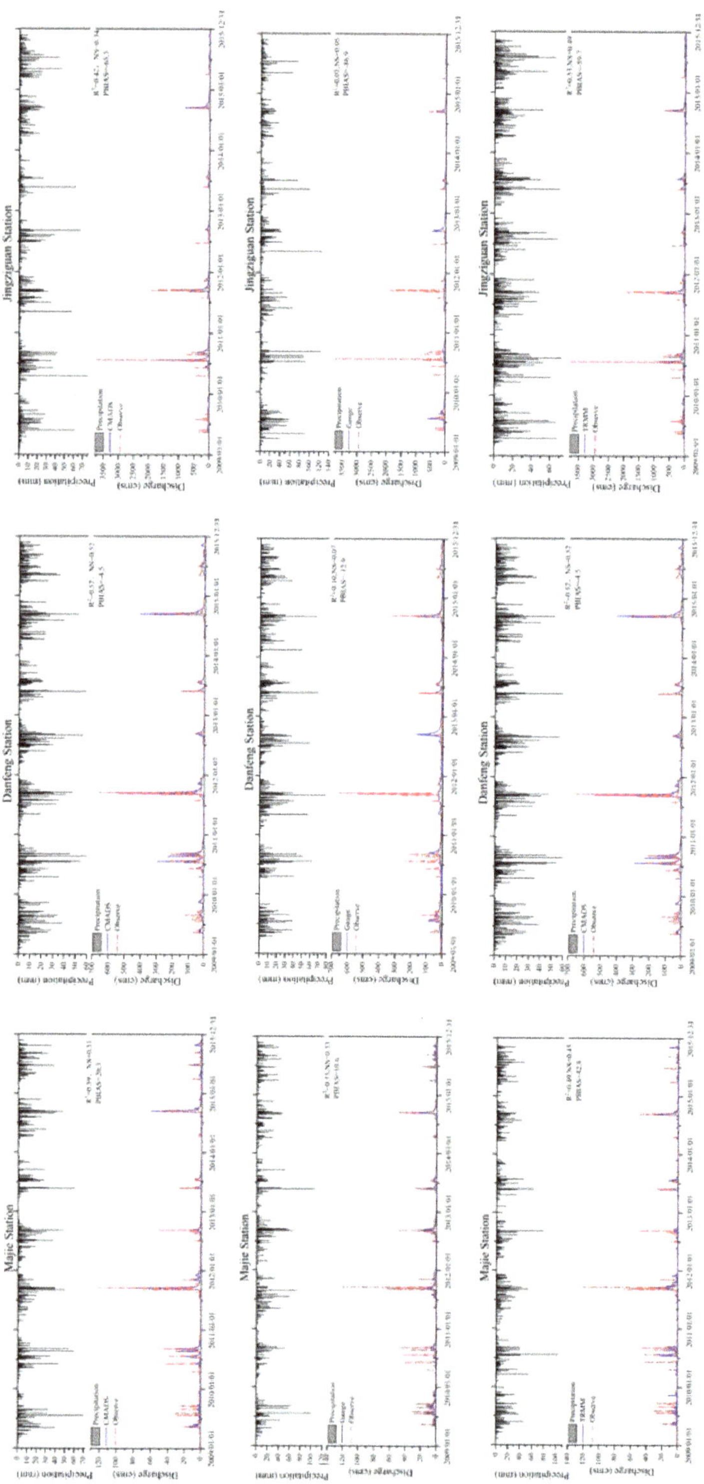


**Figure 9.** Observed and simulated discharges at Majie Station, Danfeng Station, and Jingziguan Station of the DRB at a daily scale using inputs of Gauge, CMADS, and TRMM. Note that this figure is given as year/month/day.

## 4. Discussion

Precipitation inputs play an important role in runoff simulation, and the errors can influence the accuracy of the hydrographical outputs [52]. Generally, precipitation inputs are evaluated on the basis of their predictable performances with hydrological parameters at the watershed scale, which avoids the scale difference found when using ground-based observations for validation [53]. This study evaluated the performance of RO and satellite-based precipitation datasets (CMADS and TRMM) in driving the SWAT model to simulate streamflow in the DRB on both monthly and daily scales. All modeling scenarios were calibrated and validated against runoff data measured at Majie Station, Danfeng Station, and Jingziguan Station. The SWAT-Cup's SUFI-2 algorithm was used for calibration and validation. Indices, including NSE, $R^2$, and PBIAS, were selected to evaluate the efficiency of simulation runoff outcomes.

We found that compared with rainfall gauge observations, TRMM tended to underestimate the precipitation on both monthly and daily scales, while CMADS tended to overestimate the rainfall on the monthly scale but understate the rainfall on the daily scale. These findings are similar with previous studies. For example, Jiang et al. [21] found that TRMM underestimated precipitation on a daily scale in the Mishui Basin, Jiang et al. [47] found an average bias of −20.5% for CMADS over Xixian Basin, and Guo et al. [38] calculated an average bias of −28.7% for CMADS over Jinhua River Basin. The reason why CMADS underestimated the precipitation was the underestimation of the background field CMORPH data [54]. Additionally, the ability of TRMM and CMADS to identify the torrential rain events was worse than that of Gauge. In summary, the performance of CMADS in precipitation simulation was better than that of TRMM, maybe because the correct process of TRMM was simpler than that of CMADS [55]. The spatial distribution of the precipitation varied from dataset to dataset, namely, the rainfall of CMADS increased from the center to the surroundings and its rainfall in the central north was the highest. TRMM increased from upstream to downstream and the highest rainfall occurred in the east, but there is no obvious spatial distribution pattern with the rainfall of Gauge. This result can be explained by the different distributions of the rain gauge. In addition, all meteorological data were categorized into each sub-basin by the "nearest-distance" principle in the SWAT model [49], which contributed to the difference in the precipitation data from CMADS, TRMM, and Gauge as well. Moreover, the similarity between the rainfall of CMADS and Gauge was higher than that of the TRMM on both the monthly and daily scales, which is consistent with Wang's [56] research in the Ganjiang River Basin, where the area and elevation are similar to the DRB. Wang et al. [56] found that CMADS performed better than TRMM in precipitation estimation because of the different development processes of these two products. Only 500 stations were used to correct the TRMM data, while 2421 stations were used to correct the CMADS data [57,58]. Song et al. [59] conducted research on the Qujiang River Basin, (38,900 km$^2$) finding that the spatial distribution of CMADS and TRMM was different from that of Gauge, which is consistent with our study.

Pre-calibration results showed that CMADS and TRMM were reliable enough to estimate runoffs on the monthly scale at Majie Station and Jingziguan Station, while they performed unsatisfactorily in simulating streamflow at Danfeng Station. The performances of Gauge in estimating runoff on the monthly scale in the middle stream and downstream were both unreliable, and only its performances in runoff simulation at the Majie Station was satisfactory. The performances of that on the daily scale, however, were all unsatisfactory. Moreover, all three datasets seriously overestimated the runoff in the middle stream and downstream on the monthly scale. Moreover, underestimation is probably better attributed to poor representation of the spatial variability of precipitation patterns in the middle and downstream, thereby causing the low ratio of streamflow to precipitation. According to Vu et al.'s [60] research, the underestimation can be attributed to the spatiotemporal uncertainty of the precipitation inputs. Previous studies indicated that the spatiotemporal uncertainty of the catchment rainfall was one of the main sources of uncertainty in runoff simulation using rainfall–runoff models [61–63]. Additionally, satellite-based precipitation estimations have their own uncertainties [13]. This means that the satellite-based rainfall estimation affects the runoff simulation significantly [64]. Moreover, Rivera's research [65] found that previous conditions were more important before extreme floods, while previous conditions had little effect on conditions after extreme floods. It explains the result that the average observed streamflow was lower than the average simulated runoff and the average measured runoff was higher than the average observed runoff throughout the year in the upstream, except for September, because the underestimation in September in the Majie Station is the continuous precipitation before an extreme flood occurred in September.

When it comes to the post-calibration results, all three products tended to overestimate the runoff in the upper and middle reaches and underestimate the downstream at the monthly scale, while all three products overestimate the streamflow upstream and underestimate the runoff in the lower and middle streams at the daily scale. The overestimation at the monthly scale and the underestimation at the daily scale may be due to the overall inaccurate estimation of precipitation with the CMADS and TRMM data. In most cases, RO performed better than satellite precipitation in runoff simulation, even in sparsely gauged areas, when SWAT was used for modeling both monthly and daily scales, such as Vu et al.'s [60] research. Namely, they tested the accuracy of four satellite precipitation products, including TRMM 3B42 V7, PERSIANN, PERSIANN-Climate Data Record (PERSIANN-CDR), and CMADS, by using these four products to drive the SWAT model and comparing the runoff simulation results with the runoff simulated by gauged rainfall data in the Han River Basin in South Korea. Their results illustrated that the application of TRMM and CMADS in runoff simulation was worse than that of the gauges. However, our results vary from theirs. It is found in this paper that CMADS-SWAT was superior to the other two precipitation products in both monthly and daily runoff simulations, but Gauge-SWAT performed the worst in both monthly and daily streamflow simulations. This finding is not uncommon; for example, Song et al.'s [59] research on the Qujiang River Basin (38,900 km$^2$) proved that the CMADS-SWAT performed best in the whole basin, followed by TRMM-SWAT and Gauge-SWAT, which performed the worst. That was mainly because of the non-uniform distribution of the gauges. According to Wang et al.'s [66] research, when the number of the stations is similar, the more uniform distribution of rainfall stations, the greater the NSE. In this study, the distribution of the Gauge was the most nonuniform, causing the performance of Gauge-SWAT to be the least satisfactory among the three products. Moreover, the gauge data only represent the observed rainfall at a specific station, whereas the CMADS and TRMM data represent precipitation averages over a large area [67]. For the variation of the topography that causes the precipitation variations over a short distance, the heterogeneity of the landscape of the weather forecasts by CMADS and TRMM is better. In addition, CMADS is a combination of the gauge and satellite data; therefore, its accuracy is higher than that of Gauge and TRMM.

At present, few studies have compared the performance of CMADS and TRMM data by using the SWAT model, because CMADS only covers East Asia and is a newly released dataset. Additionally, most studies are focused on evaluating the performance of CMADS, CHIPRS, and CPC data or only studying the applicability of CFSR data by using the SWAT model [35,68]. Using different models or inputting different parameters will cause different results [69]. Therefore, the satellite or satellite-based products (including CMADS and TRMM) can be applied to the SWAT model or other models in the future to ensure that their replications of runoff are accurate and their predictions of rainfall are credible. Nonetheless, the datasets evaluated in this study can serve as viable alternatives in watersheds similar to the DRB where the observed precipitation data are unavailable.

## 5. Conclusions

In this study, CMADS and TRMM were evaluated on the basis of the measured records of the DRB using the SWAT model, and the main conclusions are as follows:

(1) On the monthly scale, the precipitation measurements of CMADS and TRMM are similar to the rain gauge data. However, the rainfall data derived from TRMM and

CMADS have a different pattern from the precipitation of Gauge at the daily scale. Both TRMM and CMADS underestimate the precipitation, especially TRMM data. Moreover, the ability of CMADS and TRMM to simulate extreme precipitation (e.g., torrential rain) is worse than that of Gauge. The CMADS and TRMM data are also different from the Gauge data on the spatial scale. The rainfall data derived from CMADS tend to increase from the middle to the surroundings and the rainfall data derived from TRMM tend to decrease from upstream to downstream, while the precipitation of Gauge has no clear pattern.

(2) The performance of CMADS-SWAT and TRMM-SWAT is consistent with the observed data from upstream to downstream at a monthly scale, while they both underestimate the runoff. However, Gauge-SWAT only performs satisfactorily in the upstream and its performance in the midstream and downstream is unsatisfactory. The ability of Gauge-SWAT to simulate extreme floods is poor, and the runoff is underestimated by Guage-SWAT as well. However, only CMADS-SWAT performs satisfactorily in the whole basin at a daily scale, while both TRMM-SWAT and Gauge-SWAT performed unsatisfactorily in the middle and lower reaches. CMADS-SWAT, TRMM-SWAT, and Gauge-SWAT have all underestimated the runoff at a daily scale.

(3) Among the three precipitation products, the performance of CMADS-SWAT is the best, followed by TRMM-SWAT. Gauge-SWAT had the worst performance, whether on the monthly scale or the daily scale.

**Author Contributions:** Conceptualization, Y.G. and W.D.; Y.G. and W.D.; software, Y.G.; validation, W.X. and X.Z.; formal analysis, Y.G.; investigation, Y.G.; resources, Y.G.; data curation, Y.G.; writing—original draft preparation, Y.G.; writing—review and editing, W.X., X.Z., W.T and X.W.; visualization, W.D.; supervision, W.D.; project administration, W.D.; funding acquisition, W.D. and W.T. All authors have read and agreed to the published version of the manuscript.

**Funding:** This research was funded by the National Key R&D Program of China (2021YFE0111900, 2019YFC1510705-05) and the National Natural Science Foundation of China (No.52109002). This paper was also funded by the basic scientific research business fee of scientific research institutes of the Changjiang Academy of Sciences, grant number CKSF2019185/TB.

**Institutional Review Board Statement:** Not applicable.

**Informed Consent Statement:** Not applicable.

**Data Availability Statement:** Not applicable.

**Acknowledgments:** We thank the editor and reviewers for their useful feedback that improved this paper. This study was supported by the National Key R&D Program of China (No. 2021YFE0111900, No. 2019YFC1510705-05), the National Natural Science Foundation of China (No.52109002), and the basic scientific research business fee of scientific research institutes of Changjiang Academy of Sciences (CKSF2019185/TB).

**Conflicts of Interest:** The authors declare no conflict of interest.

## References

1. Lobligeois, F.; Andréassian, V.; Perrin, C.; Tabary, P.; Loumagne, C. When does higher spatial resolution rainfall information improve streamflow simulation: An evaluation using 3620 flood events. *Hydrol. Earth Syst. Sci.* **2014**, *18*, 575–594. [CrossRef]
2. Galván, L.; Olías, M.; Izquierdo, T.; Cerón, J.C.; Fernández de Villarán, R. Rainfall estimation in SWAT: An alternative method to simulate orographic precipitation. *J. Hydrol.* **2014**, *509*, 257–265. [CrossRef]
3. Roth, V.; Lemann, T. Comparing CFSR and conventional weather data for discharge and sediment loss modelling with SWAT in small catchments in the Ethiopian Highlands. *Hydrol. Earth Syst. Sci. Discuss.* **2015**, *12*, 2113–2153. [CrossRef]
4. Cornelissen, T.; Diekkrüger, B.; Bogena, H.R. Using high-resolution data to test parameter sensitivity of the distributed hydrological model HydroGeoSphere. *Water* **2016**, *8*, 202. [CrossRef]
5. Mileham, L.; Taylor, R.; Thompson, J.; Todd, M.; Tindimugaya, C. Impact of rainfall distribution on the parameterisation of a soil-moisture balance model of groundwater recharge in equatorial Africa. *J. Hydrol.* **2008**, *359*, 46–58. [CrossRef]
6. Remesan, R.; Holman, I.P. Effect of baseline meteorological data selection on hydrological modelling of climate change scenarios. *J. Hydrol.* **2015**, *528*, 631–642. [CrossRef]

7. Bohnenstengel, S.I.; Schlünzen, K.H.; Beyrich, F. Representativity of in situ precipitation measurements – A case study for the LITFASS area in North-Eastern Germany. *J. Hydrol.* **2011**, *400*, 387–395. [CrossRef]
8. Liu, J.; Kummerow, C.D.; Elsaesser, G.S. Identifying and analysing uncertainty structures in the TRMM microwave imager precipitation product over tropical ocean basins. *Int. J. Remote Sens.* **2016**, *38*, 23–42. [CrossRef]
9. Zhu, H.; Li, Y.; Liu, Z.; Shi, X.; Fu, B.; Xing, Z. Using SWAT to simulate streamflow in Huifa River basin with ground and Fengyun precipitation data. *J. Hydroinf.* **2015**, *17*, 834–844. [CrossRef]
10. Musie, M.; Sen, S.; Srivastava, P. Comparison and evaluation of gridded precipitation datasets for streamflow simulation in data scarce watersheds of Ethiopia. *J. Hydrol.* **2019**, *579*, 124168. [CrossRef]
11. Belete, M.; Deng, J.; Wang, K.; Zhou, M.; Zhu, E.; Shifaw, E.; Bayissa, Y. Evaluation of satellite rainfall products for modeling water yield over the source region of Blue Nile Basin. *Sci. Total Environ.* **2020**, *708*, 134834. [CrossRef] [PubMed]
12. Oreggioni Weiberlen, F.; Báez Benítez, J. Assessment of satellite-based precipitation estimates over Paraguay. *Acta Geophys.* **2018**, *66*, 369–379. [CrossRef]
13. Duan, Z.; Liu, J.; Tuo, Y.; Chiogna, G.; Disse, M. Evaluation of eight high spatial resolution gridded precipitation products in Adige Basin (Italy) at multiple temporal and spatial scales. *Sci. Total Environ.* **2016**, *573*, 1536–1553. [CrossRef]
14. Luo, X.; Wu, W.; He, D.; Li, Y.; Ji, X. Hydrological Simulation Using TRMM and CHIRPS Precipitation Estimates in the Lower Lancang-Mekong River Basin. *Chinese Geogr. Sci.* **2019**, *29*, 13–25. [CrossRef]
15. Cecinati, F.; Moreno-Ródenas, A.M.; Rico-Ramirez, M.A.; ten Veldhuis, M.C.; Langeveld, J.G. Considering Rain Gauge Uncertainty Using Kriging for Uncertain Data. *Atmosphere* **2018**, *9*, 446. [CrossRef]
16. Peleg, N.; Ben-Asher, M.; Morin, E. Radar subpixel-scale rainfall variability and uncertainty: Lessons learned from observations of a dense rain-gauge network. *Hydrol. Earth Syst. Sci.* **2013**, *17*, 2195–2208. [CrossRef]
17. Hwang, Y.; Clark, M.P.; Rajagopalan, B. Use of daily precipitation uncertainties in streamflow simulation and forecast. *Stoch. Environ. Res. Risk Assess.* **2011**, *25*, 957–972. [CrossRef]
18. Alijanian, M.; Rakhshandehroo, G.R.; Mishra, A.K.; Dehghani, M. Evaluation of satellite rainfall climatology using CMORPH, PERSIANN-CDR, PERSIANN, TRMM, MSWEP over Iran. *Int. J. Climatol.* **2017**, *37*, 4896–4914. [CrossRef]
19. Sun, Q.; Miao, C.; Duan, Q.; Ashouri, H.; Sorooshian, S.; Hsu, K.L. A Review of Global Precipitation Data Sets: Data Sources, Estimation, and Intercomparisons. *Rev. Geophys.* **2018**, *56*, 79–107. [CrossRef]
20. Hur, J.; Raghavan, S.V.; Nguyen, N.S.; Liong, S.Y. Evaluation of High-resolution Satellite Rainfall Data over Singapore. *Procedia Eng.* **2016**, *154*, 158–167. [CrossRef]
21. Jiang, S.; Ren, L.; Hong, Y.; Yang, X.; Ma, M.; Zhang, Y.; Yuan, F. Improvement of Multi-Satellite Real-Time Precipitation Products for Ensemble Streamflow Simulation in a Middle Latitude Basin in South China. *Water Resour. Manag.* **2014**, *28*, 2259–2278. [CrossRef]
22. Duncan, J.M.A.; Biggs, E.M. Assessing the accuracy and applied use of satellite-derived precipitation estimates over Nepal. *Appl. Geogr.* **2012**, *34*, 626–638. [CrossRef]
23. Yan, R.; Gao, J.; Huang, J. WALRUS-paddy model for simulating the hydrological processes of lowland polders with paddy fields and pumping stations. *Agric. Water Manag.* **2016**, *169*, 148–161. [CrossRef]
24. Gokhan Yilmaz, A.; Alam Imteaz, M.; Ogwuda, O. Accuracy of HEC-HMS and LBRM Models in Simulating Snow Runoffs in Upper Euphrates Basin. *J. Hydrol. Eng.* **2011**, *17*, 342–347. [CrossRef]
25. Wu, J.; Chen, X.; Yu, Z.; Yao, H.; Li, W.; Zhang, D. Assessing the impact of human regulations on hydrological drought development and recovery based on a 'simulated-observed' comparison of the SWAT model. *J. Hydrol.* **2019**, *577*, 123990. [CrossRef]
26. Bhuiyan, M.A.E.; Nikolopoulos, E.I.; Anagnostou, E.N.; Polcher, J.; Albergel, C.; Dutra, E.; Fink, G.; Martínez-De La Torre, A.; Munier, S. Assessment of precipitation error propagation in multi-model global water resource reanalysis. *Hydrol. Earth Syst. Sci.* **2019**, *23*, 1973–1994. [CrossRef]
27. Solakian, J.; Maggioni, V.; Lodhi, A.; Godrej, A. Investigating the use of satellite-based precipitation products for monitoring water quality in the Occoquan Watershed. *J. Hydrol. Reg. Stud.* **2019**, *26*, 100630. [CrossRef]
28. Price, K.; Purucker, S.T.; Kraemer, S.R.; Babendreier, J.E.; Knightes, C.D. Comparison of radar and gauge precipitation data in watershed models across varying spatial and temporal scales. *Hydrol. Process.* **2014**, *28*, 3505–3520. [CrossRef]
29. Wang, H.; Sun, F.; Xia, J.; Liu, W. Impact of LUCC on streamflow based on the SWAT model over the Wei River basin on the Loess Plateau in China. *Hydrol. Earth Syst. Sci.* **2017**, *21*, 1929–1945. [CrossRef]
30. Qiu, J.; Yang, Q.; Zhang, X.; Huang, M.; Adam, J.C.; Malek, K. Implications of water management representations for watershed hydrologic modeling in the Yakima River basin. *Hydrol. Earth Syst. Sci.* **2019**, *23*, 35–49. [CrossRef]
31. Li, D.; Christakos, G.; Ding, X.; Wu, J. Adequacy of TRMM satellite rainfall data in driving the SWAT modeling of Tiaoxi catchment (Taihu lake basin, China). *J. Hydrol.* **2018**, *556*, 1139–1152. [CrossRef]
32. Huang, Y.; Bárdossy, A.; Zhang, K. Sensitivity of hydrological models to temporal and spatial resolutions of rainfall data. *Hydrol. Earth Syst. Sci.* **2019**, *23*, 2647–2663. [CrossRef]
33. Dhanesh, Y.; Bindhu, V.M.; Senent-Aparicio, J.; Brighenti, T.M.; Ayana, E.; Smitha, P.S.; Fei, C.; Srinivasan, R. A Comparative Evaluation of the Performance of CHIRPS and CFSR Data for Different Climate Zones Using the SWAT Model. *Remote Sens.* **2020**, *12*, 3088. [CrossRef]

34. Al-Falahi, A.H.; Saddique, N.; Spank, U.; Gebrechorkos, S.H.; Bernhofer, C. Evaluation the performance of several gridded precipitation products over the highland region of yemen for water resources management. *Remote Sens.* **2020**, *12*, 2984. [CrossRef]
35. Mararakanye, N.; Le Roux, J.J.; Franke, A.C. Using satellite-based weather data as input to SWAT in a data poor catchment. *Phys. Chem. Earth, Parts A/B/C* **2020**, *117*, 102871. [CrossRef]
36. Dao, D.M.; Lu, J.; Chen, X.; Kantoush, S.A.; Van Binh, D.; Phan, P.; Tung, N.X. Predicting tropical monsoon hydrology using CFSR and CMADS data over the Cau river basin in Vietnam. *Water* **2021**, *13*, 1314. [CrossRef]
37. Gao, Z.; Long, D.; Tang, G.; Zeng, C.; Huang, J.; Hong, Y. Assessing the potential of satellite-based precipitation estimates for flood frequency analysis in ungauged or poorly gauged tributaries of China's Yangtze River basin. *J. Hydrol.* **2017**, *550*, 478–496. [CrossRef]
38. Guo, M.; Zhang, T.; Li, J.; Li, Z.; Xu, G.; Yang, R. Reducing nitrogen and phosphorus losses from different crop types in the water source area of the Danjiang river, China. *Int. J. Environ. Res. Public Health* **2019**, *16*, 3442. [CrossRef] [PubMed]
39. Gu, W.; Shao, D.; Jiang, Y. Risk Evaluation of Water Shortage in Source Area of Middle Route Project for South-to-North Water Transfer in China. *Water Resour. Manag.* **2012**, *26*, 3479–3493. [CrossRef]
40. Dong, Z.; Yan, Y.; Duan, J.; Fu, X.; Zhou, Q.; Huang, X.; Zhu, X.; Zhao, J. Computing payment for ecosystem services in watersheds: An analysis of the Middle Route Project of South-to-North Water Diversion in China. *J. Environ. Sci.* **2011**, *23*, 2005–2012. [CrossRef]
41. Hu, S.; Qiu, H.; Yang, D.; Cao, M.; Song, J.; Wu, J.; Huang, C.; Gao, Y. Evaluation of the applicability of climate forecast system reanalysis weather data for hydrologic simulation: A case study in the Bahe River Basin of the Qinling Mountains, China. *J. Geogr. Sci.* **2017**, *27*, 546–564. [CrossRef]
42. Shan, Z.B.Z.Z.R. The Impact of Land Use and Agricultural Management on Non-point Source Nitrogen Pollution in Dan River Watershed. *J. Soil Water Conserv.* **2020**, *34*, 135–141.
43. Meng, X.; Sun, Z.; Zhao, H.; Ji, X.; Wang, H.; Xue, L.; Wu, H.; Zhu, Y. Spring flood forecasting based on the WRF-TSRM mode. *Teh. Vjesn.* **2018**, *25*, 141–151. [CrossRef]
44. Meng, X.; Wang, H.; Wu, Y.; Long, A.; Wang, J.; Shi, C.; Ji, X. Investigating spatiotemporal changes of the land-surface processes in Xinjiang using high-resolution CLM3.5 and CLDAS: Soil temperature. *Sci. Rep.* **2017**, *7*, 1–14. [CrossRef] [PubMed]
45. Keikhosravi Kiany, M.S.; Masoodian, S.A.; Balling, R.C.; Montazeri, M. Evaluation of the TRMM 3B42 product for extreme precipitation analysis over southwestern Iran. *Adv. Sp. Res.* **2020**, *66*, 2094–2112. [CrossRef]
46. Huffman, G.J.; Adler, R.F.; Bolvin, D.T.; Nelkin, E.J. The TRMM Multi-Satellite Precipitation Analysis (TMPA). *Satell. Rainfall Appl. Surf. Hydrol.* **2010**, 3–22. [CrossRef]
47. Jiang, S.; Liu, R.; Ren, L.; Wang, M.; Shi, J.; Zhong, F.; Duan, Z. Evaluation and Hydrological Application of CMADS Reanalysis Precipitation Data against Four Satellite Precipitation Products in the Upper Huaihe River Basin, China. *J. Meteorol. Res.* **2020**, *34*, 1096–1113. [CrossRef]
48. Duan, Z.; Tuo, Y.; Liu, J.; Gao, H.; Song, X.; Zhang, Z.; Yang, L.; Mekonnen, D.F. Hydrological evaluation of open-access precipitation and air temperature datasets using SWAT in a poorly gauged basin in Ethiopia. *J. Hydrol.* **2019**, *569*, 612–626. [CrossRef]
49. Pang, J.; Zhang, H.; Xu, Q.; Wang, Y.; Wang, Y.; Zhang, O.; Hao, J. Hydrological evaluation of open-access precipitation data using SWAT at multiple temporal and spatial scales. *Hydrol. Earth Syst. Sci.* **2020**, *24*, 3603–3626. [CrossRef]
50. Zhang, B.; Xu, X.; Liu, W.; Chen, T. Dynamic changes of soil moisture in loess hilly and gully region under effects of different yearly precipitation patterns. *Chinese J. Appl. Ecol.* **2008**, *19*, 1234–1240.
51. Azarnivand, A.; Camporese, M.; Alaghmand, S.; Daly, E. Simulated response of an intermittent stream to rainfall frequency patterns. *Hydrol. Process.* **2020**, *34*, 615–632. [CrossRef]
52. Knoche, M.; Fischer, C.; Pohl, E.; Krause, P.; Merz, R. Combined uncertainty of hydrological model complexity and satellite-based forcing data evaluated in two data-scarce semi-arid catchments in Ethiopia. *J. Hydrol.* **2014**, *519*, 2049–2066. [CrossRef]
53. Bai, P.; Liu, X. Evaluation of Five Satellite-Based Precipitation Products in Two Gauge-Scarce Basins on the Tibetan Plateau. *Remote Sens.* **2018**, *10*, 1316. [CrossRef]
54. Zhou, Z.; Gao, X.; Yang, Z.; Feng, J.; Meng, C.; Xu, Z. Evaluation of hydrological application of CMADS in Jinhua River Basin, China. *Water* **2019**, *11*, 138. [CrossRef]
55. Sun, R.; Yuan, H.; Liu, X.; Jiang, X. Evaluation of the latest satellite–gauge precipitation products and their hydrologic applications over the Huaihe River basin. *J. Hydrol.* **2016**, *536*, 302–319. [CrossRef]
56. Wang, Q.; Xia, J.; Zhang, X.; She, D.; Liu, J.; Li, P. Multi-scenario integration comparison of cmads and tmpa datasets for hydro-climatic simulation over ganjiang river basin, china. *Water* **2020**, *12*, 3243. [CrossRef]
57. Liu, S.; Yan, D.; Qin, T.; Weng, B.; Li, M. Correction of TRMM 3B42V7 Based on Linear Regression Models over China. *Adv. Meteorol.* **2016**, *2016*, 3103749. [CrossRef]
58. Chen, S.; Hong, Y.; Cao, Q.; Gourley, J.J.; Kirstetter, P.E.; Yong, B.; Tian, Y.; Zhang, Z.; Shen, Y.; Hu, J.; et al. Similarity and difference of the two successive V6 and V7 TRMM multisatellite precipitation analysis performance over China. *J. Geophys. Res. Atmos.* **2013**, *118*, 13060–13074. [CrossRef]
59. Song, Y.; Zhang, J.; Meng, X.; Zhou, Y.; Lai, Y.; Cao, Y. Comparison study of multiple precipitation forcing data on hydrological modeling and projection in the qujiang river basin. *Water* **2020**, *12*, 2626. [CrossRef]

60. Vu, T.T.; Li, L.; Jun, K.S. Evaluation of multi-satellite precipitation products for streamflow simulations: A case study for the Han River Basin in the Korean Peninsula, East Asia. *Water* **2018**, *10*, 642. [CrossRef]
61. Hromadka, T.V.; McCuen, R.H. Uncertainty estimates for surface runoff models. *Adv. Water Resour.* **1988**, *11*, 2–14. [CrossRef]
62. Maskey, S.; Guinot, V.; Price, R.K. Treatment of precipitation uncertainty in rainfall-runoff modelling: A fuzzy set approach. *Adv. Water Resour.* **2004**, *27*, 889–898. [CrossRef]
63. Jones, P.D.; Lister, D.H.; Wilby, R.L.; Kostopoulou, E. Extended riverflow reconstructions for England and Wales, 1865-2002. *Int. J. Climatol.* **2006**, *26*, 219–231. [CrossRef]
64. Andréassian, V.; Perrin, C.; Michel, C.; Usart-Sanchez, I.; Lavabre, J. Impact of imperfect rainfall knowledge on the efficiency and the parameters of watershed models. *J. Hydrol.* **2001**, *250*, 206–223. [CrossRef]
65. Solano-Rivera, V.; Geris, J.; Granados-Bolaños, S.; Brenes-Cambronero, L.; Artavia-Rodríguez, G.; Sánchez-Murillo, R.; Birkel, C. Exploring extreme rainfall impacts on flow and turbidity dynamics in a steep, pristine and tropical volcanic catchment. *CATENA* **2019**, *182*, 104118. [CrossRef]
66. Xu, Q.W.Y.C.X.Z. Influence of Rain Gauges Network Configuration on the Accuracy of Rainfall Spatial Interpolation and Hydrological Modeling. *J. Yangtze River Sci. Res. Inst.* **2019**, *4*, 19–26.
67. Fuka, D.R.; Walter, M.T.; Macalister, C.; Degaetano, A.T.; Steenhuis, T.S.; Easton, Z.M. Using the Climate Forecast System Reanalysis as weather input data for watershed models. *Hydrol. Process.* **2014**, *28*, 5613–5623. [CrossRef]
68. Tuo, Y.; Duan, Z.; Disse, M.; Chiogna, G. Evaluation of precipitation input for SWAT modeling in Alpine catchment: A case study in the Adige river basin (Italy). *Sci. Total Environ.* **2016**, *573*, 66–82. [CrossRef] [PubMed]
69. Nash, J.E.; Sutcliffe, J.V. River flow forecasting through conceptual models part I — A discussion of principles. *J. Hydrol.* **1970**, *10*, 282–290. [CrossRef]

*water*

MDPI

*Communication*

# An Impacts-Based Flood Decision Support System for a Tropical Pacific Island Catchment with Short Warnings Lead Time

**Shaun Williams [1,*], James Griffiths [1], Bernard Miville [1], Emarosa Romeo [2], Mafutaga Leiofi [2], Michael O'Driscoll [1], Malaki Iakopo [2], Silipa Mulitalo [3], Josephina Chan Ting [4], Ryan Paulik [1] and Graham Elley [1]**

[1] NIWA Taihoro Nukurangi, Christchurch 8011, New Zealand; James.Griffiths@niwa.co.nz (J.G.); Bernard.Miville@niwa.co.nz (B.M.); michael.odriscoll@niwa.co.nz (M.O.); ryan.paulik@niwa.co.nz (R.P.); graham.elley@niwa.co.nz (G.E.)
[2] Water Resources Division, Ministry of Natural Resources and Environment, Apia WS1338, Samoa; emarosa.romeo@mnre.gov.ws (E.R.); mafutaga.leiofi@mnre.gov.ws (M.L.); malaki.iakopo@mnre.gov.ws (M.I.)
[3] Meteorology Division, Ministry of Natural Resources and Environment, Apia WS1338, Samoa; silipa.mulitalo@mnre.gov.ws
[4] Disaster Management Office, Ministry of Natural Resources and Environment, Apia WS1338, Samoa; josephina.chanting@mnre.gov.ws
* Correspondence: shaun.williams@niwa.co.nz

**Abstract:** Early warnings decision support systems are recognized as effective soft adaptation tools to prepare for the impacts of imminent flooding and minimize potential injuries and/or loss of life in flood-prone regions. This paper presents a case study of a pilot project that aimed to establish an impacts-based flood monitoring, early warnings, and decision support system for the Vaisigano River which flows through Apia, the capital of Samoa. This river is located in a characteristic short and steep catchment with rapid critical flood peak durations following periods of intense rainfall. The developed system integrates numerical weather prediction rainfall forecasts, real-time rainfall, river level and flow monitoring data, precomputed rainfall-runoff simulations, and flood inundation estimates of exposure levels and threat to human safety at buildings and on roads for different return period events. Information is ingested into a centralized real-time, web-based, flood decision support system portal that enables hydrometeorological officers to monitor, forecast and alert relevant emergency or humanitarian responders of imminent flooding with adequate lead time. This includes nowcasts and forecasts of estimated flood peak time, magnitude and likely impacts of inundation. The occurrence of three distinct extreme rainfall and flood events over the 2020/2021 tropical cyclone season provided a means to operationally test the system. In each case, the system proved adequate in alerting duty officers of imminent flooding in the Vaisigano catchment with up to 24 h warnings and response lead time. Gaps for improvement of system capabilities and performance are discussed, with recommendations for future work suggested.

**Keywords:** flood monitoring; forecasting; hazard exposure; emergency response; Vaisigano River; Samoa

**Citation:** Williams, S.; Griffiths, J.; Miville, B.; Romeo, E.; Leiofi, M.; O'Driscoll, M.; Iakopo, M.; Mulitalo, S.; Ting, J.C.; Paulik, R.; et al. An Impacts-Based Flood Decision Support System for a Tropical Pacific Island Catchment with Short Warnings Lead Time. *Water* **2021**, *13*, 3371. https://doi.org/10.3390/w13233371

Academic Editors: Xiekang Wang, Philippe Gourbesville and Changjun Liu

Received: 20 October 2021
Accepted: 26 November 2021
Published: 29 November 2021

**Publisher's Note:** MDPI stays neutral with regard to jurisdictional claims in published maps and institutional affiliations.



## 1. Introduction

Flood-related events have been the most frequent natural hazard disaster type between 2000–2019, incurring the largest average annualized losses at a global scale compared with any other natural hazard [1–4]. Soft adaptation approaches such as early warnings systems are considered effective tools that enhance the safety and resilience of people to the impacts of flood events (e.g., [5–8]). Indeed, it has become widely accepted that real-time monitoring and early warnings systems provide necessary information tools that enable people living in exposed areas to respond accordingly in order to minimize potential injuries, loss of life and/or livelihoods (e.g., [9–11]).

Flash flooding in particular poses significant early warning challenges due to short catchment response times of <2–3 h compared with the time required for making informed decisions on emergency response and resourcing e.g., [12,13]. These challenges are exacerbated in ungauged catchments where people, lifelines and infrastructure are exposed e.g., [13]. While much progress has been made in weather forecasting over the last few decades [12], challenges remain in forecasting flash floods which are influenced by a combination of geomorphic, hydrological, soil moisture capacity and antecedent trigger conditions [14–16].

In the Pacific Small Islands Developing States (SIDS) region, weather forecasting and early warning systems (EWS) have typically taken a hazard-centric approach in terms of tracking, estimating and alerting affected people of what a specific event 'might be' in terms of magnitude and extent (with several hours to days lead time). Over the last decade, however, Pacific SIDS are increasingly integrating impacts-based approaches into EWSs (i.e., what a specific weather event 'might do' in terms of exposure and losses) [16]. These shifts reflect responses by Pacific SIDS in delivering on global and regional framework targets such as the 2015–2030 Sendai Framework targets [17], and the Framework for Resilient Development in the Pacific [18].

Extreme rainfall-induced fluvial and pluvial flooding are an annual risk faced by many populated centres in high volcanic islands across the tropical Pacific SIDS region (e.g., [19]). Many catchments in these settings are typically characterized as short and steep, with rapid critical rainfall to flood peak durations. This results in extremely limited observations-based (or nowcast) early warnings and response lead times, and motivates the shortening of alerting lead times through the incorporation of forecast-based techniques. For example, the Vaisigano River catchment in Samoa, has an extreme critical peak duration of approximately 1-h resulting in very short warnings and response lead times.

In this paper, we present a case study of a pilot impacts-based flood early warning decision support system developed for the Vaisigano River which flows through the Samoan capital of Apia. The aim of the pilot was to implement a context-specific, real-time, automated river monitoring and forecast-based flood early warnings decision support system capable of alerting hazard monitoring and emergency responders of imminent flooding with up to 24–48 h lead time. We describe the methods and analysis used in the development of different system components, including a web-based information portal which enables interaction with the decision support information tools. We discuss the practical implications of the developed system based on operational testing during the 2020/2021 tropical cyclone season in Samoa, with recommendations offered for future research.

## 2. The Vaisigano Catchment

The Vaisigano catchment is located in the central north of Upolu Island and is characterized by a short, steep, funnel-shaped drainage morphology covering an area of approximately 33 km$^2$ [20] (Figure 1). The catchment extends approximately 12 km from sea level to an elevation of 1158 m along Upolu's volcanic spine [20,21]. The catchment comprises three main tributaries that converge into a single channel at the Alaoa confluence approximately 4.5 km from the coastline, exiting at the Vaisigano bridge in Apia (Figure 1). It is worth noting that Apia itself is generally less than 1 m above mean sea level and was developed in a deltaic/wetland environment which multiple rivers east and west of the catchment drain through.

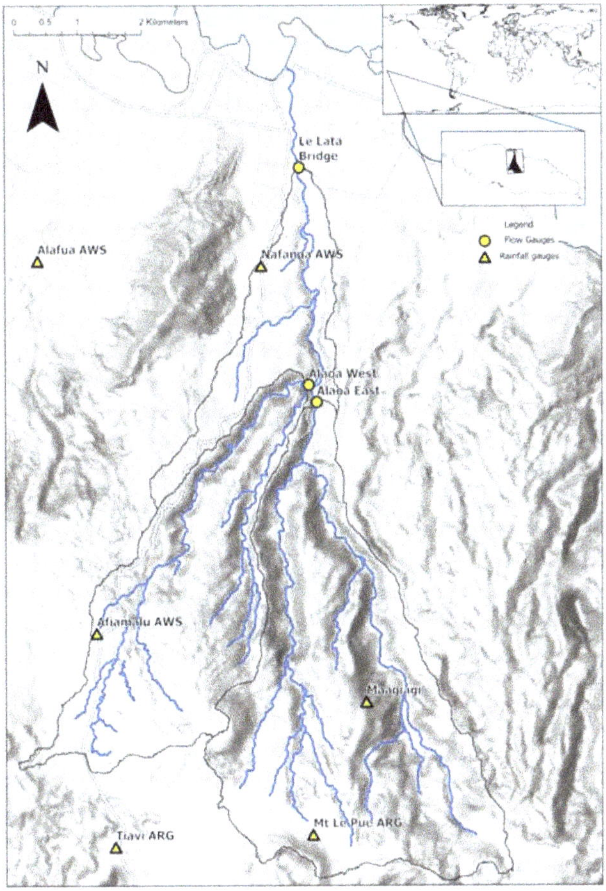


**Figure 1.** Monitored sub-catchments in the Vaisigano River catchment boundary showing perennial and ephemeral tributaries in relation to the downstream urban areas of Apia. The distribution of rainfall and river level/flow monitoring stations used in this study are shown.

The average annual rainfall in the catchment is between 3000–6000 mm, with about 70% occurring during the wet or tropical cyclone season from November to April. There is an approximate 1500 mm difference between annual rainfall in the upper reaches of the catchment (e.g., Afiamalu), compared with the lower plains (e.g., Nafanua).

Morphological drainage characteristics of the catchment favour rapid-rising floods during periods of intense rainfall, with rapid critical storm peak durations of approximately 1 h which can activate ephemeral stream channels in the wider floodplain [22,23]. This will be particularly exacerbated by ongoing climate change where rainfall intensities are estimated to be more severe in the Samoa region by the year 2100 [24].

Urban growth accompanied by residential and commercial development in the floodplains over the last century have rendered a significant proportion of infrastructure and up to several thousand people exposed to the potential impacts of flooding. Indeed, this was demonstrated by several flood events which occurred this millennium: the 2001 [21], 2012 [25], and 2018 [26] flood events. Up to 14 people lost their lives in the 2012 event, with up to US$204 million in damage and losses [27]. The 2012 disaster triggered significant national investment in hard and soft adaptation measures aimed at enhancing the safety and resilience of assets within the catchment, and minimize the impacts of future events in the light of changing climate (e.g., [27,28]). A key soft adaptation investment was to enhance real-time river monitoring and early warnings systems which can support emergency decisions in response to imminent flooding with sufficient lead time.

## 3. System Development: Methodologies and Analysis

The combined physical and exposure characteristics of the Vaisigano catchment results in short flood early warnings response lead times using observation-based monitoring (or nowcasting) systems only. Hence, the incorporation of a real-time forecast-based solution was required to enable longer lead times. In addition, the inclusion of accurate precomputed hazard exposure representations depicting the likely magnitude and extent of forecast flooding as well as areas/assets that could be affected, would provide necessary response planning tools that enable prioritization of emergency resources (including the identification of potential recovery needs).

The necessity for a precomputed impacts forecast approach recognizes the current technological and financial limitations in rapidly simulating accurate two-dimensional (2D) flood inundation models with only a few hours of warning lead time. Here, we describe the data, methodologies and analysis undertaken to develop the Vaisigano flood decision support system (FDSS) framework which encompasses these various capabilities (Figure 2).

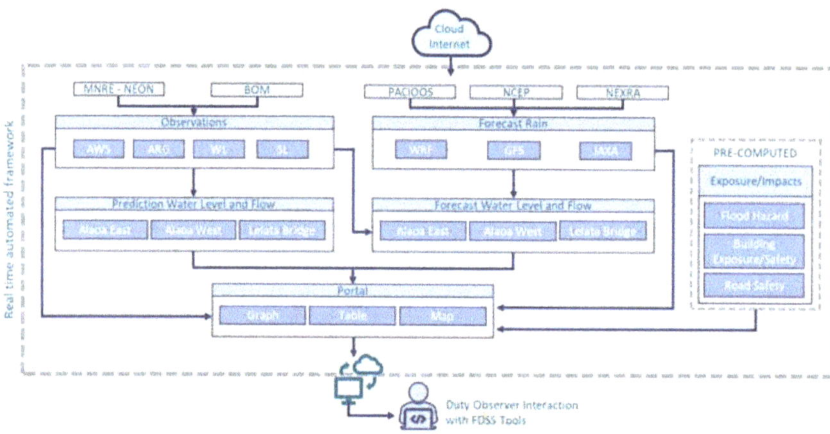


**Figure 2.** Simplified flood decision support system (FDSS) schema showing key data components and information workflow. Real-time observations (or nowcast) and forecast data are integrated with a library of precomputed flood hazard and impacts data, which are accessible to local operators as decision-ready tools on a web-based portal. This relatively low-cost FDSS was developed between February 2020 and April 2021.

### 3.1. Observations and Forecast Data

Near real-time rainfall and river stage/flow monitoring data from gauges located within and around the Vaisigano catchment (Figure 1) provided a basis for developing the observations-based elements of the system. Five (5) minute rainfall data are telemetered from each rain gauge every 10 min via dual mobile (primary) and satellite (backup) communications networks, which are accessible through a web-based telemetry management system (MNRE-NEON) administered by the Samoa Ministry of Natural Resources and Environment (MNRE) (Figure 2) [29]. Similarly, 5-min stage and flow data from each river gauge are telemetered via dual communications, although data packets are sent every 5 min. These data were used along with historical hindcast analysis, described in the sections below, to formulate empirical-based flood thresholds and predictive nowcasting tools capable of alerting flood monitoring and emergency responders of imminent flooding with up to 3-h lead time.

The Weather Research and Forecasting (WRF) numerical weather prediction (NWP) 3 km rainfall model, available for Samoa through the Pacific Islands Ocean Observing System (PACIOOS), provided a means to enable the forecast-based elements of the system. Forecasts of hourly rainfall intensities up to 48 h in advance are updated every 24 h, with site-specific raw data (csv files) obtained via a THREDDS Data Server (TDS) [30]. Integration of these data with the real-time observation baselines using similar empirical relationships as those developed for the nowcasting tools enabled the development of automated and manual tools capable of providing forecast-based alerts with up to 48-h lead time.

Real-time sea level data (1-min) from the Apia tide gauge accessible via the Climate and Oceans Support Program for the Pacific (COSPPac) enabled the tidal state and residual coinciding with a nowcast/forecast flood event to be monitored. This provides a proxy for understanding the effect this might have in exacerbating surface flooding in the coastal, lowest reaches of the catchment.

### 3.2. Flood Frequency Analysis

Rainfall and river flow thresholds for different return period events were required to produce representative real-time forecasts of flood magnitude (up to 48 h lead time), and corresponding precomputed hazard exposure.

#### 3.2.1. Rainfall Analysis

Five (5) and ten (10) minute historical rainfall data held within the Samoa Tideda hydrological and Climate Database for the Environment (CliDE) archives were used to assess and/or validate previously established rainfall return period thresholds (Table 1) [22].

**Table 1.** Historic rainfall and flow timeseries for the Vaisigano catchment used in this study.

| Parameter | Monitoring Station | 2009 | 2010 | 2011 | 2012 [1] | 2013 | 2014 | 2015 | 2016 | 2017 | 2018 [2] | 2019 | 2020 |
|---|---|---|---|---|---|---|---|---|---|---|---|---|---|
| Rainfall | Lake Lanotoo | | | ✓ | ✓ | ✓ | ✓ | ✓ | ✓ | ✓ | ✓ | ✓ | ✓ |
| | Mt. Le Pue | | | | | ✓ | ✓ | ✓ | ✓ | ✓ | ✓ | ✓ | |
| | Tiavi | | | | | | | ✓ | ✓ | ✓ | ✓ | ✓ | |
| | Alaoa West | | | | | ✓ | ✓ | ✓ | ✓ | ✓ | ✓ | ✓ | ✓ |
| | Afiamalu AWS | | ✓ | ✓ | ✓ | ✓ | ✓ | ✓ | ✓ | ✓ | ✓ | ✓ | ✓ |
| | Nafanua AWS | | ✓ | ✓ | ✓ | ✓ | ✓ | ✓ | ✓ | ✓ | ✓ | ✓ | |
| River Level | Lelata Bridge | | | | | | | | | | ✓ | ✓ | ✓ |
| | Alaoa West | | | | | | | | | | ✓ | ✓ | |
| | Alaoa East | ✓ | ✓ | ✓ | ✓ | ✓ | ✓ | ✓ | ✓ | ✓ | ✓ | ✓ | ✓ |

[1] Tropical cyclone Evans impacted Samoa between 11–15 December 2012. [2] Tropical cyclone Gita impacted Samoa on 8–9 February 2018.

Both daily and sub-daily records for Afiamalu, Nafanua, Alaoa, Le Pue and Tiavi sites were particularly crucial as they represented the spatial variability of available data within the catchment (Figure 1), with maximum daily rainfall extremes of over 400 mm recorded at Afiamalu, Nafanua, and Alaoa. The most significant large rainfall event since 2010 was associated with tropical cyclone Evans in 2012. Overall rainfall records suggest that the event was between a 20- to >100-year return period event (5% to <1% annual exceedance probability), and was compounded by antecedent rainfall of 174 mm at Afiamalu and 120.5 mm at Alaoa during the preceding 24-h period. This caused multiple landslips within the catchment [25]. On average it was estimated that up to 206 mm of rainfall fell within a 2-h period [31]. The 24-h rainfall recorded at Afiamalu was 413.8 and 467 mm at the Alaoa gauge. Other gauges within the catchment recorded less (e.g., 144 mm at Apia and 207 mm at Nafanua), illustrating the high spatial variability of rainfall even for large events. Maximum hourly rainfall for the on 13 December 2012 event varied from 56 mm/h between 12 and 1pm to 85 mm/h from 2 to 3pm at Afiamalu, and from 12.2 to 156.8 mm/h at the Alaoa gauge [31]. Again, this illustrates the local spatial variability of rainfall in the catchment.

Data from seven operational rain gauges located within or near the catchment (i.e., Afiamalu, Nafanua, Alaoa, Mt Le Pue, Tiavi, Maagiagi and Togitogiga) were used to produce area-weighted catchment averaged rainfall and rainfall-surface runoff relationships (detailed in Section 3.2.2), to enable reliable estimation of real-time nowcast and forecast peak flow and level for river gauges at Alaoa West, Alaoa East and Lelata Bridge (Figure 1). Thiessen polygons were drawn around each of the above rainfall gauges with only the gauges at Nafanua, Afiamalu, Maagiagi and Le Pue intersecting with the polygons (Figure 3). This enabled calculation of the area associated with each rainfall polygon (Equation (1)) to underpin the estimation of area-weighted average rainfall for each sub-catchment (Equations (2)–(4)):

$$[(Area'0' \times Rain@Nafanua) + (Area'2' \times Rain@Afiamalu) + (Area'3' \times Rain@LePue) + (Area'4' \times Rain@Tiavi) + (Area'5' \times Rain@Togitogiga)] + (Area'6' \times Rain@Maagiagi)/Totalareaofcatchment \tag{1}$$

$$Lelatacatchmentaveragerainfall = [(6.53 \times Rain@Nafanua) + (8.39 \times Rain@Afiamalu) + (5.31 \times Rain@LePue) + (0.67 \times Rain@Tiavi) + (0.17 \times Rain@Togitogiga) + (11.03 \times Rain@Maagiagi)]/32.11 \tag{2}$$

$$AlaoaEastcatchmentaveragerainfall = [(0.93 \times Rain@Nafanua) + (0.17 \times Rain@Afiamalu) + (5.27 \times Rain@LePue) + (0.02 \times Rain@Tiavi) + (0.17 \times Rain@Togitogiga) + (10.48 \times Rain@Maagiagi)]/17.04 \tag{3}$$

$$AlaoaWestcatchmentaveragerainfall = [(1.39 \times Rain@Nafanua) + (7.94 \times Rain@Afiamalu) + (0.05 \times Rain@LePue) + (0.65 \times Rain@Tiavi)] + (0.55 \times Rain@Maagiagi)/10.57 \tag{4}$$

If one or more rain gauges breaks down or becomes faulty, then the following assumptions are applied:

- If Tiavi data are missing then use data from the next closest gauge (which in order of priority) Le Pue, Afiamalu, Togitogiga, Maagiagi, Alafua, or Nafanua;
- If Le Pue data are missing then use data from the next closest gauge (which in order of priority) Tiavi, Togitogiga, Afiamalu, Maagiagi, Nafanua, or Alafua;
- If Togitogiga data are missing then use data from the next closest gauge (which in order of priority) Le Pue, Tiavi, Maagiagi, Afiamalu, Nafanua, or Alafua;
- If Afiamalu data are missing then use data from the next closest gauge (which in order of priority) Maagiagi, Le Pue, Tiavi, Alafua, Nafanua or Togitogiga; and
- If Nafanua data are missing then use data from the next closest gauge (which in order of priority) Alafua, Afiamalu, Maagiagi, Le Pue, Tiavi or Togitogiga.

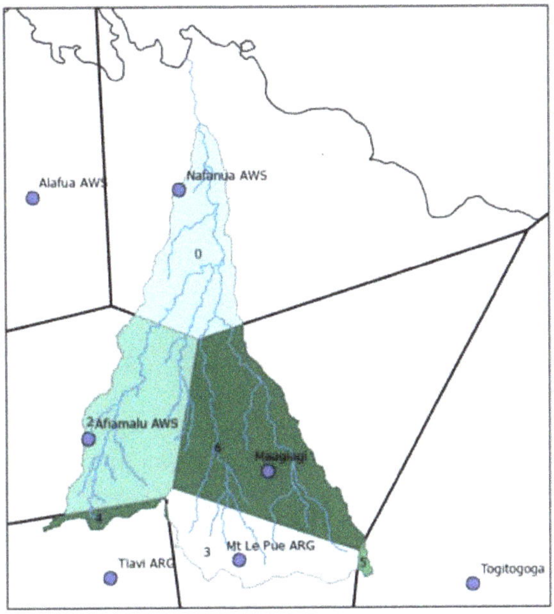


**Figure 3.** Thiessen polygons for automatic rain gauges (ARG) and automatic weather stations (AWS) in or near the Vaisigano catchment.

Rainfall thresholds for design event durations were identified using rainfall intensity-duration-frequency (IDF) curves previously established at Afiamalu (Table 2) [22,23,32]. These were used to identify thresholds required to trigger early warnings for a given real-time nowcast or forecast event. For example, 1-h rainfall intensity of 156.8 mm observed at Afiamalu during the 2012 flood represents an event greater than a 100-year return period event (Table 2).

**Table 2.** Intensity-Duration-Frequency (IDF) threshold values used in this study.

| Duration | Rainfall Intensity (mm/h) for Different Return Period Events [1] | | | | |
|---|---|---|---|---|---|
| | 1 in 5 Year | 1 in 10 Year | 1 in 25 Year | 1 in 50 Year | 1 in 100 Year |
| 5-min | 313 | 365 | 412 | 450 | 493 |
| 10-min | 242 | 283 | 320 | 350 | 382 |
| 15-min | 204 | 240 | 270 | 295 | 323 |
| 30-min | 142 | 166 | 187 | 204 | 223 |
| 1-h | 90 | 105 | 119 | 129 | 142 |
| 2-h | 57 | 66 | 72 | 80 | 87 |
| 3-h | 40 | 47 | 50 | 57 | 62 |

[1] Thresholds for additional design event durations (up to 60-day duration and 1000-year return period) were adapted from American Samoa based on similar catchment characteristics identified in [33].

### 3.2.2. Flood Stage and Flow Thresholds

Stage and flow data from the operational gauges at Alaoa West, Alaoa East and Lelata were used to identify/monitor flow thresholds that are expected to be exceeded based on real-time observations/forecasts of rainfall. Long-term records at these sites along with a rating curve developed in [22] for flows up to 15 $m^3/s$ at the Alaoa East gauging station were used. The associated rating equations for different stage-flow ranges at Alaoa East up to a stage of 3.2 m and discharge of 75 $m^3/s$ are shown in and are summarized in Table 3.

**Table 3.** Rating curve equations for different stage-flow ranges at Alaoa East river gauge.

| Stage (m) | Discharge Equation ($m^3/s$) |
|---|---|
| 0 to 0.125 | Discharge = 0.470 × Stage$^{1.232}$ |
| >0.125 to ≤0.375 | Discharge = 0.470 × Stage$^{1.233}$ |
| >0.375 to ≤0.6 | Discharge = 0.470 × Stage$^{1.234}$ |
| >0.6 to ≤0.825 | Discharge = 2.799 × Stage$^{3.807}$ |
| >0.825 to ≤1.0625 | Discharge = 2.374 × Stage$^{3.345}$ |
| >1.0625 to ≤1.3125 | Discharge = 2.364 × Stage$^{3.261}$ |
| >1.3125 to ≤1.5625 | Discharge = 2.421 × Stage$^{3.114}$ |
| >1.5625 to ≤1.825 | Discharge = 2.455 × Stage$^{3.076}$ |
| >1.855 to ≤2.1125 | Discharge = 2.424 × Stage$^{3.101}$ |
| >2.1125 to ≤2.375 | Discharge = 2.409 × Stage$^{3.110}$ |
| >2.375 to ≤2.85 | Discharge = 2.710 × Stage$^{2.965}$ |
| >2.85 | Discharge = 4.360 × Stage$^{2.446}$ |

These stage-discharge data, along with preliminary analysis by [23], provided a basis to derive flood frequency distributions from Alaoa East (Table 4). For design events up to the 10-year return period (10% annual exceedance probability), the flow magnitudes were within 10% of previous flow estimates as detailed in [33]. For higher magnitude design events, flows have increased from those previously available (e.g., the 100-year return period event increased from 331 to 542 $m^3/s$) (Table 5) [23].

**Table 4.** Design event flows for Alaoa East station adapted from [23].

| Return Period (Years) | Design Event Flows at Alaoa East Gauge (Eastern Catchment) | | Design Flows at Electric Power Corporation Weir (Eastern, Central and Western Catchments) ($m^3/s$) |
|---|---|---|---|
| | Flow ($m^3/s$) | Level (m Gauge Datum) | |
| 1 | 17 | 1.7 | 29 |
| 2 | 41 | 2.6 | 69 |
| 5 | 82 | 3.2 | 139 |
| 10 | 131 | 3.6 | 222 |
| 20 | 204 | 4.0 | 346 |
| 50 | 358 | 4.8 | 608 |
| 100 | 542 | 5.4 | 921 |
| 200 | 814 | 6.4 | 1384 |
| 500 | 1382 | 7.8 | 2350 |
| 1000 | 2052 | 9.3 | 3489 |

Modelled flow levels at Lelata for the 5-, 20-, 100- and 1000-year return period events were derived using the area weighted sub-catchment areas and rescaling of peak flows to Alaoa East and Alaoa West (Table 5). Compared to the values shown in Table 5, the rescaled flows for Alaoa East are higher for the 5- and 20-year return period conditions but are lower for the 100- and 1000-year return periods. Similar trends are reflected at the Lelata gauge site [23].

**Table 5.** Predicted critical rainfall characteristics for 5-, 20-, 100- and 1000-year return period and associated predicted flows at Lelata Bridge and re-scaled critical flows for Alaoa East and Alaoa West using catchment areas.

| | | Return Period Event | | | | Source |
|---|---|---|---|---|---|---|
| | | 5-Year | 20-Year | 100-Year | 1000-Year | |
| Rainfall | Total (mm) | 93.5 | 127.0 | 133.7 | 337.3 | [33] |
| | Duration (mins) | 45.0 | 60.0 | 45.0 | 45.0 | [22] |
| | Intensity (mm/h) | 133.2 | 127.0 | 190.5 | 240.4 | [32], estimated from 30 and 60 min for the 45 min critical duration events |
| Flow ($m^3/s$) | Lelata Bridge | 550.0 | 690.0 | 862.0 | 1137.0 | [22] |
| | Alaoa East | 292.0 | 366.0 | 458.0 | 604.0 | Scaled by area from the Lelata Bridge model in [22] |
| | Alaoa West | 181.0 | 227.0 | 284.0 | 374.0 | Scaled by area from the Lelata Bridge model in [22] |

### 3.3. Peak Flow Estimation

Forecasts of peak flows estimated using the relationships described in Sections 3.1 and 3.2 for the Alaoa East gauged site were used to provide short-term warnings of potential high flows at this location; based on observed/nowcast rainfall and stage-flow as well as forecast rainfall sourced from the WRF model (see Figure 2). This site is particularly important due to the absence of reliable gauged data at the Lelata and Alaoa West sites. A unit hydrograph approach was used to predict the flood hydrograph at this site using the procedure adapted from [34]. This approach was used because of the lack of detailed soil or vegetations maps at the site that could be used to inform a more physically based model. Also, as the model was to be hosted by a web-based service, it was more convenient to use an empirical-based model that could be hard-coded into the system. The key steps involved in developing the model included:

- Identifying observed flow data for the given catchment. These data should be available in the required temporal resolution (duration) of the unit hydrograph (e.g., 10-min, 1 h or 3 h);
- Identifying an appropriate flow event (i.e., where river flow starts and ends at a common baseflow value and exhibits a standard hydrograph shape);
- Plotting the direct runoff hydrograph (i.e., observed flow/baseflow) and then calculate the area under the hydrograph curve to determine the volume of runoff for the event (trapezoidal method);
- Calculating the rainfall excess (mm) by dividing the volume of runoff by the catchment area ($m^2$); and
- Calculating each temporal ordinate of the unit hydrograph by dividing each observed discharge ordinates (t1, t2, t3, ... , *tx*) by the excess rainfall figure described above.

To use the unit hydrograph to predict the direct runoff resulting from each 1 mm of rainfall, the following steps were used:

- Subtract the estimated 'rainfall loss' from the rainfall timeseries to get 'effective rainfall';
- Multiply each rainfall value by the unit hydrograph to obtain multiple direct runoff hydrographs (one for each rainfall increment, and each lagged by a successive time increment); and

- Calculate the sum of all the direct runoff hydrographs and add the baseflow component of observed data to obtain the flow timeseries.

Implementation of the unit hydrograph method in the FDSS allowed the starting condition (baseflow) to be determined from observed conditions. In addition, the rainfall loss factor was made dependent on antecedent rainfall conditions, thus acting as a surrogate for the extent of soil saturation at the start of each storm event.

Flow event data, which captured the extreme flood in 2012, were used to calculate the 10-min unit hydrograph at Alaoa East, as both the rainfall and flow data were available in 10-min increments during this period. For each identified event, a baseflow value was calculated from the flow before rainfall commenced. The 10-min data were re-scaled to create a 1-h unit hydrograph which we use to predict flow from real-time forecast rainfall data.

### 3.4. Alert Thresholds

Flood thresholds used to trigger an alert were primarily based on the rainfall and water-level thresholds described in Sections 3.2 and 3.3. In addition, rainfall thresholds identified from IDF curves derived for American Samoa [22,33] were used to compensate for data gaps or short available records within the Vaisigano catchment. A simplified schema showing the analytical data flow from real-time observations and forecasts to alert triggers are shown in Figure 4.

Observed water level alerts are determined directly from in-catchment water level gauges at Lelata, Alaoa East and Alaoa West. The thresholds (5-, 20-, 100-, 1000-year return periods) at which an alert is signalled are based on those identified within the most recent flood modelling of the catchment presented in [23]. Thresholds for the Alaoa East and Alaoa West site were derived by rescaling the estimates for the Lelata using flow per unit area.

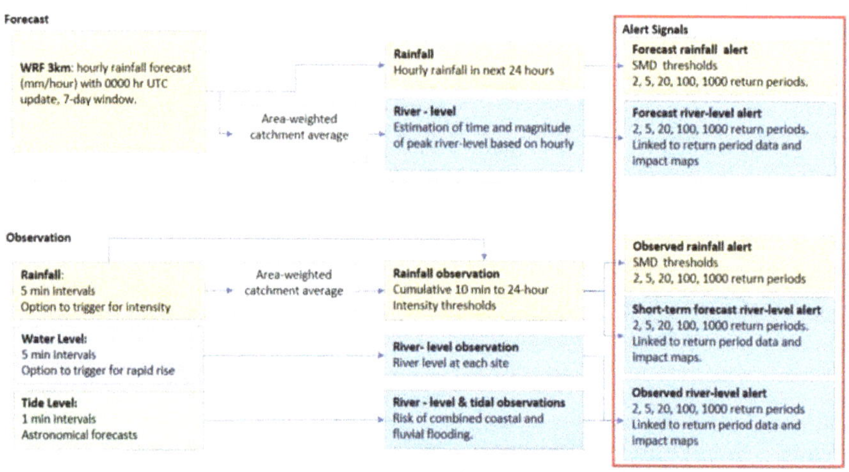


**Figure 4.** Schematic of alerting criteria used to indicate increased flood risk. This is pre-configured and displayed on the predicted 'Flood Status' and 'Peak Expected Time' tools on the flood decision support system (FDSS) web-portal.

Observed rainfall alerts are based on information taken from [22], i.e., from data that have been derived for American Samoa [33]. Rainfall thresholds can be either a rainfall intensity or a total amount of rain over a specified period. Observed rainfall is also used to predict water level at each river gauging station, enabling advanced warning of up to 1 h or more depending on rainfall intensity, of the likelihood of water level thresholds being exceeded at those locations.

Two NWP sources provide medium-term rainfall forecasts:

- Weather Research and Forecasting Model (WRF) (3 km resolution) 1-h rainfall, available at 00 UTC, for a 7-day window; and
- NOAA Global forecast systems (GFS) forecast (~20 km resolution) 3-h rainfall, available at 00, 06, 12, 18 UTC, for a 10 day window.

Alert signal thresholds for forecast rainfall in addition to observed rain and water levels, include:

- A threshold of 5 mm/10-min to indicate risk of flooding;
- A threshold of 25 mm/12-h for two consecutive timeframes is also used to indicate increased flood risk;
- A threshold of 75 mm/24-h used to indicate increased flood risk.

Flood warnings issued by MNRE are conveyed via available communications outlets (e.g., website, national media stations, social media), and guided by the MNRE Forecast Operations Plan [35].

### 3.5. Flood Hazard and Impacts Analysis

To facilitate the rapid representation of inundation hazard and exposure for a given forecast flood magnitude, it was necessary to compile a library of precomputed scenarios corresponding to the given forecast flood event. This was particularly crucial due to the flashy nature of the Vaisigano catchment and current computational limitations in simulating real-time, on-the-fly, hydrodynamic simulations of flood hazard intensity. To this end, the multi-hazard model framework supported by RiskScape software [36] was configured to analyse building and road exposure to flooding hazards. The RiskScape software engine combines spatial layers representing hazards and exposures (i.e., elements at risk) with vulnerability functions in a model workflow to quantify the impacts from hazard events. Here, we used flood hazard layers represented by available TuFlow 2D flood inundation modelled depth and velocity hazard intensities (2 m grid resolution) for the 2-, 5-, 20-, 100- and 1000-year return period flood scenarios [22]. These were intersected with exposure layers representing building polygon and road polyline features to calculate metrics on the exposure and threat to safety at building and road locations (Figure 5).

The hydraulic model was configured by Filer et al. (2019) [22] using a rectangular grid domain (5.3 × 7.5 km), with base topography and nearshore bathymetry derived using the 5 m LiDAR digital elevation model (DEM) produced in [37]. DEM modifications enabled hydraulic features such as river channels and infrastructure (e.g., road crests, bridges, culverts) to be represented in the model outputs. Variable roughness was applied to represent different land uses and ranged from Manning's coefficients of 0.02 (roads, car parks) to 0.3 (buildings). The model was calibrated against the 2012 and 2018 flood events, with the results showing good modelled-to-observed water levels within a desirable tolerance of 0.25 m [22].

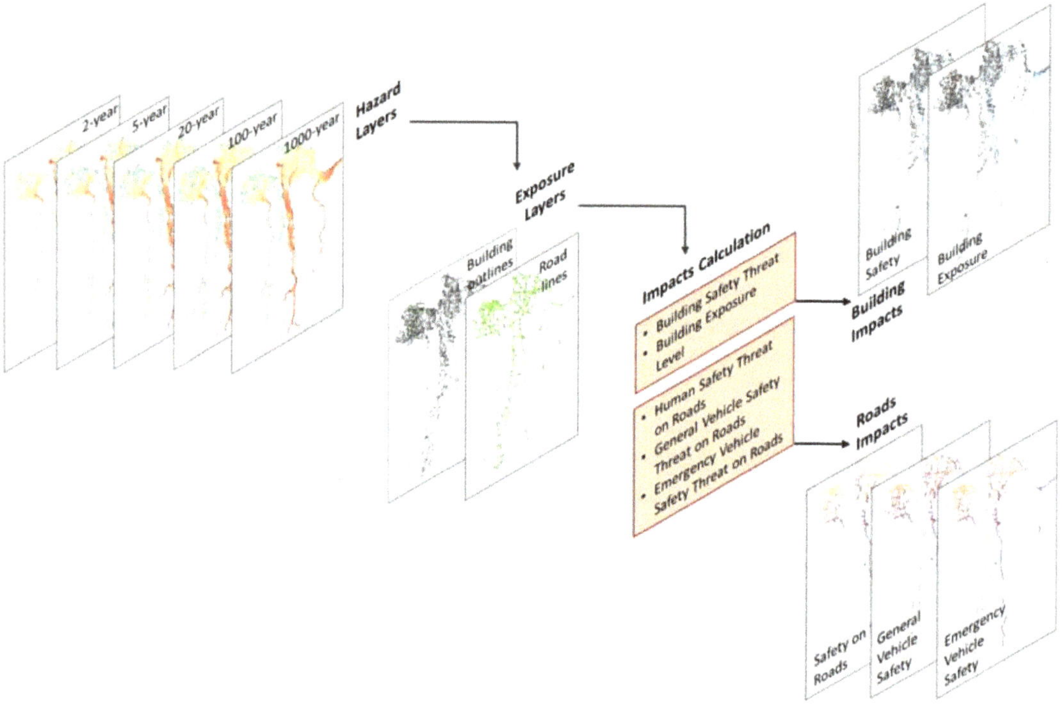


**Figure 5.** Conceptual representation of the impact model workflow in RiskScape. Here, flood inundation depth (D) and depth-velocity (DV) represent hazard layers, building and road objects represent exposure layers, with impacts calculated from vulnerability functions relating D and DV hazard with building and roads exposure and threat to safety levels.

The flood hazard layers are representative of existing mitigation works in the Vaisigano catchment (e.g., levee segments) as well as the rainfall and river stage-flow relationships described in Section 3.2 and 3.3. In addition, each hazard layer represents the estimated flood hazard intensities coinciding with storm/high tide to provide conservative indications of expected impacts to a forecast flood event. Digital building outlines used in this analysis were obtained from the 2009 Pacific Catastrophe Risk Assessment and Financing Initiative (PACRAFI) database available for Apia [38]. This polygon dataset was manually updated in QGIS software using 2020 Google satellite base imagery to include outlines of new buildings constructed since 2009, and exposed to the 1000-year maximum available scenario. Road polylines were obtained from the Open Street Map (OSM) database [39], and split into 10 m segments. These provided the input exposure layers for RiskScape to calculate flood exposure metrics (building count and road length) and human threat to safety at building and road locations (Figure 5).

Threat to safety functions represent the human stability response to maximum flood depth-velocity (DV) at building locations [40]. Here, ordinal categories (e.g., none, low, medium high, extreme), describe the relative threat to a person's safety at ground-level for each building location and 10-m road segment. Similar categories determine the safety of persons located in a vehicle (e.g., sedan, truck, fire-engine) at the time of maximum flood DV on roads [41]. Human threat to safety for each flood event was reported at feature- and village-scale web-supported GIS files for ingestion and display as decision-ready maps and tables on the FDSS portal.

## 4. Flood Decision Support System (FDSS): Operations Portal

The FDSS Portal provides a centralized web-based information interface/dashboard that incorporates multiple information tools, enabling MNRE and other authorized operators to monitor and predict/forecast imminent flood events in real-time. This includes rapid access to decision-ready tools for determining the potential impacts in advance. The portal is built within the Climate Data for the Environment Services application Client (CliDEsc) framework [42] and provides a flexible platform for future MNRE data visualization needs.

The portal is cloud-based and is password-protected. It can be configured to enable additional or new data/information sources required to support hydrometeorological or climate operational decision-making as these needs arise in future. Information ingested and displayed within the Vaisigano FDSS includes (Figure 6), but is not limited to:

- Real time rainfall and river stage data from monitoring stations within and surrounding the Vaisigano catchment;
- Real time sea-level data from the Apia tide gauge;
- Nowcast (observations-based) flood prediction tool;
- Real time NWP (WRF 3 km) rainfall forecasts and flood forecast tools; and
- Pre-computed flood hazard and exposure/impacts maps and data tables for available flood scenarios.

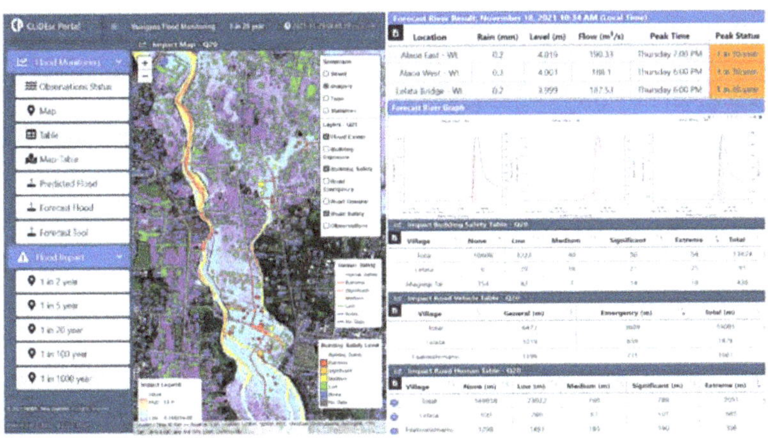


**Figure 6.** Vaisigano flood decision support system (FDSS) web-portal interface tools (example forecast event only). In this example, time series of forecast flood peak and magnitude at river gauge sites along with corresponding maps/tables of flood hazard and human safety at building/road locations are displayed.

## 5. System Performance

Significant rainfall events which occurred during the 2020–2021 tropical cyclone season provided a basis to operationally test the system and identify ongoing gaps for future development. In particular, the rainfall events on 18 December 2020, 7 January and 22 February 2021 proved useful in assessing the efficiency of the prediction model. Figure 7 illustrates the river levels recorded at Alaoa East and Lelata flow gauging stations from the 15 to 19 December 2020 [43], associated with the 18 December 2020 flooding. Higher rainfall was observed at Afiamalu rainfall station than Le Pue during this period highlighting the high local spatial variability in the catchment (as previously described in Section 3.2). Whilst the 5-, 10-, and 60-min rainfall intensities at Afiamalu were all less than 1-year return period flood, the 24-h rainfall accumulation of 392 mm from 2 pm on the 17 to 18 December was greater than a 25-year event (Table 5). For the same period, 316 mm of rainfall was recorded at the Le Pue site, equivalent to between a 5- to 10-year event [33].

(a)

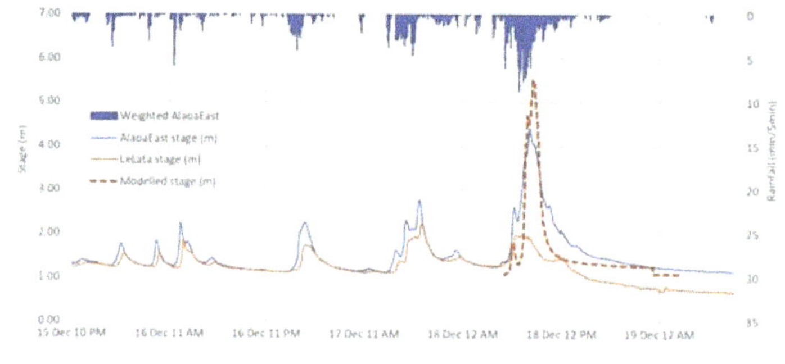


(b)

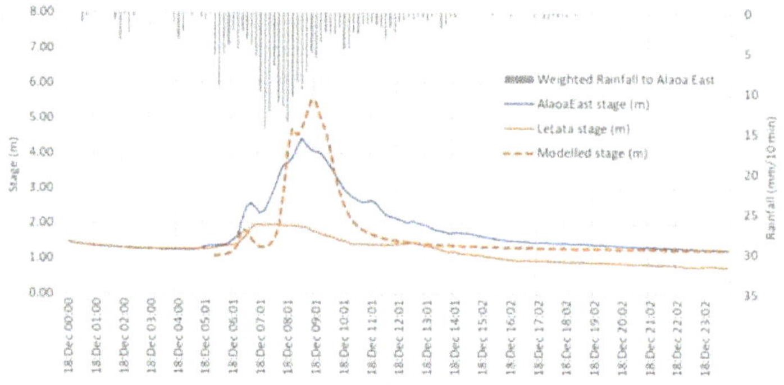


(c)

**Figure 7.** (**a**) Forecast rainfall for 9am on 18th December 2020 local Samoan time (WRF model issued at 0000 UTC on 17th December 2020). (**b**) Observed 5-min and (**c**) 10-min river stage at Lelata and Alaoa East river gauging sites, and weighted rainfall for Alaoa East associated with the 18th December 2020 flooding.

The data at this and subsequent events on 7 January 2021 and 22 January 2021 have been used to validate and improve the prediction accuracy of the FDSS system by increasing the amount of data that the 10-min and 1-h unit hydrographs are based upon. That is, these recent extreme events have extended the pool of data from which the hydrographs are derived.

Modelled stage at Alaoa East are also shown in Figure 7. It should be noted that the river stage is based on area-weighted rainfall for the Alaoa East catchment and correlation of modelled and observed flows yielded an $R^2$ value of 0.87 for this event.

*Uncertainty*

The current system was developed under conditions of data scarcity and discontinuity. Uncertainty surrounding the validity of rating curves for each streamflow dataset presented a particular challenge. The uncertainty is produced because of limited opportunity to measure river flows during high flow conditions, and ongoing changes in channel morphology during such conditions without associated changes to the rating curve information. For example, it can be seen in Figure 7 that the flood peak at the Lelata site on 18th December 2020 was not recorded because the river over-topped the monitored channel at just below 2 m and resulted in a flat hydrograph. Limited data for the development of river rating curves mean that assumptions made during construction of the unit hydrographs will impact the performance of the model in predicting the magnitude of peak flows.

A second source of uncertainty lies within the area-weighted rainfall estimates used by the model for each catchment, which despite being informed by new rain-gauges within the catchment will be susceptible to errors for spatially heterogeneous storm events. The impact of rainfall estimation uncertainty on model performance at this site will be reported in a subsequent publication.

To improve model performance, the model was calibrated using the rainfall loss factor which was dynamically set to reflect different antecedent soil moisture at the start of each rainfall event. It should be noted, however, that the model was less accurate at predicting lower peak flows.

## 6. Practical Implications

This pilot has demonstrated the suitability of integrating real-time forecast with observations of rainfall, river level/flow and precomputed representations of flood magnitude and exposure to warn responders of imminent flooding with adequate alert lead times in a short, steep, flashy tropical Pacific catchment. Indeed, this was exemplified by the extreme rainfall and subsequent flooding on 18 December 2020. In this case, responders were alerted up to 24 h lead time, enabling an evidence-based evacuation of affected residents several hours before the Vaisigano River overtopped its channel.

While the developed FDSS tools have proven effective in helping to inform evacuation response decisions which minimize the threat of injuries and loss of life, key challenges to guide future work include, but are not limited to:

- Implementing a regular programme of flow gaugings, in particular during and after high flow events to improve existing rating curves and unit hydrograph calibrations used in the FDSS predictive models;
- Statistical downscaling of NWP models for Samoa to finer grid resolutions, as well as more frequent real-time updates consistent with the GFS/ECMWF forecast frequencies (e.g., [44,45]). This includes the refinement of rainfall IDF's at key sites and ingestion of additional forecast models where available for Samoa;
- Application of new impacts/exposure forecasting techniques using statistical, precomputed, metamodeling approaches (e.g., [46–48]); and
- Assessing community perspectives in relation to warnings communications and response behaviour to progress towards the integration of a more people-centred approach to early warnings systems (e.g., [49–51]).

Nevertheless, the FDSS provides a first order solution towards impacts-based forecasting for the Samoa region which is consistent with Pacific [16,18] and international [17] efforts. Future extension of the framework to build on the capabilities developed in this pilot would help to safeguard the effectiveness of the FDSS in minimizing flood threats to injury, lives and livelihoods in Samoa, with potential applications to similar SIDS contexts.

## 7. Conclusions

This paper aimed to describe a pilot flood decision support system (FDSS) for the Vaisigano River, Samoa, which is comprised of a suite of integrated rainfall/river observations and forecast tools that are linked with precomputed hazard exposure scenarios to enable real-time flood-impacts warnings forecasting for the Vaisigano catchment.

A number of challenges face the development of such systems in data sparse areas. In particular, the availability of sufficient and consistent historic rainfall and flow data is required to design, calibrate and validate even parsimonious runoff models such as those used in the developed system. Despite such challenges, this study illustrates that it is possible to develop an early warning system by identifying specific thresholds of hazard (flood) occurrence and then targeting prediction capability on those thresholds. Although, the absence of soil and vegetation data as well as the need for model parsimony were pivotal in the decision to use the unit hydrograph approach for peak flow prediction.

The effectiveness of the developed FDSS was demonstrated in the lead up to the 18th December 2020 extreme rainfall and flooding, whereby residents in affected floodplains were successfully warned and evacuated several hours prior to river channel overtopping. This was achieved through the use of a multi-layered warning system that provided 1-h resolution rainfall and river level forecasts (30-h into the future); 10-min resolution prediction of rainfall and river level (24-h into the future); and real-time rainfall and river level monitoring.

While this pilot provides a benchmark for future work and extension to other catchments in the Samoa region, the findings offer a conceptual framework for developing similar impacts-based flood decision support systems in comparable Pacific Islands and SIDS contexts.

**Author Contributions:** Conceptualization, S.W., J.G., B.M. and M.I.; methodology, S.W., B.M., J.G., M.O., E.R., R.P., M.L. and G.E.; software, B.M., J.G., S.W., G.E., M.O. and R.P.; validation, J.G., B.M., M.O., E.R. and R.P.; formal analysis, E.R., M.L., M.I., S.M. and J.C.T.; investigation, B.M., J.G., S.W., M.O., R.P., E.R., M.L., S.M. and J.C.T.; resources, G.E., S.W., M.I., E.R., M.L., S.M. and J.C.T.; data curation, B.M., E.R., J.C.T., M.O. and S.W.; writing—original draft preparation, S.W. and J.G.; writing—review and editing, B.M., R.P., E.R., M.L. and M.I.; visualization, B.M., J.G. and S.W.; supervision, M.I., G.E. and S.W.; project administration, S.W. and M.L.; funding acquisition, M.I., G.E. and S.W. All authors have read and agreed to the published version of the manuscript.

**Funding:** This pilot project was enabled through the Government of Samoa's Green Climate Fund, Vaisigano Catchment Project No: MNRE-2019/001 implemented by the Samoa Ministry of Natural Resources and Environment (E.R.; M.L.; M.I.; S.M. and J.C.T.), and NIWA Taihoro Nukurangi Project No: MNR20901 and CARH2206 (S.W.; J.G.; B.M.; M.O.; R.P. and G.E.). The APC was funded by NIWA Taihoro Nukurangi, Project No:MNR20901.

**Institutional Review Board Statement:** Not applicable.

**Informed Consent Statement:** Not applicable.

**Data Availability Statement:** Hydrometeorological and climate data used in this study are available via formal request to the Samoa Ministry of Natural Resources and Environment, and were sourced from the MNRE-NEON, BOM CliDE and MNRE-NIWA Tideda databases.

**Acknowledgments:** The authors would like to thank Tuaifaiva Titimanu Simi, Ta'inau Mulipola Ausetalia Titimaea, Amituana'i Lameko Simanu, Alan Porteous and Doug Ramsay for the helpful advice provided throughout this research. The Australian Bureau of Meteorology, Tides Unit through the Pacific Sea Level & Geodetic Monitoring project as part of the Climate and Oceans Support Program in the Pacific (COSPPac), is thanked for facilitating the integration of the Apia tide gauge data. BMT Ltd. are thanked for providing the flood inundation layers via the Samoa Ministry of Natural Resources and Environment. Three anonymous reviewers are thanked for their helpful comments which improved the paper.

**Conflicts of Interest:** The authors declare no conflict of interest.

## References

1. Centre for Research on the Epidemiology of Disasters (CRED); United Nations Office for Disaster Risk Reduction (UNSIDR). *Human Cost of Disasters: An Overview of the Last 20 Years (2000–2019)*; UNDRR: Geneva, Switzerland, 2020.
2. Marsh McLean. *Sunk Costs: The Socioeconomic Impact of Flooding*; Rethinking Flood Series, Report 1; Marsh and McLean Companies Inc.: New York, USA, USA, 2021.
3. Xu, W.; Cong, J.; Proverbs, D.; Zhang, L. An Evaluation of Urban Resilience to Flooding. *Water* **2021**, *13*, 2022. [CrossRef]
4. Armal, S.; Porter, J.R.; Lingle, B.; Chu, Z.; Marston, M.L.; Wing, O.E.J. Assessing Property Level Economic Impacts of Climate in the US, New Insights and Evidence from a Comprehensive Flood Risk Assessment Tool. *Climate* **2020**, *8*, 116. [CrossRef]
5. United Nations Development Programme (UNDP). *Five Approaches to Build Functional Early Warning Systems*; United Nations Development Programme: New York, NY, USA, 2018.
6. The European Climate Adaptation Platform (Climate-ADAPT). Establishment of Early Warning Systems. Available online: https://climate-adapt.eea.europa.eu/metadata/adaptation-options/establishment-of-early-warning-systems (accessed on 1 October 2021).
7. World Meteorological Organization (WMO). *2020 State of Climate Services: Risk Information and Early Warning Systems*; WMO-No. 1252; World Meteorological Agency: Geneva, Switzerland, 2020.
8. Duminda, P.; Ousmane, S.; Jetal, A.; Hamid, M.; Mohamed, R. Challenges and Technical Advances in Flood Early Warning Systems (FEWSs). In *Flood Impact Mitigation and Resilience Enhancement*; Huang, G., Ed.; Intech Open: London, UK, 2020. [CrossRef]
9. Chen, G.; Hou, J.; Zhou, N.; Yang, S.; Tong, Y.; Su, F.; Huang, L.; Bi, X. High-Resolution Urban Flood Forecasting by Using a Coupled Atmospheric and Hydrodynamic Flood Models. *Front. Earth Sci.* **2020**, *8*, 545612. [CrossRef]
10. Schroeter, S.; Richter, H.; Arthur, C.; Wilke, D.; Dunford, M.; Wehner, M.; Ebert, E. Forecasting the impacts of severe weather. *Aust. J. Emerg. Manag.* **2021**, *36*, 76–83. [CrossRef]
11. Handmer, J.; Nalau, J. Understanding Loss and Damage in Pacific Small Island Developing States. In *Loss and Damage from Climate Change. Climate Risk Management, Policy and Governance*; Mechler, R., Bouwer, L., Schinko, T., Surminski, S., Linnerooth-Bayer, J., Eds.; Springer: Cham, Switzerland, 2019. [CrossRef]
12. Bucherie, A.; Werner, M.; van den Homberg, M.; Tembo, S. Flash Flood warning in context: Combining local knowledge and large-scale hydro-meteorological patterns. *Nat. Hazards Earth Syst. Sci. Discuss.* **2021**, preprint. [CrossRef]
13. Li, Z.; Zhang, H.; Singh, V.P.; Yu, R.; Zhang, S. A Simple Early Warning System for Flash Floods in an Ungauged Catchment and Application in the Loess Plateau, China. *Water* **2019**, *11*, 426. [CrossRef]
14. Drobot, S.; Parker, D.J. Advances and challenges in flash flood warnings. *Environ. Hazards* **2007**, *7*, 173–178. [CrossRef]
15. Young, A.; Bhattacharya, B.; Zevenbergen, C. A rainfall threshold-based approach to early warnings in urban data-scarce regions: A case study of pluvial flooding in Alexandria, Egypt. *J. Flood Risk Manag.* **2021**, *14*, e12702. [CrossRef]
16. United Nations Economic and Social Commission for Asia and the Pacific (UN-ESCAP). *Seasonal Outlook to Socio-Economic Impact Based Forecasting: Proposed Methodology and Key Results*; Asia-Pacific Disaster Report 2021, Working Paper 02; ICT and Disaster Risk Reduction Division, UN-ESCAP: Bangkok, Thailand, 2020.
17. UNISDR (United Nations Office for Disaster Risk Reduction). *Sendai Framework for Disaster Risk Reduction 2015–2030*; UNISD: Geneva, Switzerland, 2015.
18. SPC; SPREP; PIFS; UNDP; UNISDR; USP. *Framework for Resilient Development in the Pacific: An integrated Approach to Address Climate Change and Disaster Risk Management (FRDP)*; Pacific Community (SPC): Suva, Fiji, 2017; ISBN 978-982-00-1034-5.
19. Yeo, S.; Esler, S.; Taaffe, F.; Jordy, D.; Bonte-Grapentin, M. *Urban Flood Risk Management in the Pacific: Tracking Progress and Setting Priorities*; UFCOP, World Bank Group: Washington, DC, USA, 2017.
20. Baisyet, P.M. *Vaisigano River Watershed Management Plan*; Food and Agricultural Organization (FAO): Rome, Italy, 1990.
21. Yeo, S. *A Review of Flooding in Apia, Samoa*; SOPAC Technical Report 338; South Pacific Applied Geosciences Commission (SOPAC): Suva, Fiji, 2001.
22. Filer, B.; Dearnley, C.; Buchanan, M.; Caddis, B. *Review of the Interdependence of Flood Mitigation Options for the Vaisigano River Catchment*; Prepared for United Nations Development Programme, Client Reference GCF SAM 001/17, BMT Document Reference: R.B22946.006.01.; BMT Eastern Australia Pty Ltd.: Brisbane, Australia, 2019.

23. Water Technology Pty Ltd.; UNDP. *Updated Flood Study of the Vaisigano River–Volume 1, 2 and 3*; Prepared for the Ministry of Natural Resource and Environment, Water Technology; Water Technology Pty Ltd.: Melbourne, Australia, 2014.
24. Intergovernmental Panel on Climate Change (IPCC). AR6 Climate Change 2021: The Physical Science Basis. Available online: https://www.ipcc.ch/report/ar6/wg1/ (accessed on 30 August 2021).
25. The World Bank. *Samoa Post-Disaster Needs Assessment Cyclone Evans 2012*; Report No. ACS4432; The World Bank: Washington, DC, USA, 2013.
26. Flood List. Samoa–Tropical Cyclone Gita Causes Landslides and Flooding. Available online: https://floodlist.com/australia/samoa-tropical-cyclone-gita-causes-landslides-and-flooding (accessed on 1 October 2021).
27. Green Climate Fund. Strengthened Adaptive Capacity and Reduced Exposure to Climate Risks of Vulnerable Livelihoods and Infrastructure in the Vaisigano River Catchment, Environmental and Social Management Framework and Management Plan. Available online: https://pims.undp.org/attachments/5919/214510/1700637/1700918/FP-UNDP-021116-5919-Annex%20VI%20(b).pdf (accessed on 1 October 2021).
28. The World Bank. Draft Early Warning System Needs Assessment and Investment Plan for the Pacific Islands–Samoa. Available online: https://documents1.worldbank.org/curated/en/983251584942847169/pdf/Early-Warning-System-Needs-Assessment-and-Investment-Plan-for-the-Pacific-Islands-Samoa.pdf (accessed on 30 August 2021).
29. Williams, S.; Griffiths, J.; Miville, B.; O'Driscoll, M.; Kinsman, M.; Porteous, A.; Paulik, R.; Elley, G. *Vaisigano Flood Monitoring, Early Warning and Decision Support System: Final Report*; Vaisigano Catchment Project No. MNRE-2019-001, NIWA Client Report No. 2021070EI; NIWA: Christchurch, New Zealand, 2021.
30. Samoa PacIOOS. WRF Simulations at UHM. Available online: http://www.soest.hawaii.edu/MET/Faculty/wrf/arwGS/arw_samoa_3km.html (accessed on 1 October 2021).
31. MNRE; GHD Ltd. *Vaisigano Flood of December 2012 Report on Damage and Possible Causes*; GHD Ltd.: Christchurch, New Zealand, 2013.
32. Lumbroso, D.; Titimaea, A.; Penaia, A.; Bonte-Grapentin, M. *Samoa Technical Report 13th July–3rd August 2006, WS 3.5.4 Capacity Building in Flood Risk Management*; EU-SOPAC Project Report 69c; SOPAC Commission: Suva, Fiji, 2008.
33. Perica, S.; Dietz, S.; Heim, S.; Hiner, L.; Maitaria, K.; Martin, D.; Pavlovic, S.; Roy, I.; Trypaluk, C.; Unruh, D.; et al. Point Precipitation Frequency Estimates 2015. Available online: https://hdsc.nws.noaa.gov/hdsc/pfds/pfds_printpage.html?lat=34.4297&lon=-117.4522&da (accessed on 30 August 2021).
34. Sherman, L. Stream Flow from Rainfall by the Unit Graph Method. *Eng. News Rec.* **1932**, *108*, 501–505.
35. Ministry of Natural Resources and Environment (MNRE). *Samoa Meteorological Services Forecasting Operational Plan*, 3rd ed.; Ministry of Natural Resources and Environment: Apia, Samoa, 2020.
36. Schmidt, J.; Matcham, I.; Reese, S.; King, A.; Bell, R.; Henderson, R.; Smart, G.; Cousins, J.; Smith, W.; Heron, D. Quantitative multi-risk analysis for natural hazards: A framework for multi-risk modelling. *Nat. Hazards* **2011**, *58*, 1169–1192. [CrossRef]
37. FUGRO. *Report of Survey: Airborne Lidar Bathymetric and Topographic Survey of Samoa 2015*; Contract: ECRCR-LIB 2.1, Fugro LADS Document Reference No: TLCS00.047.008, Issue No: 1.00; Fugro LADS Corporation Pty Ltd.: Adelaide, Australia, 2015.
38. Pacific Community (SPC). Pacific Risk Information System. Available online: http://pcrafi.spc.int/ (accessed on 1 October 2021).
39. Australia and Oceania OpenStreetMap Data. Available online: https://download.geofabrik.de/australia-oceania.html (accessed on 30 October 2020).
40. Cox, R.J.; Shand, T.D.; Blacka, M.J. *Australian Rainfall and Runoff Revision Project 10: Appropriate Safety Criteria for People*; Water Research Laboratory, Stage 1 Report P10/S1/006; Australian Rainfall and Runoff (AR&R): Manly Vale, Australia, 2010.
41. Paulik, R. *Methods for Estimating Vehicle Impacts from Flood Hazards in RiskScape*; NIWA Client Report No: 2016008WN; NIWA: Wellington, New Zealand, 2016.
42. CliDEsc. Available online: https://clidesc.readthedocs.io/en/latest/ (accessed on 18 October 2021).
43. Flood List. Samoa–Damaging Floods and Landslides Strike After Heavy Rain. Available online: https://floodlist.com/australia/samoa-floods-landslides-december-2020 (accessed on 1 October 2021).
44. Global Forecast System. Available online: https://www.ncei.noaa.gov/products/weather-climate-models/global-forecast (accessed on 1 October 2021).
45. ECMWF Forecasts. Available online: https://www.ecmwf.int/en/forecasts (accessed on 1 October 2021).
46. Dazzi, S.; Vacondio, R.; Mignosa, P. Flood Stage Forecasting Using Machine-Learning Methods: A Case Study on the Parma River (Italy). *Water* **2021**, *13*, 1612. [CrossRef]
47. Ivanov, V.; Xu, D.; Dwelle, C.; Sargsyan, K.; Wright, D.; Katopodes, N.; Kim, J.; Tran, V.; Warnock, A.; Fatichi, S.; et al. Breaking Down the Computational Barriers to Real-Time Urban Flood Forecasting. *Geophys. Res. Lett.* **2021**, *48*, e2021GL093585. [CrossRef]
48. Merz, B.; Kuhlicke, C.; Kunz, M.; Pittore, M.; Babeyko, A.; Bresch, D.N.; Domeisen, D.I.V.; Feser, F.; Koszalka, I.; Kreibich, H.; et al. Impact forecasting to support emergency management of natural hazards. *Rev. Geophys.* **2021**, *58*, 4, e2020RG000704. https://doi.org/10.1029/2020RG000704. 10.1029/2020RG000704
49. Practical Action. *Transforming Lives Through Ingenuity: Practical Action and Early Warning Systems*; Practical Action Publishing: Rugby, UK, 2020.

50. Marchezini, V.; Horita, F.E.A.; Matsuo, P.M.; Trajber R.; Trejo-Rangel, M.A.; Olivato, D. A Review of Studies on Participatory Early Warning Systems (P-EWS): Pathways to Support Citizen Science Initiatives. *Front. Earth Sci.* **2018**, *6*, 184. https://doi.org/10.3389/feart.2018.00184.
51. Zhang, X.; Zhong, Q.; Zhang, R.; Zhang, M. People-centered early warning systems in China: A bibliometric analysis of policy documents. *Int. J. Disaster Risk. Reduct.* **2020**, *51*, 101877. https://doi.org/10.1016/j.ijdrr.2020.101877.

MDPI
St. Alban-Anlage 66
4052 Basel
Switzerland
Tel. +41 61 683 77 34
Fax +41 61 302 89 18
www.mdpi.com

*Water* Editorial Office
E-mail: water@mdpi.com
www.mdpi.com/journal/water

www.ingramcontent.com/pod-product-compliance
Lightning Source LLC
Chambersburg PA
CBHW040840250926
51850CB00007BA/1368
* 9 7 8 3 0 3 6 5 7 8 1 8 7 *